Clinical Aspects of Blood Viscosity and Cell Deformability

Royal College of Physicians and Surgeons of Glasgow

Clinical Aspects of Blood Viscosity and Cell Deformability

Edited by

G.D.O. Lowe, J.C. Barbenel and C.D. Forbes

Foreword by George P. McNicol

With 79 Figures and 20 Tables

Springer-Verlag
Berlin Heidelberg New York 1981

G.D.O. Lowe, MB, ChB, MRCP
Lecturer in Medicine
and Honorary Senior Registrar
University Department of Medicine
Royal Infirmary, 86 Castle Street
Glasgow G4 0SF, Scotland

J.C. Barbenel, BDS, BSc, MSc, PhD, LDS, RCS, MInstP
Senior Lecturer
Bioengineering Unit, University of Strathclyde
Glasgow G4 0NW, Scotland

C.D. Forbes, MD, FRCP
Senior Lecturer in Medicine
and Honorary Consultant Physician
University Department of Medicine
Royal Infirmary, 86 Castle Street
Glasgow G4 0SF, Scotland

ISBN-13:978-1-4471-3107-6 e-ISBN-13:978-1-4471-3105-2
DOI: 10.1007/978-1-4471-3105-2

Softcover reprint of the hardcover 1st edition 1981

Typeset by Photo-Graphics, Stockland, Honiton, Devon, England.

2128/3916-543210

Foreword

After many years of relative neglect, the importance of study of factors governing blood flow has at last achieved recognition; in this volume are documented many of the techniques, and the basic scientific and clinical observations, which have helped to open up understanding of this highly important aspect of human physiology and pathology in recent years. The text is logically divided into five sections beginning with blood cell deformability, then moving on to theoretical consideration of blood rheology, followed by accounts of the interrelationships between rheology, blood flow and vascular occlusion. The final two sections deal with blood rheology in clinical practice and therapeutic aspects of the study of blood flow.

As regards blood cell deformability (Section A), the basic problem is set out by Kiesewetter and colleagues in the first paragraph of chapter 1 (p.3), in which they point out that whereas human erythrocytes at rest have a diameter of approximately 7.5 μm, nutritive capillaries have diameters ranging from 3–5 μm, and chapters in section A give an account of the ways in which the red cell can undergo deformation to permit capillary perfusion and the maintenance of the microcirculation. In the opening paper, Kiesewetter and his associates review the fundamental problems inherent in attempts to measure red cell deformability and conclude that the multiplicity of techniques at present in use demonstrates the difficulties involved in obtaining reproducible data; they point out how important it is to define which specific facet of erythrocyte structure and function is actually being measured in various test systems. In chapter 2 (p.9) Sirs gives an account of his personal work on the relationships between erythrocyte flexibility and whole-blood viscosity and he points out that his observations demonstrate that the packing rate of the red cells in a centrifuge gives a direct measure of erythrocyte flexibility and in his view is a suitable index of their flow properties. He concludes his discussion with the provocative statement that there is no quantitative physical explanation of how the deformability of an erythrocyte lowers blood viscosity at high shear rates. In the following chapter Kiesewetter and colleagues (p.19) provide an account of two sophisticated methods for measuring red cell deformability, involving in the first instance monitoring of the ability of red cells to drop through pores in filter membranes under the driving pressure generated by their own weight at very low shear stresses; the second technique involves the measurement of the passage times of individual erythrocytes through a single pore with an optical-electronic device. It is possible that the first, simpler method may have some routine applications, but both methods should yield useful information needed for the basic understanding of the alterations in erythrocyte deformability found in health and disease. Bagge and colleagues (p.27) then draw attention to the relatively neglected but intriguing area of white cell deformability, of obvious clinical importance in patients with elevated white cell counts. They point out how important it is to recognise that white cells are potential blockers of nutritive flow in a low-flow state associated with hypotension.

Barbenel (p.37) presents a valuable review chapter on the measurement of red blood cell deformability in which he concludes by pointing out that the nature of the response of red blood cells to applied forces depends on three primary factors of cell shape, the mechanical properties of the cell membrane and cell content. Although occasionally one of these factors may be changed in isolation, more commonly two or more changes occur together. After noting that there is a wide variety of methods available for the measurement of red cell deformability, he observes that the ingenuity in producing new test methods has not been mirrored by parallel mathematical analysis of the tests or demonstration of relevance for routine clinical use.

Section B presents a series of chapters on general aspects of blood rheology, beginning with an account from Schmid-Schönbein (p.49) of the interactions between haemodynamic factors and blood rheology in which he suggests, perhaps somewhat pessimistically, that *in vitro* measurement of viscosity is of extremely limited value because it can either underestimate or overestimate the viscous resistance to the motion of blood in the microvasculature in disease. In the following chapter Dormandy (p.67) takes a slightly more optimistic view, and concludes that there are generally accepted techniques for measuring whole-blood viscosity using commercially available viscometers which give reproducible results with physiological validity and that measurement of viscosity has a place in clinical medicine in the investigation and treatment of circulatory disorders. Plasma viscosity measurement is discussed by Harkness (p.79), who points out that measurement of plasma viscosity is simple, speedy and accurate. The values obtained are an index of plasma protein concentrations and measurement of plasma viscosity he feels may be of value in diagnosis, prognosis and treatment. His chapter includes a useful account of variations in plasma viscosity in health with age, sex, diet, exercise, menstruation and pregnancy.

Section C examines the important relationships between blood rheology and vascular occlusions. In the opening chapter in this section, Charlesworth (p.91) sets out the relationships between blood rheology and blood flow and points out the potential hazards of reducing blood viscosity in patients with arteriosclerosis; the decrease in blood viscosity may well produce loss of pulsatile energy and hence outweigh the advantages which might result from reduction in peripheral resistance. On the other hand, in the following chapter (p.97), Pearson and colleagues point out that there is unequivocal evidence of an association between a high venous haematocrit, for example in primary proliferative polycythaemia, and the risk of occlusive vascular disease, particularly in the cerebral circulation. There is also some evidence of an increased risk of vascular occlusion at high normal compared with low normal venous haematocrit values. They conclude that it is now appropriate to initiate clinical trials of venesection in patients with high normal venous haematocrit values in the hope of reducing the incidence of vascular occlusion.

An obvious area of association between disorders of rheology and vascular occlusion is sickle cell disease where the evidence is reviewed by Stuart and Kenny (p.109). While more evidence is needed, the data at present available suggest that changes in erythrocyte deformability may precede changes in plasma protein levels and plasma viscosity.

The effect of white cells on blood flow has in the past been relatively neglected, but Preston (p.123) points out that when whole-blood viscosity is measured, although patients with leukaemia may have normal values because the increase due to the elevated white cell count is compensated for by the reduced haematocrit, blood transfusion may carry considerable potential hazards. Following transfusion the increase in haematocrit may raise the whole-blood viscosity sufficiently to produce coma or even death. The increased viscosity associated with the high white cell count can be treated by leucopheresis, and plasmapheresis is rapidly effective in reducing viscosity in patients with paraproteinaemias. Lowe and colleagues (p.133) examine the relationships between occlusive arterial disease and blood rheology and conclude that there is evidence that blood viscosity and its major determinants — haematocrit and plasma fibrinogen — are associated with clinical arterial events and possibly with the extent of arterial occlusion. There is some evidence that increased viscosity may reduce blood flow in patients, contributing to the clinical outcome, but there is at present no evidence that steps taken to alter blood rheology affect the natural history of arterial disease.

Section D gives an account of blood rheology in medicine and surgery. The first chapter (p.151), by Barnes, discusses relationships between blood viscosity and the clinical manifestations of diabetes. Evidence is presented that in diabetes mellitus, blood flow properties are abnormal, with increased whole-blood viscosity at low shear rates in long-standing diabetics, particularly in those with complications. There is also strong evidence to suggest that erythrocyte deformability is reduced in diabetes. The aetiology of these changes is complex and not yet fully elucidated.

Matthews (p.163) reviews the evidence for an association between alterations in blood haemorheology and thrombosis post-operatively and suggests, although the evidence is by no means conclusive, that it is logical to suppose that blood viscosity may influence the development of arterial and venous thrombosis in post-operative patients.

Turning to obstetrics and gynaecology, Buchan and Macdonald (p.175) present a large body of experimental evidence on changes in blood viscosity in obstetrics and gynaecology. Fibrinogen levels change during pregnancy, as does erythrocyte deformability although not quite in parallel. Packed cell volume is an even more important determinant of whole-blood viscosity than erythrocyte deformability and Buchan and McDonald point out that packed cell volume tends to be higher in pill-users in any stage of the menstrual cycle. The present incomplete data suggests that in heavy smokers and patients with intrauterine growth retardation there may be important abnormalities in blood rheology. They conclude that the field requires intensive study which might reveal previously unsuspected therapeutic avenues of value.

Walker (p.193) points out that in the newborn, incidence of hyperviscosity is estimated at 3%–5% of all newborns and may be of importance in the development of neonatal symptoms and long-term sequelae. He suggests that estimation of the haematocrit should be routine in the ill newborn and that haemodilution should be considered in the treatment of any baby when venous haematocrit exceeds 0.65. Further studies are needed to evaluate the utility of this intervention.

Section E is devoted to therapeutic aspects of blood rheology. Schmid-Schönbein and Rieger (p.211) report on a study of 25 patients with arterial occlusive disease in whom therapeutic haemodilution was produced by repeated venesection followed by reinfusion of the autologous plasma and low molecular weight dextran. There was suggestive evidence of therapeutic benefit from the study which, however, was not case-controlled or blind. Dodds (p.227) examines the benefits of plasma exchange, and concludes that although this form of therapy is both novel and dramatic it is rather crude in concept because it removes all components of plasma rather than selectively removing what is required. He feels that it must remain in many conditions an experimental tool with, however, the potential for a considerable impact in medicine, particularly in the malignant paraproteinaemias and patients with digital ischaemia.

Plasma exchange is further considered by Allan (p.235), who discusses its value in Waldenström's macroglobulinaemia, and feels that it is of real utility in rapid and effective treatment of the hyperviscosity syndrome, and may also be helpful in patients who are precluded from having chemotherapy by concomitant illnesses, or whose tumour has proved unresponsive to chemotherapy.

Lowe and colleagues (p.241) discuss the effect of two defibrinating snake enzymes, ancrod and batroxobin, which convert fibrinogen to fibrin by splitting off fibrinopeptide A only, unlike thrombin, which splits fibrinopeptides A and B. The resulting fibrin micro-clots rapidly cleared from the circulation, resulting in a fall in circulating fibrinogen. As a consequence blood and plasma viscosity is reduced and there is some evidence that blood flow to the periphery may be increased. It is suggested that this is a safe and effective therapeutic approach to venous thrombosis, but controlled trials do not yet support the use of defibrination in patients with arterial disease.

The final chapter in this section, from Dormandy (p.251), discusses evidence that drugs can modify erythrocyte deformability and suggests that none of the evidence is conclusive. However, he feels that it is plausible that therapeutic improvement in red cell deformability can be achieved by drug intervention and that further clinical studies, particularly in the microcirculation, are well warranted.

This is perhaps a good concluding sentiment. The Royal College of Physicians and Surgeons of Glasgow, and the organisers of the Symposium on which this book is based, are to be warmly congratulated for an imaginative and timely venture. This balanced, useful and informative book indicates that blood rheology has now come of age; there is presented an impressive account of basic scientific data, clinical observation and indeed clinical successes. It is, I think, a milestone in the development of the field, justifying further intensive efforts in a new, rapidly growing and exciting area of applied biology, with potentially very large applications both to the understanding of disease and to its treatment.

December 1980

Professor George P. McNicol, MD, PhD
University Department of Medicine
The General Infirmary
Leeds, LS1 3EX, U.K.

Preface

Blood viscosity, rheology and red cell deformation has become a major area of scientific research over the past few years and was originally stimulated in Glasgow by Dr. Leopold Dintenfass while he was a Visiting Professor. The editors felt that the scientific techniques now available for measurement of various facets of viscosity and cell deformation had been widely used by individual clinicians in a variety of diverse disorders. The purpose of this Symposium was to compare methods and techniques and attempt to correlate results with the clinical findings in patients. In addition we hoped to encourage clinicians to use viscosity and rheology measurements in everyday clinical practice. There is no doubt that newer and better methods will become available, in particular for deformability measurement, and the information in these pages will be rapidly outdated.

The editors wish to thank Mrs. Moira Hargreaves for her infinite patience in retyping manuscripts and checking references and the Royal College of Physicians and Surgeons of Glasgow for arranging the Symposium.

Glasgow, December 1980

G.D.O. Lowe
J.C. Barbenel
C.D. Forbes

Table of Contents

List of Contributors

T.L. Allan, MB., Ch.B., Registrar in Haematology, Stobhill General Hospital, Glasgow, G21 3UW, UK.

U. Bagge, MD., Ph.D., (Professor), Laboratory of Experimental Biology, Department of Anatomy, University of Göteborg, Göteborg, Sweden.

J.C. Barbenel, BDS., MSc., Ph.D., Senior Lecturer, Bioengineering Unit, University of Strathclyde, Glasgow G4 0NW, UK.

A.J. Barnes, MB., MCRP., Lecturer, Department of Medicine, Royal Postgraduate Medical School, Hammersmith Hospital, Du Cane Road, London, W12 0HS, UK.

P-I Brånemark, MD., Ph.D., (Professor), Laboratory of Experimental Biology, Department of Anatomy, University of Göteborg, Göteborg, Sweden.

P.C. Buchan, BSc., MB., MRCOG., Senior Registrar, Department of Obstetrics and Gynaecology, St. James' University Hospital, Leeds LS9 7TF, UK.

D. Charlesworth, MD., FRCS., Reader in Surgery, University Hospital of South Manchester, Nell Lane, West Disbury, Manchester M20 8LR, UK.

U. Dauer, Dr. Ing., Department of Physiology, Medical Faculty, Rhine-Westphalia Technical High School, Melatenerstrasse 211, 5100 Aachen, West Germany.

A.J. Dodds, MB., BS., FRACP., FRCPA., Research Fellow and Honorary Research Registrar, Department of Haematology, St. George's Hospital Medical School, London SW17 0RE, UK.
Currently – Haematologist, St. Vincent's Hospital, Darlinghurst, New South Wales 2010, Australia.

J.A. Dormandy, FRCS., FRCS. (Ed.)., Consultant Surgeon, St. James' Hospital, Sarsfeld Road, Balham, London, SW12 8HW, UK.

M.M. Drummond, Biochemist, Bioengineering Unit, University of Strathclyde, Glasgow G4 0NW, UK.

C.D. Forbes, MD., FRCP., (Glas., Edin., Lond.), Senior Lecturer, Department of Medicine, Royal Infirmary, Glasgow G4 0SF, UK.

J. Harkness, BSc., MD., FRCPath., Consultant Haematologist, Department of Clinical Pathology, Musgrove Park Hospital, Taunton, Somerset, UK.

P.R.D. Humphrey, MA., BM., B.Ch., MRCP., National Hospital for Nervous Diseases, Queen Square, London, WC1, UK.

M.W. Kenny, MB., Ch.B., MRCP., MRC(Path), Senior Registrar, Department of Haematology, Queen Elizabeth Hospital, Birmingham B15 2TH, UK.

H. Kiesewetter, Dr.Ing., Dr.Med., Department of Physiology, Medical Faculty, Rhine-Westphalia Technical High School, Melatenerstrasse 211, 5100 Aachen, West Germany.

G.D.O. Lowe, MB., MRCP., Lecturer, Department of Medicine, Royal Infirmary, Glasgow G4 0SF, UK.

H.N. Macdonald, Ph.D., MB., FRCOG., Consultant, Department of Obstetrics and Gynaecology, St. James' University Hospital, Leeds LS9 7TF, UK.

P.N. Matthews, FRCS., FRCS. (Ed)., Senior Surgical Registrar, St. James' Hospital, Sarsfeld Road, Balham, London SW12 8HW, UK.

K. Mussler, Dipl. Ing., Department of Physiology, Medical Faculty, Rhine-Westphalia Technical High School, Melatenerstrasse 211, 5100 Aachen, West Germany.

T.C. Pearson, MD., MRCPath., Consultant, Department of Haematology, St. Thomas' Hospital, London, SE1 7EH, UK.

C.R.M. Prentice, MD., FRCP. (Glasg. & Lond.), Reader, Department of Medicine, Royal Infirmary, Glasgow G4 0SF, UK.

F.E. Preston, MD., MRCPath., Consultant Haematologist, Hallamshire Hospital, Sheffield, S10 2JF, UK.

H. Rieger, Priv. Doz. Dr. Med., Department of Physiology, Medical Faculty, Rhine-Westphalia Technical High School, Melatenerstrasse 211, 5100 Aachen, West Germany.

H. Schmid-Schönbein, Professor, Department of Physiology, Medical Faculty, Rhine-Westphalia Technical High School, Melatenerstrasse 211, 5100 Aachen, West Germany.

D. Seiffge, Dr. Med. Vet., Department of Physiology, Medical Faculty, Rhine-Westphalia Technical High School, Melatenerstrasse 211, 5100 Aachen, West Germany.

J.A. Sirs, BSc., Ph.D., Reader and Head of Department, Department of Biophysics, St. Mary's Hospital Medical School, University of London, Paddington, London, W2, UK.

R. Skalak, BS., CE., Ph.D., (Professor), Department of Civil Engineering and Engineering Mechanics, Columbia University, New York, U.S.A.

R. Spohr, Dr. Rer. Nat., Department of Physiology, Medical Faculty, Rhine-Westphalia Technical High School, Melatenerstrasse 211, 5100 Aachen, West Germany.

J. Stuart, MD., FRCP., FRC(Path)., Professor, Department of Haematology, Queen Elizabeth Hospital, Birmingham B15 2TH, UK.

P. Teitel, Priv. Doz. Dr. Med., Department of Physiology, Medical Faculty, Rhine-Westphalia Technical High School, Melatenerstrasse 211, 5100 Aachen, West Germany.

D.J. Thomas, MD., Ch.B., National Hospital for Nervous Diseases, Queen Square, London, WC1, UK.

C.H.M. Walker, MD., FRCPE., FACC., DCH., Consultant Paediatrician, Department of Child Health, Ninewells Hospital, Dundee, DD1 9SY, UK.

G. Wetherley-Mein, BA., MD., FRCP., Professor, Department of Haematology, St. Thomas' Hospital, London, SE1 7EH, UK.

Editors' Note

Units of Measurement

To facilitate conversion of traditional to SI (Système Internationale) units, the following table should be consulted:

	Traditional units	*Equivalent in SI units*
Force	1 dyne	10^{-5} Newtons (N)
Shear stress	1 dyne/cm^2	10^{-1} Pascals (Pa) = 10^{-1} N/m^2
Viscosity	1 Poise (P)	10^{-1} Pascal. second (Pa.s)
$= \frac{\text{shear stress}}{\text{shear rate}}$	1 centiPoise (cP)	1 milliPascal. second (mPa.s)

Section A
Blood Cell Deformability

1 Problems of Measurement of Red Cell Deformability

H. Kiesewetter, H. Schmid-Schönbein, D. Seiffge and P. Teitel

1.1 Introduction

Human erythrocytes at rest have a diameter of approximately 7.5 μm, whereas nutritive capillaries have diameters ranging from 3 to 5 μm. Following from this simple geometrical boundary condition, it is obvious that the ability of the red blood cells (RBCs) to undergo deformation is of paramount importance for normal fluidity of blood and also for adequate perfusion of the nutritive capillaries of all organs.

Recent years have witnessed detailed fundamental studies of causes and effects of the deformability of RBCs, and these support the theory that the fluidity of blood flowing rapidly in the terminal microcirculation depends critically upon this micromechanical peculiarity of the RBC. All nutritive capillaries, and in particular the sinuses of the spleen, are constantly filtering out from the active circulation RBCs with impaired deformability, so that the life span of such an RBC is considerably reduced. This fact suggests that only measurements with a high sensitivity will be able to detect minor changes in the normal deformability of the RBC. As far as possible, such methods should simulate both the geometrical and the hydrodynamical conditions prevailing in the microcirculation.

1.2 Measurement Techniques

The most common measurement techniques currently utilised will be listed and briefly discussed.

1.2.1 Viscosity Measurements in Rotational and Capillary Viscometers

For viscosity measurements suspensions of washed RBCs in buffered (pH 7.4) isotonic saline solutions must be used since the deformability of the cells is measured only indirectly. If the RBCs are suspended in autologous plasma or in other fluids containing aggregation-enhancing components, the measurement of viscosity will yield data not only on deformability of the RBC but also on the phenomenon of cell aggregation, which considerably influences the viscosity of such suspensions.

In these measurements the effective shear stresses are usually so high that only the behaviour of RBCs in rapid flow through large vessels is stimulated. Under these conditions, the RBCs behave like fluid droplets, the shape of which is adapted to the streamlines; the RBC's shape in the microcirculation, however, is an adaptation to a well-defined geometrical constraint. Therefore the deformability of RBCs in nutritive capillaries can be assessed only crudely by viscosity measurements.

1.2.2 Determination of RBC Geometry in Well-defined Shear Fields

Determination of RBC geometry in well-defined shear fields has been undertaken in many different ways, but none of these actually mimics the conditions in the microcirculation. In one of these methods, the shape taken up by erythrocytes in a shear field induced in a highly viscous suspension fluid (30 Pa.s) is recorded as a stationary picture (Fischer, 1978). For such investigations, the haematocrit of the suspension must be kept low. Recent studies in our department have disclosed that diamide-treated RBCs, the elongation ability of which is strongly curtailed in the shear field of a "rheoscope", do not show a marked impairment in their flow behaviour either in the living microcirculation (Haest et al., 1978) or in the newly designed "filtrometer" (Fig. 1.1.)

An approach similar to the rheoscope has been utilised in the "ectacytometer" of Bessis and Mohandas (1978); here, the elongation ability of RBCs suspended in highly viscous suspending fluids and submitted to a shear field in a rheoscope chamber is assessed by analysis of a diffraction pattern produced by a laser beam. The approach utilised by Hochmuth et al. (1973) makes use of fixation of RBCs at the bottom wall of a flow chamber mounted on a microscope stage and measurement of the elongation resulting from the influence of well-defined shear stresses. This approach, however, has the drawback that the immobile (fixed) surface of the RBC membrane is not well defined and therefore poorly suited to quantitate deformability.

By the same token, the approach described by Bull and Brailsford (1976), whereby RBCs hanging on fibrin strands and submitted to defined shear stresses have their elongation measured optically, also implies poorly defined conditions of both the geometry of the contact surface and the type of interaction between the fibrin strands and the RBC membranes. In another approach, RBC geometry is studied in the shear field produced inside the aperture of a Coulter counter; the parameter measured here is the disturbance of the electrical field produced by the

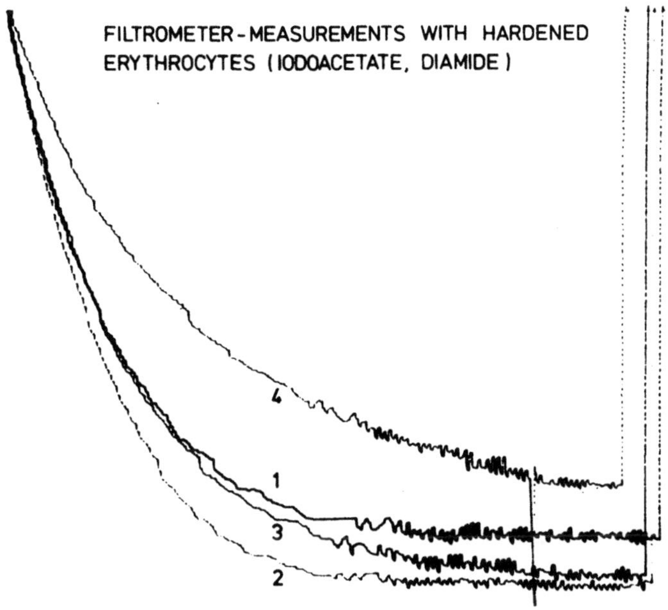

Fig.1.1. Filtrometer filtration curves. Filtered volume (ordinate) as a function of time (abscissa)
1. Concentrated RBC 2. Concentrated RBC after treatment with iodoacetate 3. Concentrated RBC after treatment with iodoacetate and diamide 0.05 mM/litre 4. Concentrated RBC after treatment with iodoacetate and diamide 0.5 mM/litre

passage of an erythrocyte, which allegedly is nearly proportional to the volume of the cell. The great drawback of this approach is that one has also to determine a form-factor, since a sphere and an ellipsoid can have the same volume.

1.2.3 Aspiration of Part of, or the Entire Erythrocyte in a Micropipette

Another group of research workers partially or completely aspirated RBCs into micropipettes. In channels of approximately 1 μm in diameter, the negative pressures are unphysiologically high (5×10^4 Pa). For diameters of approximately 3 μm, the corresponding pressures are within the physiological range (30 Pa), yet the procedure is cumbersome and extremely time consuming for measuring a representative number of individual cells from a given cell population. Therefore, this approach appears to be inappropriate for routine investigations.

1.2.4 Determination of the Packing Density of RBCs Submitted to Defined Forces

In a further approach, the packing density of RBC sediments obtained by centrifugation is quantitated. Here, too, the applied shear stresses are unphysiologically high. A further and basically erroneous notion here is an attempt to draw conclusions on the deformability of the RBCs from their sedimentation rate; the latter is mainly determined by the extent of aggregation, which is

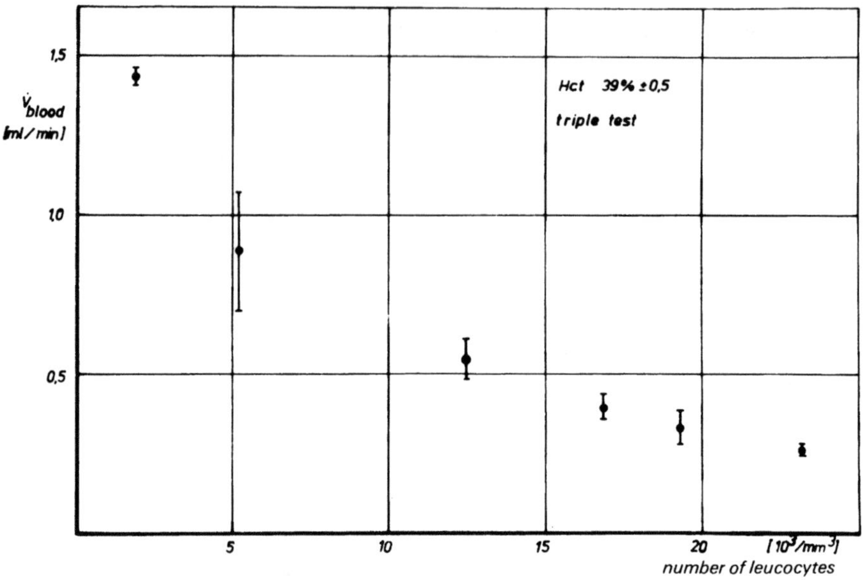

Fig.1.2. Measurements of filtration with whole blood (anti-coagulated with EDTA) to which has been added various numbers of leucocytes (mean ± SD).
$\dot{V}_{blood}$ = filtered volume per minute

considerably increased in the presence of high concentrations of aggregating macromolecules, particularly of fibrinogen.

1.2.5 Filtration of Erythrocytes in Well-defined Boundary Conditions

Filtration methods are commonly used to asses RBC deformability and several parameters have been quantitated (driving pressure, volume of filtered suspension, number of filtered cells, flow rate, etc.). The nature of the filtering membrane (Nuclepore, Millipore, sieves of sintered metal powders, glass capillary arrays) and also the dimension of the narrow channels in the filtering membrane (diameter and length) vary greatly from one method to another. A common feature is that most of them use too high a driving pressure, and consequently too high a shear stress. A further fundamental error is utilisation of whole blood, since the less deformable leucocytes and platelets may plug a great many of the pores. Thus, in addition to deformability of the RBC, the time-dependent reduction in the total number of available and still permeable pores will be assessed. The results obtained, therefore, reflect a complex, undesirable, artefact. This assertion has been substantiated by determinations made in our department, in which suspensions of RBC have been contaminated with various amounts of leucocytes and platelets (Fig. 1.2).

The only method exempt from these criticisms is the procedure described by Teitel and Mussler (1979), which has recently been further developed in our department and which will be described elsewhere (see Chapter 3).

1.2.6 Measurement of Passage Time of Individual Erythrocytes Through Single-pore Membranes

The drawback of all filtration procedures is the fact that they can only make an integrative assessment. In other words, for the same measured parameter (flow rate, i.e. first derivative of the filtered volume with respect to time), it is not possible to tell if the findings result from a relatively small impairment of the deformability of all cells from the sample analysed, or whether it results from gross changes in only some of the cells.

The only procedure which makes such differentiation possible, while maintaining the required geometrical and hydrodynamic boundary conditions, is determination of the deformability of individual erythrocytes flowing through single-pore membranes. This procedure consists in determination of the passage time of each individual red cell. The result appears as a histogram from which can be extrapolated the deformability of all cells in the population analysed, the distribution of this micromechanical feature within this population, and also the deformability of the individual RBCs (see Chapter 3).

1.3 Conclusions

The multiplicity of techniques available to assess quantitatively the deformability of RBCs shows how difficult it is to obtain comparative and reproducible data; most techniques fulfil this aim only partially. With these difficult measurements, one should always be aware of which specific "property" of the erythrocytes is actually being measured and what kind of factors influence the parameters measured.

1.4 References

Bessis M, Mohandas N (1978) Laser diffraction patterns of sickle cells in fluid shear fields. Blood Cells *3*: 225-232

Bull B S, Brailsford J D (1976) Red cell membrane deformability: New data. Blood *48*: 663-667

Fischer T M (1978) Mechanisches Verhalten von Erythrozyten in Scherfeld zäher Lösungen. Dissertation from the Rhine-Westphalia Technical School, Aachen

Haest C W M, Driesen G, Heldtmann H, Kamp D, Fischer T M, Stohr-Liesen M (1978) Is deformability a parameter for the rate of elimination of erythrocytes from the circulation? Abstracts of 50th Meeting Deutsche Physiologische Gesellschaft (Göttingen). Pfluegers Arch Suppl. to vol. *377*

Hochmuth R M, Mohandas N, Blackshear P L (1973) Measurement of elastic modulus for red cell membrane using a fluid mechanical technique. Biophys J *13*: 747-762

Teitel P, Mussler K (1979) A new automatic instrument to investigate the rheological behaviour of red blood cells. Thromb Haemostas *42*: 105

2 Erythrocyte Flexibility and Whole-blood Viscosity

J. A. Sirs

2.1 Introduction

One of the remarkable properties of blood is its ability to flow with a viscosity only a few times greater than water, though approximately half its volume is composed of cells. If the cellular components were rigid spheres, blood would have the consistency of a solid (Taylor et al., 1965). By hardening cells with formalin or acetaldehyde, Kuroda et al. (1958) and Seaman (1966) found that the viscosity was increased by an order of magnitude. It has also been demonstrated (Chien et al., 1967) that the whole-blood viscosity is Newtonian in these circumstances. The viscosity of normal whole blood is non-Newtonian with a magnitude of about 5 cP (mPa. s) at 37°C and high shear rates. On this basis it has become accepted that the flexibility of red blood cells and the plasma viscosity are the main factors which determine the viscosity of blood at high shear rates, though quantitative evidence for this hypothesis is lacking in the range of viscosity between these two extremes.

A theoretical analysis of the viscosity of a dilute suspension of rigid spheres was derived by Einstein (1906; 1911) and extended by Jeffery (1922) to ellipsoidal particles. Taylor (1932) developed a theory that can be applied to dilute suspensions of immiscible fluid drops. Various extensions of these theories (Roscoe, 1952; Brinkman, 1952; Oldroyd, 1953) have been adapted by Dintenfass (1968a) to derive an equation that could be applied to whole blood. Though qualitative use has been made of this equation, several discrepancies have become apparent (Dintenfass, 1975). It has also been applied by Dupont and Sirs (1977) qualitatively to demonstrate that variations of the internal viscosity, η_i, of the red blood cell, calculated from the whole-blood viscosity, η_b, and plasma viscosity,

η_p, correlate with measurements of erythrocyte flexibility using the centrifuge technique (Sirs, 1968; 1970). In this paper a more systematic study has been made of this problem by measuring blood viscosity, plasma viscosity and erythrocyte flexibility under a variety of conditions.

The measurement of erythrocyte flexibility using the centrifuge method is simple, requires the minimum quantity of whole blood and results can be obtained within several minutes. Attempts to correlate such measurements, using freshly drawn blood samples, with results obtained using the filtration technique, have not been satisfactory. While qualitative agreement can be obtained when the cells are hardened by various physico-chemical methods, discrepancies occur in measuring the more subtle differences present *in vivo*. This is in part due to the difficulty of obtaining reproducible results with the filtration technique (Dintenfass, 1968b; Schmid-Schönbein et al., 1973) and partly because erythrocyte flexibility is not the only factor involved in determining the rate of filtration. A consideration of other methods, to obtain independent checks, disclosed a number of difficulties. Techniques such as sucking cells into pipettes, which make measurements on individual cells, require a large number of measurements to obtain a representative sample. Another problem is to avoid additional artefacts introduced by manipulation of the blood sample. An objective of any measurement of flexibility should be to provide an index that reflects the changes occurring in whole blood *in vivo*. In this respect it was apparent that the best evidence to confirm the validity of the centrifuge method would be to compare the results with estimations of erythrocyte flexibility made from measurements of η_b and η_p.

2.2 Materials and Methods

Blood samples were collected from the ante-cubital veins of two subjects, using heparin (12.5 iu/ml). It was kept at room temperature following collection, and measurements of viscosity and erythrocyte flexibility were made within 2-3 h. Various investigations and studies over the past 10 years have been made on the blood of the two donors, so that information on several parameters was available. In particular they were both of blood group A Rh + ve and their erythrocytes contained haemoglobin HbA, with identical reaction kinetics for oxygen and carbon monoxide. The total protein, fibrinogen, albumin, globulin, alkaline phosphatase, aspartate transaminase, and bilirubin in the plasma were determined and found to be normal.

Whole-blood viscosity and plasma viscosity were measured using a Wells-Brookfield cone-on-plate LVT viscometer at shear rates in the range 1.15-230 s^{-1} and between 10° and 38°C. The viscometer was calibrated with standard fluids, and the relative accuracy of the measurements was ±5% at shear rates below 23 s^{-1} and ±2% at high shear rates. Care was taken to avoid clotting and any sample which gave erratic readings was discarded.

The flexibility of the red blood cells was determined using the stop-start and automatic centrifuge methods (Sirs, 1968; 1970). The initial packing rate (%/min), of the length of the red-cell column relative to the total initial length, gives an index of erythrocyte flexibility. The rate depends on the centrifugal acceleration, the haematocrit, the density difference between cells and plasma, and the flexibility of the cells (Sirs, 1968). In the automatic recording centrifuge the centrifugal acceleration is constant at 200 *g*. With the stop-start procedure

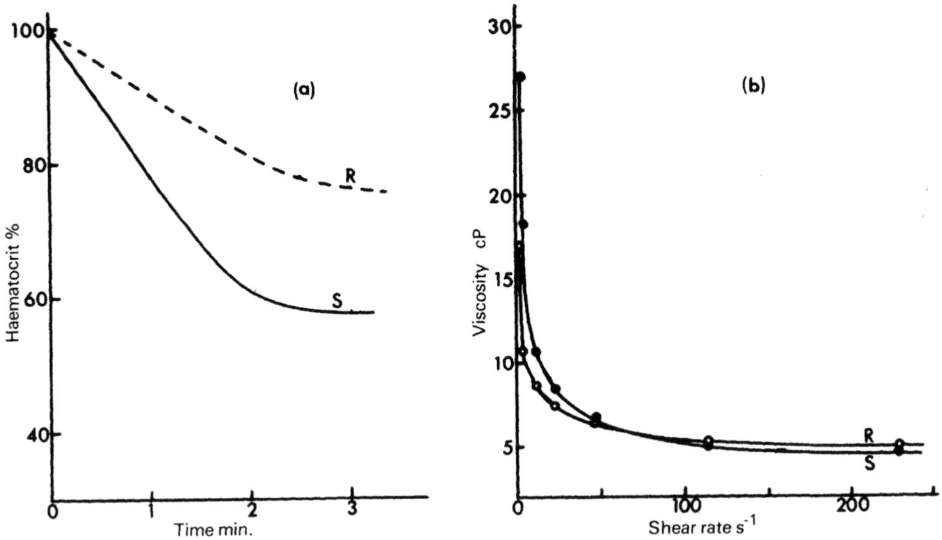

Fig.2.1. Erythrocyte flexibility and blood viscosity of subjects R and S: (a) packing rate with time; subject S, *solid curve* at a haematocrit of 44.3%; Subject R, *broken curve,* 43.3%; (b) whole-blood viscosity at 37.3°C and 45% haematocrit with variation of shear rate. Subject S, *solid circles;* Subject R, *open circles*
1 cP (centipoise) = 1 mPa.s.

relative estimations are made, to allow for small variations of the speed, by centrifuging two or more samples simultaneously. The small differences between cell-to-plasma densities in the blood samples have a negligible effect on the packing rate (Sirs, 1968). At haematocrits greater than 35% and when the centrifugal acceleration is more than 200 *g*, rouleaux formation and aggregation do not alter the rate of packing (Rampling and Sirs, 1972). The initial packing rate values were standardised to a packed cell volume (PCV) of 40% or 45% at 13,000 *g*, using a calibration curve obtained by diluting the blood in its own plasma. The index of packing rate has an accuracy of ±0.5%.

2.3 Results

The two subjects, R and S, investigated in this study, were chosen because R has an erythrocyte flexibility representative of the mean value of a healthy population, whereas S is one of 1.5% of the total population whose erythrocyte flexibility was found to be more than three standard deviations from the mean (Rampling and Sirs, 1976). The rate of packing and blood viscosity over a range of shear rates at 37.3°C are shown in Fig. 2.1. The more flexible cells of S significantly reduce the whole-blood viscosity at high shear rates, but increase the viscosity at low shear rates. The respective plasma viscosities of R and S at 37.3°C were 1.42 and 1.49 cP respectively. These results are similar to those previously reported using a capillary viscometer (Sirs, 1975). The viscosity of blood should increase at low shear rates, as the cell flexibility becomes significant relative to the applied force, until it behaves as if the cells were rigid. The shear rate at which this occurs would, however, fall with an increase of the cell flexibility. In Fig. 2.1

the reverse is the case. This is in accord with the accepted hypothesis that aggregation and rouleaux formation also contribute to the increased blood viscosity at low shear rates. The formation of rouleaux involves a change from a biconcave to a flattened cell profile (Chien et al., 1971; Rowlands and Skibo, 1972) and the more flexible cells facilitate the formation of aggregates.

2.3.1 Internal Viscosity of the Erythrocyte

Dintenfass (1968) has derived an equation relating whole-blood viscosity, η_b, plasma viscosity, η_p, and the internal viscosity of the erythrocytes, η_i, in the form:

$$\eta_r = (1 - HT)^{-2.5} \quad (1)$$

where η_r is the relative viscosity, η_b/η_p, H the haematocrit, and T, Taylor's factor, as given by

$$T = (p + 0.4)/(p + 1) \quad (2)$$

where p is equal to η_i/η_p. Though this equation had previously been applied with apparent success to data relating η_i to the plasma fibrinogen concentration (Dupont and Sirs, 1977), it was decided to test its validity further by varying the haematocrit. Samples of blood of varying haematocrit were obtained by diluting the whole blood with its own plasma. The latter was obtained by centrifuging a portion of the blood sample at 3000 g for 10 min. The results are shown in Fig. 2.2 as a semi-logarithmic plot of the blood viscosity, at 37.3°C and a shear rate of 230 s^{-1}, with haematocrit. The viscosity at zero haematocrit is that obtained using plasma. With these data the internal viscosity of the red cells, η_i, was calculated using equation (1). The results are shown in Table 2.1. Though it is possible that small differences in η_i could occur by diluting whole blood with plasma, obtained by centrifugation, the large differences obtained using equations (1) and (2) are not acceptable. There are several possible explanations for this discrepancy. The haematocrit, H, is an approximation only and should in its correct form be replaced by KC, where K is a coefficient of immobilisation of the continuous fluid, and C the concentration of the disperse phase. Another factor is that the constant of 2.5 may not be applicable to biconcave discs. There are also theoretical difficulties in taking into account mutual interactions which disturb both the flow round each cell and within the cell, and involve direct cell-to-cell contact.

2.3.2 Alternative Estimation of the Internal Viscosity

In order to minimise these complications it was decided to revert to the formula for dilute suspensions derived by Taylor:

$$\eta_r = 1 + KCT \quad (3)$$

where C is the volume fraction of the disperse phase and K is equal to 2.5 for spheres. This equation is valid for C equal to 0.01 or less, equivalent to a haematocrit of 1%. Equation (3) applies to immiscible fluid drops that are spherical. Its application to biconcave erythrocytes, consisting of a concentrated haemoglobin solution enclosed by a membrane, requires justification. Consider

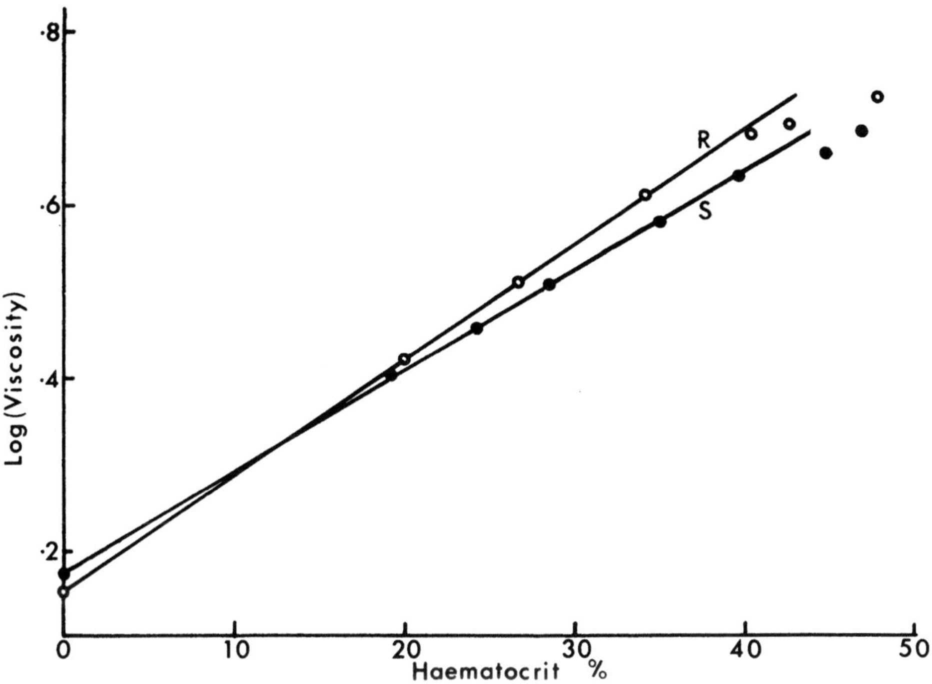

Fig.2.2. A plot of the variation of log η with haematocrit, at 37.3°C and 230 s^{-1} shear rate. Subject S, *solid circles;* Subject R, *open circles*

Table 2.1 Calculated values of η_i with haematocrit

Haematocrit	(%):	44.6	39.6	34.9	28.4	24.2	19.1
η_i	(cP):	4.27	5.61	7.06	12.75	− 152.6	

first the situation when the suspended particles have a high internal viscosity, η_i, and the value of T in equation (2) is equal to unity. In these circumstances equation (3), with T equal to one, represents the relation of viscosity to the concentration of the dilute suspension of inflexible particles, and K is a constant whose value depends on the particle shape. For spherical particles K is equal to 2.5, as derived by Einstein, and for asymmetric shapes is greater than 2.5. It can be applied to a dilute suspension of inflexible erythrocytes by using an appropriate value for K. Now consider that the erythrocytes are made progressively more deformable, with a consequent lowering of η_r. To allow for this effect we can, in a similar manner to Taylor's derivation for spherical fluid drops, multiply K by a cell flexibility factor, T_c. There is no theoretical analysis available inter-relating erythrocyte deformability and η_r in terms of T_c, equivalent to Taylor's treatment of immiscible fluid drops. The possibility that, as a first approximation, T_c may be given by equation (2) is discussed later. The important implication of this approach is that by measurement of η_r we can estimate KT_c, which provides an index of erythrocyte deformability directly related to the flow properties of blood. It is of course implicit in the use of equations (2) and (3) that there is no aggregation between particles, and this restricts their application to the interpretation of blood viscosity at high shear rates.

There is one further aspect to be considered before discussing the experimental determination of KT_c. Equation (3) applies to haematocrits of 1% or less. For practical purposes it would be advantageous to estimate η_{r1}, at 1% haematocrit, from measurements of η_r made on whole blood at a haematocrit, H. The effect of haematocrit on blood viscosity, for subjects R and S, is shown in Fig. 2.2. The lines fit the relationship,

$$\log \eta_b = aH + \log \eta_p \tag{4}$$

where a is the slope of the line. A logarithmic relationship of this form has been observed by several investigators (Haynes, 1960; Peric, 1964; Gregerson et al., 1965) and has a theoretical basis (Richardson, 1950). Since η_r is equal to η_b/η_p, equation (4) reduces to

$$\log \eta_r = aH \tag{5}$$

To obtain η_{r1} from η_r, equation (5) can be put in the form:

$$\eta_r = (\eta_{r1})^H \tag{6}$$

2.3.3 *Measurement of* KT_c *and* K

In principle values of KT_c were obtained by substituting experimental values of η_{r1}, at 1% haematocrit, into equation (3), using the bloods R and S, which had cells of different deformability. The erythrocyte flexibility was determined in terms of the initial packing rate, using the centrifuge technique, on the same blood samples. The value of K can then be assessed by plotting KT_c against the initial packing rate. When erythrocytes are rigid, the packing rate is zero and T_c is equal to unity. The value of K can thus be obtained from the extrapolation of the curve to determine the point of intercept with the KT_c axis.

The measurements of erythrocyte flexibility and KT_c obtained from Figs. 2.1 and 2.2 were not sufficient to assess if the relationship was linear. To obtain two additional points a sample of each blood was incubated at 48°C for 45 min. It was ascertained by microscopic observation in previous studies, and in the present case, that this procedure hardened cells with no visible effect on the erythrocyte size (Sirs, 1968), though with a maximum resolution of $\pm$ 0.1 μm it was uncertain to what degree the shape might have been affected. After hardening, the flexibility of the erythrocytes was determined in the automatic centrifuge. The blood viscosity was measured over a range of haematocrits, obtained by diluting the whole blood in its own plasma. The viscosity of the plasma was also obtained, and the data fitted a linear plot of log η_b with haematocrit. The value of η_r at 1% haematocrit was calculated from the regression line, and in turn the respective value of KT_c from equation (3). The values of KT_c for normal bloods R and S, and hardened cells of R and S, plotted against erythrocyte flexibility, in terms of the initial percentage change of haematocrit per minute (i.e. the initial slope of the curves in Fig. 2.1a), are shown in Fig. 2.3. If rigid discs or O-rings, of the same overall dimensions, are packed at random in a cylinder the respective maximum haematocrits are 54.5% $\pm$2% and 31% $\pm$2% (Sirs, 1969). The maximum

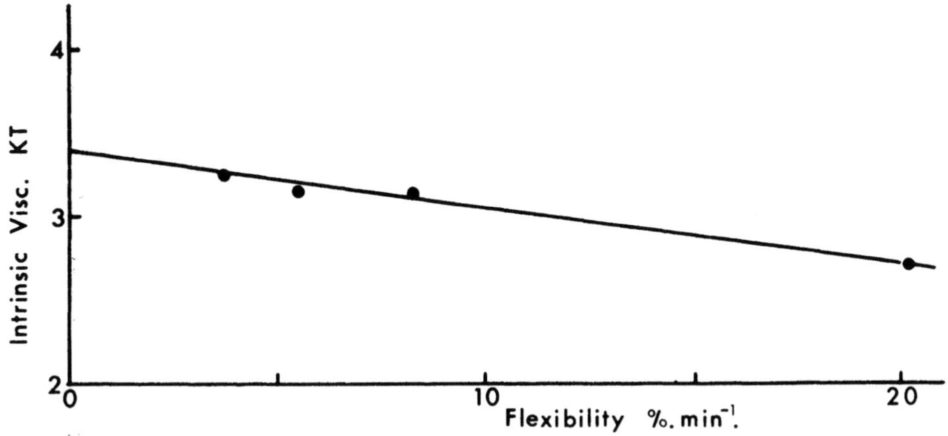

Fig.2.3. The effect of erythrocyte flexibility (packing rate, %.min^{-1}) on KT_c (intrinsic viscosity in cP or mPa.s)

haematocrit for *rigid* red blood cells can be estimated from such experiments as 35%-42%. The stack of discs or O-rings remain locked together and immobile when spun to simulate centrifugation. This indicates that with rigid cells the packing rate in a centrifuge would be zero, as can be demonstrated with blood after hardening the cells by heating at 49.6°C for 50 min. (Sirs, 1968). Extrapolating the line in Fig. 2.3 to zero packing rate gives a value for K of 3.4.

An interesting implication of this result is that the shape and volume of the cells must be the same for subjects R and S, and is not altered by heating. The presence of even a small stagnant layer of water round inflexible cells would significantly alter their effective volume, *C*, and these points would have been displaced from the line. In a similar manner differences of shape, corresponding to a variation of K by ±0.1, would have been discernible. To check this conclusion further measurements of the cell volumes were made. The PCV, measured at 13,000 *g*, was the same before and after heating, and there was no evidence of any haemolysis occurring during the hardening, which suggests the mean corpuscular volume (MCV) was also unchanged. The MCV was then measured using a Coulter Counter Model S. The red blood cell count (RBC) for S was the same before and after heating, at 4.99×10^{12} cells/litre. For blood R the RBC values were 5.11×10^{12} cells/litre and 5.07×10^{12} cells/litre, before and after hardening respectively. The MCVs for the S cells were 87 fl and 88 fl, and for R unchanged at 85 fl.

2.4 Discussion

These results show a good correlation between the measurement of erythrocyte flexibility using the rate of packing at 200-600 *g* in a centrifuge, and KT_c, estimated from blood and plasma viscosity measurements. They confirm that the packing rate gives a direct measure of erythrocyte flexibility, and is a suitable index of their flow properties. From an experimental standpoint, the measurement of the packing rate is preferred to viscosity measurements, because it is simpler, uses a relatively cheap technique, and requires the minimal quantity of blood with no preparatory procedures. An additional difficulty of the viscometry method is that the value of K, the shape factor, can vary, for example in haemolytic diseases. To what extent K can change is not at present known, and it may be that this factor is small compared to variations in T_c.

An example of the variation in K, and of the application of the viscometry method to confirm the effect of plasma fibrinogen concentration on erythrocyte flexibility, is obtained by applying this analysis to the earlier data of Dupont and Sirs (1977). Measurements of blood and plasma viscosity, plasma fibrinogen concentration and erythrocyte flexibility, using the centrifuge technique, were obtained on a patient during and following surgery for removal of the gall bladder. Following subsequent complications the patient's fibrinogen level exceeded 1000 mg/dl, and it took a month for the viscosity, flexibility and fibrinogen to return to normal and the patient to be discharged. A linear relationship was observed between erythrocyte flexibility and plasma fibrinogen concentration, as shown in Fig. 2.4(a). The variation of KT_c, calculated from the viscosity data, with packing rate is illustrated in Fig. 2.4(b). Again a linear and good correlation between these two factors occurs. From the intercept with the y-axis, the value of K was estimated to be 2.73. Unfortunately no other haematological data, such as the MCV, were obtained on this patient to indicate if the erythrocytes were in any way abnormal.

With the value of K we can calculate T_c. When the theoretical interpretation is sufficiently advanced it may be possible to estimate from T_c the relative contribution of the cell membrane and viscous properties of the internal haemoglobin to the erythrocyte deformability. A first approximation to this problem is to use Taylor's equation (2), as applied to immmiscible fluid droplets.

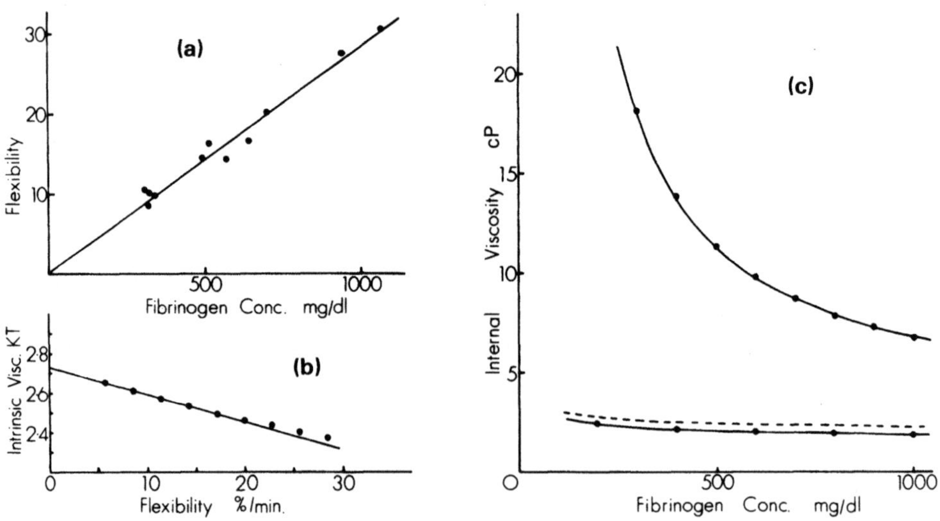

Fig.2.4. (a) The effect of plasma fibrinogen concentration on erythrocyte flexibility; (b) the variation of KT with erythrocyte flexibility; (c) calculated values of η_i with fibrinogen concentration: *solid circles* — using K equal to 2.73; *open circles* — K = 3.4. The *broken curve* is the values of η_i obtained using equation (1). All data were obtained on blood samples from a patient during and after surgery.
1 centipoise (cP) = 1 mPa.s

The calculated values of η_i for the normal bloods of R and S are 3.00 cP and 9.65 cP respectively. After hardening by heating, the respective values for R and S were 10.7 and 18.7 cP. A similar analysis of the values of η_i, with varying plasma fibrinogen concentration, is shown in Fig. 2.4(c). The lowest values of η_i, obtained with subject S, should be comparable with or greater than that of the viscosity of the intracellular haemoglobin. From the data of Cokelet and Meiselman (1968) and Schmidt-Nielsen and Taylor (1968) the viscosity of a 34% haemoglobin solution at 37°C is 8.2 ± 0.5 cP. This may be slightly high due to the presence of a small amount of other solid material in their samples. It is higher than η_i for subject S of 3.0 cP, though of comparable magnitude. Dintenfass (1971) has pointed out that within the red blood cell, haemoglobin is not in true solution but in the form of microcrystalline, liquid crystalline or globular second phase in which its viscosity may be as low as 2 cP. Another factor which would lower the haemoglobin viscosity is its concentration. If the true MCV is 100.4 fl (Sirs, 1968), and not 87 fl as accepted, the haemoglobin concentration would be reduced to 29.5%, and its viscosity to below 4.7 cP. The differences are then within the theoretical approximations and experimental errors inherent in the results. An experimental error of 3% in both the blood and plasma viscosity measurements could alter the calculated value of η_i by approximately 40%. Similarly an error of 10% in the estimate of C, in equation (3), could alter η_i by 40%, but this factor is to some extent compensated by the method of estimating K. With subject R, representative of the mean of a healthy population, and within a plasma fibrinogen concentration range of 250-400 mg/dl, the internal viscosity η_i is significantly above the haemoglobin viscosity. This suggests that the membrane normally makes a significant contribution to erythrocyte flexibility. It should perhaps be emphasised that Taylor's theory does not allow for a change of shape of the droplet, and only considers the flow currents set up inside the drop. To what extent this concept is applicable to a non-spherical cell surrounded by a membrane is not known. There is as yet no quantitative physical explanation of how the deformability of an erythrocyte lowers blood viscosity at high shear rates.

2.5 References

Brinkman R C (1952) The viscosity of concentrated suspensions and solutions. J Chem Phys *20:* 571-578

Chien S, Usami S, Dellenbeck R J, Gregerson M K (1967) Blood viscosity: Influence of erythrocyte deformation. Science *157:* 825-831

Chien S, Luse S A, Jan K M, Usami S, Miller L H, Fremont H (1971) Effects of macromolecules on the rheology and ultrastructure of red cell suspensions. In: Proc. Sixth Eur. Conf. Microcirculation. Basel, Karger, pp 29-34

Cokelet G R, Meiselman H J (1968) Rheological comparison of haemoglobin solutions and erythrocyte suspensions. Science *162:* 275-277

Dintenfass L (1968a) Internal viscosity of the red cell and a blood viscosity equation. Nature *219:* 956-958

Dintenfass L (1968b) Viscosity of blood at high haematocrits measured in microcapillary (parallel-plate) viscometers of r = 3-30 microns. In: Copley A L (ed) Hemorheology. Pergamon Press, Oxford, London, Edinburgh, New York, Toronto, Sydney, Paris, Braunschweig, pp 197-209

Dintenfass L (1971) Blood Microrheology. Butterworths, London, p 86

Dintenfass L (1975) Internal viscosity of the red cell, problems associated with definition of plasma viscosity and effective volume of red cells in the blood viscosity equation. Biorheology *12:* 253-256

Dupont P A, Sirs J A (1977) The relationship of plasma fibrinogen, erythrocyte flexibility and blood viscosity. Thromb Haemostas *28:* 660-667

Einstein A (1906) Eine neue Bestimmung der Molekuldimensionen. Ann Physik Lpz *19:* 289-306

Einstein A (1911) Berichtigung zu meiner Arbeit: "Eine neue Bestimmung der Molekuldimensionen". Ann Physik Lpz *34:* 591-592

Gregerson M I, Peric B, Chien S, Sinclair D, Chang C, Taylor H (1965) Viscosity of blood at low shear rates. In: Copley A L (ed) Symposium on Biorheology. Interscience Publ, New York, London and Sydney, pp 613-628

Haynes R H (1960) Physical basis of the dependence of blood viscosity on the tube radius. Am J Physiol *198:* 1193-1200

Jeffery G B (1922) The motion of ellipsoid particles immersed in a viscous fluid. Proc R Soc, *A102:* 161-179

Kuroda K, Mishiro Y, Wada I (1958) Relation between the viscosity of erythrocyte suspension and the shape of the erythrocyte. Tokushima J Exp Med *4:* 73-82

Oldroyd J G (1953) The elastic and viscous properties of emulsions and suspensions. Proc R Soc *A218:* 122-132

Peric B (1964) Viscosity of blood at low shear rates. Israel J Exp Med *11:* 139-149

Rampling M W, Sirs J A (1972) The interactions of fibrinogen and dextrans with erythrocytes. J Physiol *223:* 199-212

Rampling M W, Sirs J A (1976) A survey of the variation of erythrocyte flexibility within a healthy population. Biorheology *13:* 101-105

Richardson E G (1950) Dynamics of Real Fluids. Arnold, London, 119

Roscoe R (1952) The viscosity of suspensions of rigid spheres. Br J Appl Phys *3:* 267-269

Rowlands S, Skibo L (1972) The morphology of red-cell aggregates. Thromb Res *1:* 47-58

Schmidt-Nielsen K, Taylor C R (1968) Red blood cells; why or why not; Science *162:* 274-275

Schmid-Schönbein H, Weiss H, Ludwig H (1973) A simple method for measuring red cell deformability in models of the microcirculation. Blut *26:* 369-379

Seaman G V F (1966) Factors affecting the flow properties of red cell suspensions. Biorheology *3:* 169

Sirs J A (1968) The measurement of the haematocrit and flexibility of erythrocytes with a centrifuge. Biorheology *5:* 1-14

Sirs J A (1969) The respiratory efficiency and flexibility of erythrocytes stored in acid-citrate-dextrose solution. J Physiol *203:* 93-109

Sirs J A (1970) Automatic recording of the rate of packing of erythrocytes in blood by a centrifuge. Phys Med Biol *15:* 9-14

Sirs J A (1975) Erythrocyte flexibility, blood fibrinogen and surgery. In: Nicolaides A N (ed) Thromboembolism. Medical and Technical Publishing Co Ltd, Lancaster, England. 59-78

Taylor G I (1932) The viscosity of a fluid containing small drops of another fluid. Proc R Soc *A138:* 41-48

Taylor M H, Chien S, Gregerson M I, Lundberg J L (1965) Comparison of viscosity of suspensions of plastic spheres and human blood cells. Nature *207:* 77-78

3 New Methods for Red Cell Deformability Measurement

H. Kiesewetter, K. Mussler, P. Teitel, U. Dauer, H. Schmid-Schönbein and R. Spohr

3.1 Introduction

The flow behaviour of blood in general and within the microcirculation in particular critically depends upon the ability of erythrocytes to undergo deformation (change in their tridimensional configuration) when submitted to stresses. In this way the red blood cells (RBCs) are able to negotiate capillaries with diameters smaller than their own diameter at rest. Although a series of different procedures to quantitate this deformability has already been described, many of them fail to fulfil (in part or completely) the requirements to simulate the boundary conditions prevailing in the microcirculation, and specifically:

1) The small shear stresses (below 10 Pa), and

2) the small geometrical dimensions which impose mechanical constraints upon the erythrocytes.

Two methods have recently been developed in our department which go a long way towards complying with these requirements. These methods aim in the first place to detect moderate abnormalities in RBC deformability, since cells with gross curtailment of this ability are sequestered within the spleen and consequently are not to be found in the active circulation.

3.2 Methods and Apparatus

3.2.1 Large-scale Screening Filtrometer

For large-scale screening and rapid analysis of abnormalities in the flow behaviour of red blood cells (RBCs), a new instrument called a Filtrometer has

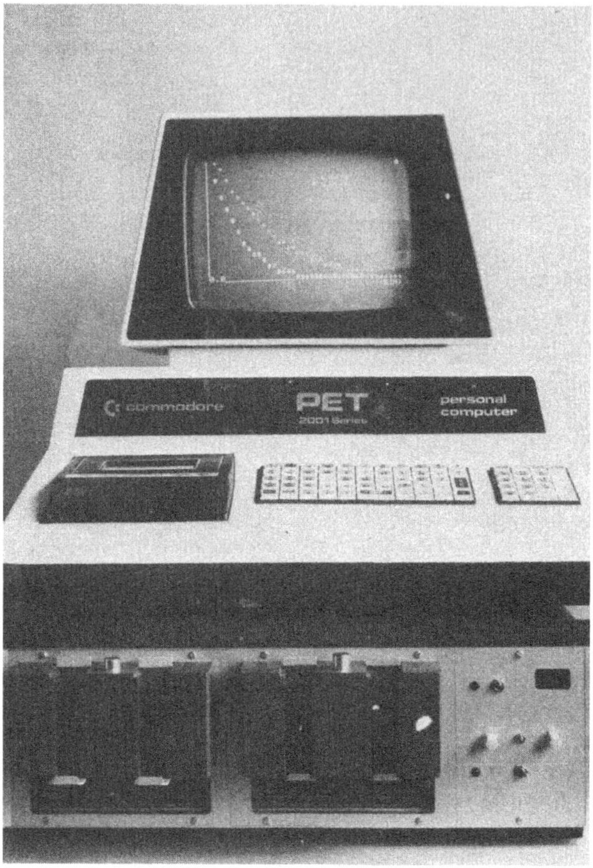

Fig.3.1. Large-scale screening Filtrometer with processing unit

been devised and constructed by Teitel and Mussler (1979) and Teitel et al., (1979) which allows complete automatation of the "polymicroviscometric" approach (Teitel, 1977). This is based on the continuous monitoring of the steadily decreasing filtration rate of concentrated RBC suspensions (volume fraction of cells = 0.80 ±0.01) through filter membranes with pores larger than the dimensions of the cells, whereby the sole driving pressure is generated by their own weight. In contradistinction to other filtrability methods, the investigation in the Filtrometer is performed at very low and non-stationary shear stresses, starting from approximately 10 Pa and then decreasing exponentially towards zero. Following the particular microrheological conditions of the measurement, it appears that the main RBC feature which is thereby quantitated is the capacity of "tank treading" of the membrane and the fluidity of the cell content.

A general view of the instrument, which is automated by a microprocessor, is shown in Fig 3.1. Four measuring channels allow the parallel investigation of four RBC samples, which takes 20-80 min; for each sample, approximately 2.0 ml of packed RBCs are necessary. Since the device has no moving parts (requiring

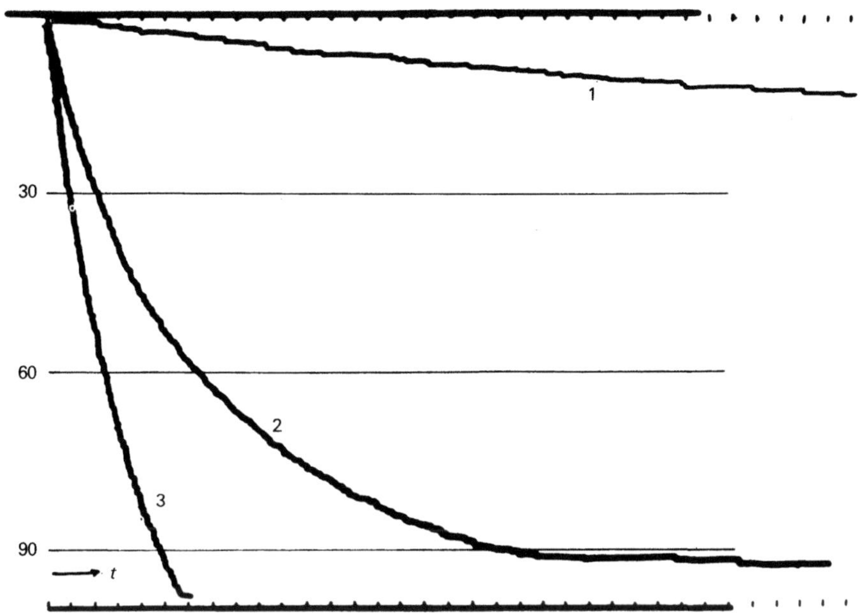

Fig.3.2. Filtration curves of different blood samples. Filtered volume (ordinate) as a function of time (abscissa)
(1) Washed red cells from a patient with chronic ulceration of the legs (2) Blood from a normal donor after 2 h at 21°C as a concentrated suspension (3) Blood from a normal donor

fine mechanical adjustments), nor a pressure-generating system, the Filtrometer is sturdy and very easy to operate. Measurement data are stored in an electronic memory unit and can be retrieved either on a potentiometric recorder [generating a "filtration curve", i.e. filtration rate ($\dot{V}$) as a function of time] or fed into a microcomputer to calculate the numerical parameters which describe (in the form of a mathematical function) the non-Newtonian flow behaviour of the sample.

Thanks to the very small shear stresses in the pores of the filter membrane and the accuracy and sensitivity of the photo-electronic measuring device, it is possible with the Filtrometer to assess very small changes in the flow behaviour of the RBCs, such as occur spontaneously when normal RBCs are kept in unphysiological conditions.

Figure 3.2 shows the following filtration curves: (1) a pathological RBC sample from a patient with chronic leg ulcers; (2) a normal RBC sample kept for 2 h at room temperature; and (3) the same sample measured fresh. The reproducibility of filtration curves obtained in a parallel run of four RBC samples from the same donor (pharmacological test) is shown in Fig. 3.3. The first parts of the curves (at the higher driving pressures) are identical; differences in flow behaviour appear only in the region of the extremely low pressures (and shear stresses). Extreme cases of impairment of the flow competence of RBCs (Teitel, 1979) are revealed by the "rheological occlusion" of the filter (arrest of the flow through the filter without actual obstruction of the individual pores), despite a finite value of shear stress (Figure 3.3).

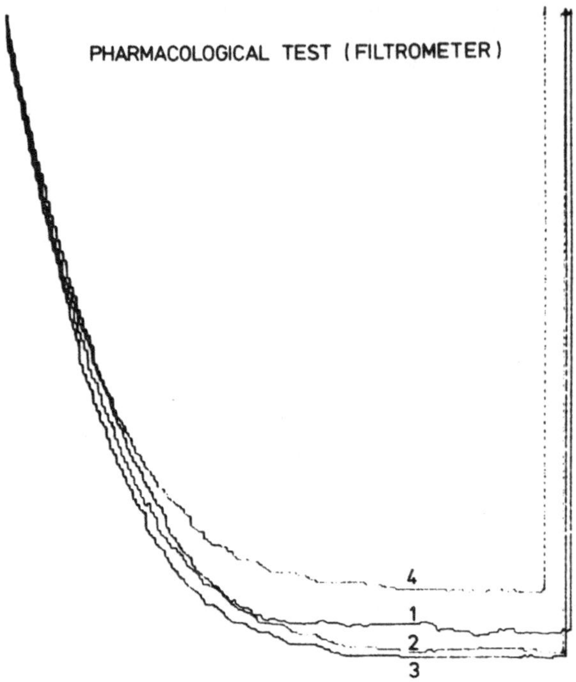

Fig.3.3. Filtration curves (pharmacological test of substance Th-926). Filtered volume (ordinate) as a function of time (abscissa) (1) Concentrated RBC (2) Concentrated RBC with Th-926 0.001 mmol/litre (3) Concentrated RBC with Th-926 0.01 mmol/litre (4) Concentrated RBC with Th-926 0.1 mmol/litre

3.2.2 Single-erythrocyte Rigidometer

A second method, with a greater power of resolution, has been developed (Mussler et al., 1978; Kiesewetter et al., 1978; Kiesewetter et al., 1979a; Kiesewetter et al., 1979b). It is based on the measurement of the passage times of individual erythrocytes through a single pore with an optico-electronic device. The main component of the "Single-erythrocyte Rigidometer" (SER) (Fig. 3.4) is a polyurethane membrane with one single measuring channel 3-5 μm in diameter and 20 μm in length, located at its geometric centre. The membrane separates the two parts of the measuring chamber, which are machined on a milling-cutter (Fig. 3.5). The upper part of the chamber is fitted with three inflow nozzles, the lower one with two outflow nozzles. The axis of the channel feeding the upper part of the chamber forms the greatest angle with its upper surface; this channel, which can be closed with a needle valve, feeds the measuring chamber with the RBC suspension, to which a driving force (pressure) is applied. This force, which can be reduced as low as 0.1 Pa, is produced by a wind-channel and measured with an accuracy of $\pm$0.05 Pa by a self-adjusting liquid manometer. The two other inflow nozzles are closed during the actual measurement and are used to fill the measuring chamber and to clean the pore in case of plugging. The two outflow nozzles of the lower part of the measuring chamber serve the same purpose; one of them is kept open during the measurement proper to allow the

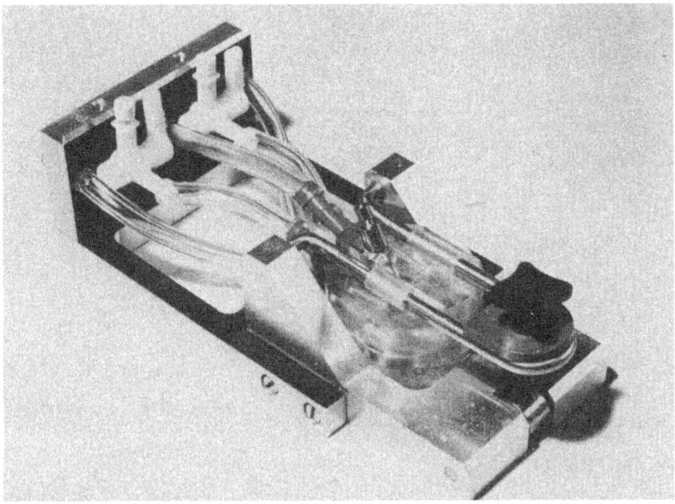

Fig. 3.4. Single-Erythrocyte Rigidometer (SER)

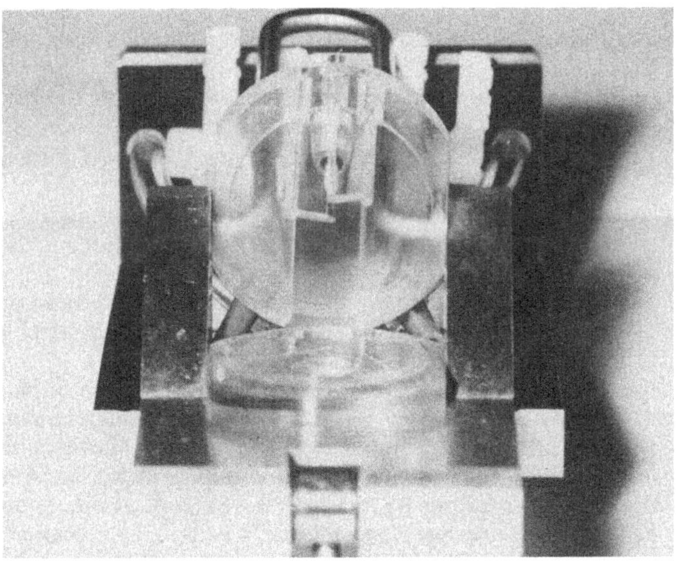

Fig.3.5. Detailed view of the SER, showing the two parts of the measuring chamber

free outflow of the cell suspension under study. All inflow and outflow nozzles are located on the same reference plane to exclude the influence of hydrostatic pressure.

The measuring chamber is mounted on the stage of a MPV-2 Leitz microscope. The light beam utilised for the measurement is focussed on the plane of the measuring pore so that it roughly corresponds to the cross-section plane of the pore. If an erythrocyte passes through the pore, the intensity of the light beam

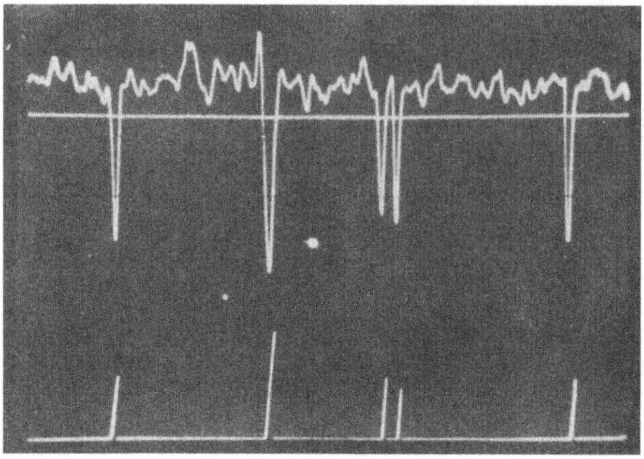

Fig.3.6. Signals obtained with the SER

will decrease. The ensuing light signal (Fig. 3.6) is fed to a photo-multiplier, amplified, and after modulation fed into one of the 256 channels of a multichannel analyser. The signals resulting from the successive passage of individual RBCs are continuously monitored on the screen of a cathode ray oscillograph.

The maximum duration of the recorded passage time is 800 ms; if the passage takes longer, a cleaning process is automatically set off. A high-frequency alternating electrical field is applied to the pore, destroying the slowly moving or fully arrested cell.

The RBCs to be investigated are first washed free of plasma and then suspended (haematrocrit 1%) in phosphate-buffered (pH 7.4) isotonic saline containing 0.5% (w/v) human albumin.

The passage of 1000 erythrocytes is measured and the histogram data stored in the multichannel analyser and then transferred to a memory and analyser unit for statistical processing.

Figure 3.7 shows the typical histogram of the frequency distribution of passage times of erythrocytes from a normal blood donor (dashed curve). The maximum (peak) of this curve is at 5 ms; the average passage time is 6.7 ms. The same figure also shows the histogram for erythrocytes from a patient with an arterial occlusive disorder who had a smoking history of 30 years duration. Here the maximum (peak) is at 15 ms; the average passage time is 24.6 ms. One can clearly see the great number of red cells with passage times greater than 30 ms. A second maximum (peak) of this histogram can be observed at 50 ms. Comparison of the two histograms in Fig. 3.7, both of which are skewed to the right, reveals that a pathological RBC sample has a prolonged average passage time and also a smaller peak of the most frequent passage time (peak passage time). Also the statistical analysis of the whole frequency distribution curve appears to be of importance. Utilising the logarithm of passage time converts the skewed frequency distribution curve into a typical symmetrical Gaussian frequency distribution curve. Figure 3.8 shows that for a normal RBC sample, such a typical Gaussian distribution curve can easily be defined, together with the corresponding parameters describing its variance. The frequency distribution

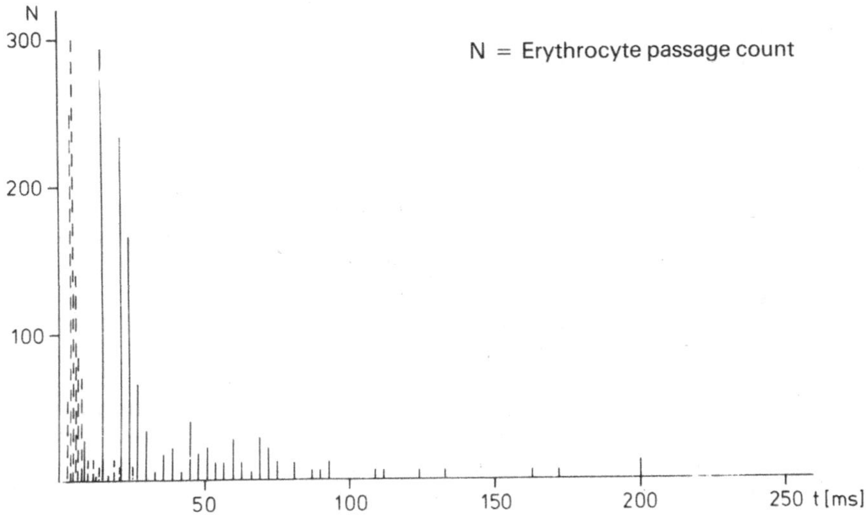

Fig.3.7. Typical histograms of the frequency distribution curve of passage times of normal RBCs and RBCs from a patient with occlusive arterial disease. See text for description

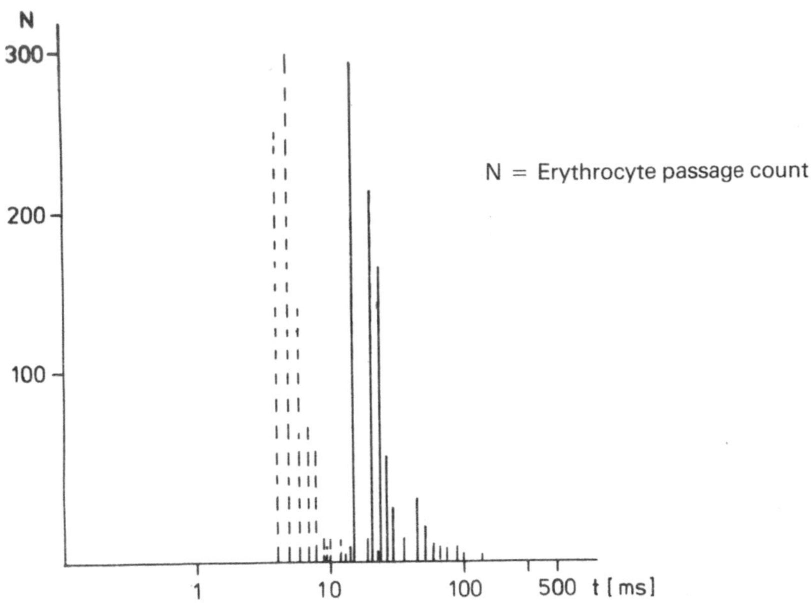

Fig.3.8. Typical histograms of the frequency distribution curve of passage times (Fig.3.7) plotted as logarithms

curves of passage times obtained with pathological or artificially altered RBCs, can now be compared with this normal distribution, and any departure from it can be construed as indicative of an abnormality.

3.3 Conclusions

We have been able to develop two new quantitative methods to investigate the ability of RBCs to deform under stress. Utilisation of the Filtrometer should be considered as a first screening approach. Those cases in which the existence of abnormal flow behaviour of the red cells is shown must then be investigated with the Single-erythrocyte Rigidometer. This enables conclusions to be drawn on the deformability of each individual erythrocyte and provides the means to assess, in various pathological conditions, whether moderate or gross impairment in the flow competence of RBCs seen in the screening test is affecting all the cells of the sample to the same extent at the same time, or only some of them. Both methods fulfil the requirements of simulating the critical geometric and hydrodynamic boundary conditions prevailing in the living microcirculation.

3.4 References

Kiesewetter H, Dauer U, Mussler K, Schmid-Schönbein H, Teitel P (1978) Method to determine the deformability (pore passage time) of single erythrocyte. Abstracts of XVII Congress of the International Society of Haematology, Paris

Kiesewetter H, Mussler K, Dauer U, Schmid-Schönbein H, Spohr R (1979a) Rigidity measurement in individual red cells. Abstracts of European Symposium "Haemorheology and Diseases", Nancy

Kiesewetter H, Schneider R, Schmid-Schönbein H (1979b) Neue methodische Konzepte für die Forschung und Praxis der Hämorheologie. 1. Gemeinsame Jahrestagung der Angiologischer Gesellschaften; Deutschlands, Österreichs und der Schweiz. Wien (1977) In: Gefässwand, Rezidivprophylaxe, Raynaud-Syndrom. Verlag Gerhard Witzstrock: Baden-Baden, Köln, New York.

Mussler K, Kiesewetter H, Schmid-Schönbein H (1978) Method to determine the deformability (pore passage time) of single erythrocytes. Abstracts of 3rd World Congress on Biorheology, La Jolla

Teitel P (1977) Basic principles of the "filterability test" (FT) and analysis of erythrocyte flow behaviour. Blood Cells, *3,* 55-70.

Teitel P (1979) Red blood cell rheology. Problems of standardising terminology and methods (plenary lecture). Abstracts of European Symposium "Haemorheology and Diseases", Nancy.

Teitel P, Mussler K (1979) A new automatic instrument to investigate the rheological behaviour of red blood cells. Thromb Haemostas *42*: 105

Teitel P, Mussler K, Hoynck V, Neumann F J, Rieger H (1979) Automatized rheological measurement for the routine haematological laboratory. II: A new instrument for screening of red cell rigidity. Microvasc Res *17*: 174

4 Measurement and Influence of White Cell Deformability

U. Bagge, P.-I. Brånemark and R. Skalak

4.1 History

Blood may be regarded as a mobile tissue. Of its cellular components the white blood cells (WBCs) and the red blood cells (RBCs) have diameters which are larger than most of the blood capillaries. Cell deformation is therefore required to let the blood corpuscles pass through those parts of the circulatory system in which blood and tissue communicate.

Quantitative analyses of the rheological properties of the individual blood cells started only about 15 years ago with micropipette studies on RBCs by Rand and Burton (1964a,b). Because of much additional effort in this area of research we have today reached a fairly good understanding of the rheological properties of the RBC and the linkage of its mechanics to the rheological behaviour of whole blood (cf. Chien, 1975).

In contrast, very little attention has been paid to the rheological properties of the WBCs. Practically all studies have been devoted to the behaviour of the WBCs in inflammatory states. Some information pointing to important differences in the ability of normal WBCs and RBCs to pass through the microcirculation was obtained, however, even in early studies of the microcirculation. Krogh (1922) noted that the resistance offered to the passage of WBCs through muscle capillaries appeared to be much greater than that for the RBCs. Sandison (1932) and Clark et al. (1936) observed, in the rabbit ear chamber, that the WBCs seemed to be stiffer than the RBCs and that the WBCs could temporarily block capillary blood flow, so-called leucocyte plugging. Nicoll and Webb (1946), reintroducing the bat wing for intravital microscopy of the microcirculation, found leucocyte plugging to be an important cause of intermittent flow in the microcirculation (see also Johnson and Wayland, 1967). More recently fairly detailed descriptions of leucocyte plugging have been presented by Palmer (1959) and Robb and Jabs (1968) from studies in the rat pancreas and the bat wing, respectively.

With the introduction of the skin tube chamber it became possible to make high-resolution observations of the microcirculation in man (Brånemark, 1971). Experiments with this technique showed that leucocyte plugging also occurs frequently in human tissue (Adell et al., 1970). This information led to a series of experiments in our laboratories, designed to evaluate more precisely the rheological properties of the WBCs and their influence on the nutritive blood flow in normal as well as pathological situations. Such information was further regarded as valuable for the understanding of the intravascular behaviour of other circulating cells of a similar topography, such as tumour cells.

4.2 Normal Rheology of White Blood Cells

The circulating WBCs are spherical in shape with diameters of 7-9 μm (granulocytes) and 6 μm (lymphocytes) (Bagge and Brånemark, 1977; Schmid-Schönbein et al., 1979). When the cells are deformed to pass through the capillaries they attain the form of cylinders capped with hemispherical ends. The deformations of WBCs take place with the volume preserved (Bagge et al., 1977b), which is intriguing in respect to the mechanical properties of the cell membrane (see below).

Typically the deformation of a WBC, for instance at the entrance of a narrow capillary, proceeds in a biphasic manner. There is an initial rapid but only partial deformation of the cell followed by a much slower final adaptation of the cell to the capillary lumen. Under normal pressure conditions and with vessel diameters of about 5 μm this visco-elastic type of deformation requires 0.5-3 s (Bagge and Brånemark, 1977). Thus, the time required for deformation of the WBCs is at least one or two orders of magnitude longer than that which RBCs need for large shape changes (Skalak and Brånemark, 1969).

Once deformed to fit into a capillary lumen the WBCs usually move on quickly, although at a somewhat lower speed than the RBCs. As a result the RBCs tend to stack up behind the WBCs, forming what is often called "train-flow". This is a phenomenon most often encountered in the long slender capillaries of skeletal muscle (Amundson et al., 1980). The slower flow of the WBCs is readily understood on the basis of the mechanics of particle flow in a tube, since a larger particle always travels more slowly (Tözeren and Skalak, 1978).

Since capillary diameters are not perfectly uniform but are locally diminished in many places by endothelial cell nuclei bulging into the lumen, WBCs may become retarded or arrested several times during their passage through a capillary. It is important to add that there is no evidence that leucocyte plugging is related to adhesion of the WBCs to the capillary walls (Bagge and Brånemark, 1977; Nicoll and Webb, 1946; Palmer, 1959). The forces involved in WBC adhesion to vessel wall, which is a venular phenomenon primarily, are high (Schmid-Schönbein et al., 1975; see also section 4.3.4) and should most probably cause a much more retarded, non-smooth passage of the WBCs through the capillaries than is actually observed *in vivo*.

After the passage through a capillary the deformed WBCs do not immediately attain a spherical shape again, unless the total duration of the deformation has been very short (Bagge and Brånemark, 1977). Many WBCs will therefore commence their venular circulation in more or less extended forms. In some cases a full recovery of the spherical shape may require half a minute or more. This mode of recovery points to a strong viscous component in the cells paralleled by a weak elastic, restoring element (for details see section 4.3.2).

In the small venules the WBCs tend to roll like wheels along the endothelium. Normally, at least, this margination of the WBCs seems to be caused not by attraction by the endothelium but by hydrodynamic interaction with the RBCs. *In vivo* observations show that the margination is initiated as soon as a vessel has a diameter which is about 50% larger than that of the WBCs. The faster-moving RBCs will then be able to pass the WBCs, which are pushed laterally in the process (Schmid-Schönbein et al., 1979). Most of the marginated WBCs will then stay marginated, at least in the small venules, although a small number of the cells are diverted into the faster axial stream at the passage over endothelial ridges (see, e.g., Vejlens, 1938) or through junctions (Bagge and Karlsson, 1979).

As was mentioned before, an intriguing aspect of WBC deformation is the relationship between the surface area and cell volume. By way of background it is important to point out that it is known for the RBC that its high surface-area-to-volume ratio, due to the disc shape, is one of the keys to the amazing deformability of the cell. Thus, the large surface-area-to-volume ratio allows the RBC to deform extensively without increasing its surface area. Deformations accompanied by an increase of the surface area require much more energy and will also, even at an increase of only about 4%, lead to membrane rupture and haemolysis (Chien et al., 1971; Evans et al., 1976).

Since the WBCs start from a spherical form, any change of shape at constant volume will require an increase of the surface area. In the deformations observed, sometimes as high as 60% increase of surface area is necessary for the WBCs to pass through the microvasculature (Bagge and Brånemark, 1977). Nevertheless, membrane rupture is not seen. The solution to this apparent paradox is to be found in the fact that the undeformed WBC has a superfluous membrane area in the form of wrinkles or projections of the membrane, which are readily seen by electron microscopy. By combining vital microscopy and electron microscopy it has been possible to identify and transfer individual deformed (cylindrical) WBCs from the mesenteric microcirculation in rabbits for ultrastructural analysis (Bagge et al., 1977a). This study shows that the membrane of deformed WBCs is stretched smooth, i.e. the superfluous membrane area is utilised during deformation. Further evidence of this mechanism is provided by the smooth appearance in SEM pictures of WBCs which have been spread out, like pancakes, on a flat surface (Bessis, 1973). In this context it deserves to be mentioned that it is this type of cell flattening, which occurs e.g. in blood smears, which is the reason for the erroneously large diameters of WBCs given in most haematology textbooks.

4.3 *In Vitro* Measurements

4.3.1 Viscometry and Sieving

The rheological properties of WBCs can be inferred to some extent from macroscopic experiments on suspensions containing WBCs in viscometers and graded filters.

Dintenfass (1971) shows some data on the bulk viscosity of packed (98% cytocrit) RBCs and WBCs in varying proportions. With 3.5% WBCs a viscosity 5-10 times greater than that of normal proportions of RBCs and WBCs is found. Increasing the WBC concentration to 80% again increases the viscosity by another factor of 5 to 10, depending on the shear rate. These experiments imply that the WBCs are considerably stiffer than the RBCs.

Another type of experiment which gives an indication of WBC properties is filtration, through a sieve of uniform pore size, usually about 6-8 μm, which is about equal to the diameter of WBCs. Tanner and Scott (1976) have used polycarbonate filters to demonstrate that filterability decreases if WBCs are hardened with formalin or if the WBC concentration is increased.

As a whole, viscometry and sieving give only a gross indication of the end result and do not allow interpretation in terms of detailed properties of the individual WBC.

4.3.2 Capillary Stenosis Experiments

Experiments on WBCs flowing through tapered glass capillaries are interesting and revealing in that they resemble constrictions which may occur in the microcirculation *in vivo* and they provide a variety of information about the WBC properties (Bagge et al., 1977b).

On passing WBCs and RBCs through the same stenosis with a minimum diameter of about 5 μm it is immediately obvious that the WBCs are much stiffer than the RBCs. The latter will pass through at very small pressure gradients, usually of a small fraction of a centimetre of H_20. But WBCs, primarily granulocytes, may take 10 cm H_20 or more to pass at the same velocity. In a typical experiment (Fig. 4.1) at a pressure of a few centimetres of H_20, there is an initial rapid compression followed by a more gradual deformation which may take many seconds before the cell is maximally deformed in the stenosis. This pattern of deformation agrees well with the observations made *in vivo,* and indicates an initial elastic behaviour followed by a viscoelastic gradual deformation.

When a deformed WBC is withdrawn from the stenosis it does not recover its spherical shape immediately. There is a small initial recovery followed by a gradual return to a spherical shape. The recovery time is dependent on the length of time and degree of initial deformation. As *in vivo,* full recovery of the WBC *in vitro* may take half a minute or more (Bagge et al., 1977b).

These qualitative and quantitative observations suggest that the appropriate model for the WBCs' mechanical properties is a viscoelastic solid, shown in Fig. 4.2. The spring K_2, in series with the dashpot, μ_3, are adequate to explain the behaviour during the compression phase. The spring K_2 allows the initial

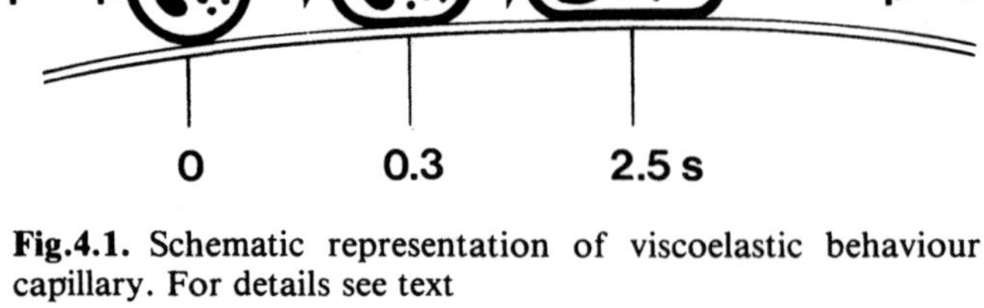

Fig.4.1. Schematic representation of viscoelastic behaviour of WBCs in a stenosed capillary. For details see text

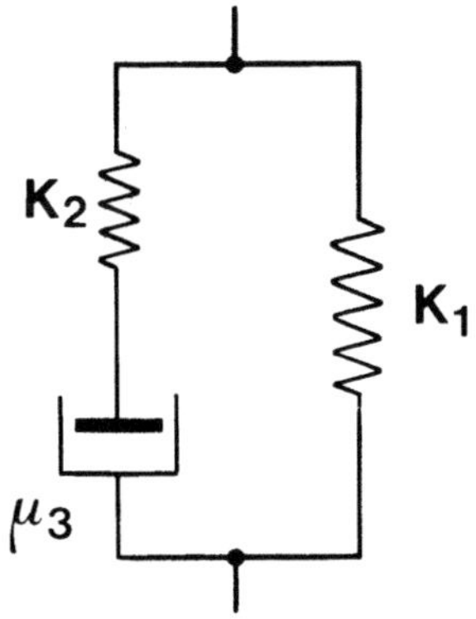

Fig.4.2. Model for the rheological properties of WBCs. For details see text

deflection and the dashpot provides the continued deformation under constant pressure. The spring K_1, parallel to the Maxwell fluid element, is required to produce the full slow restoration of the spherical shape.

The Maxwell fluid element can quite definitely be associated with the interior of the cell. The stiffness represented by K_2 is much too large to be associated with the membrane of the cell. In its normal convoluted shape, the membrane of the WBC could sustain only very small stresses. This is in contrast to RBCs in which the apparent elasticity of the cell is entirely associated with the membrane (Skalak et al., 1973).

The viscosity of the WBC, represented by the dashpot, μ_3, is also associated with the cell interior. The membrane is too thin and flexible to contribute appreciably.

The elastic element K_1 may be associated with the membrane of the cell or with the cytoplasm. The small membrane stresses acting over a sufficient period of time could restore the normal spherical shape of the WBC. It is also possible that interior structures of the WBCs, such as microfilaments which may consist of contractile proteins (cf. Murphy, 1976), somehow act to restore a spherical shape. However, since the granules and other elements participate in continuous, vigorous Brownian motions, it seems that the contents should be more nearly fluid than solid. This implies that there is no permanent cell-wide structure that could "remember" the original spherical shape.

4.3.3 Pipette Aspiration Experiments

Observations of the deformation of WBCs during aspiration into a small-diameter glass pipette have been made by a number of authors (see, e.g., Lichtman and Weed, 1972; Lichtman, 1973; Sung et al., 1979). The pipette's tip diameter is generally between 1.2 μm and 5.0 μm. The measurement made is the length of the portion of the WBC, ΔL, drawn into the pipette at a given pressure (Fig. 4.3).

In the recent experiments by Sung et al. (1979) the time history of the deformation shows a rapid increase when the load is first applied. Further, there is a gradual increase in the deformation with time. This behaviour is similar to that observed in the stenosed capillaries.

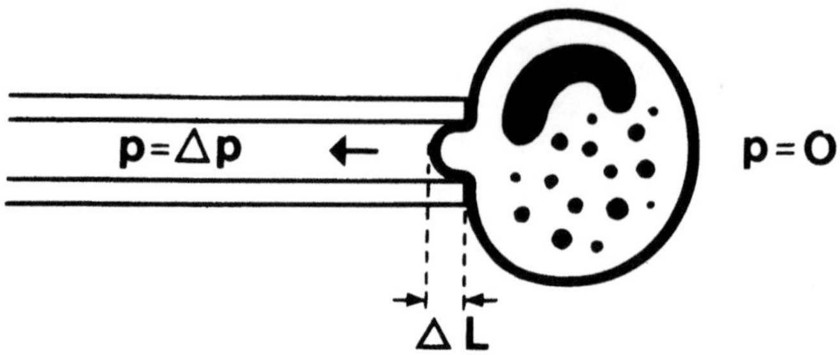

Fig.4.3. Measurement of WBC deformability by aspiration into a pipette orifice. For details see text

The pipette experiments by Sung et al. (1979) were modelled by the same idealization as shown in Fig. 4.2, but considering the parameters K_1, K_2 and μ_3 to be distributed throughout the cell. The initial shape of the cell is taken to be a sphere and the linearised theory of visco-elasticity is used to get an analytical solution in series form. The results for neutrophil granulocytes are well approximated by constants K_1 200 dyn/cm^2, K_2 2000 dyn/cm^2 and μ_3 190 dyn. s/cm^2. The time constant of the system is then about 1 s. The coefficients for the lymphocytes are similar.

A particular significance of the pipette experiment is that they are useful to check that linear theory of viscoelasticity is applicable to WBC behaviour. This is done by applying two steps of the applied pressure a short time interval apart. According to the linear theory, the resulting deformation history should be the same as for the two steps each applied separately and then summed. The experimental data by Sung et al. (1979) support this quite well.

However, as the deflection grows to about 0.1 times the cell diameter in the pipette experiments, various non-linearities come into play. Firstly, the linear theory no longer applies because of the large strains and the point of application of the pressure moves appreciably. Secondly, the boundary condition at the edge of the pipette may be questionable. The theory assumes that only normal stresses are applied. Thirdly, as the convolutions of the membrane are smoothed out, the stresses in the membrane may have substantial influence, which is neglected in the elementary theory.

The experimental data in both pipette and stenosis experiments show a decreasing slope of deflection with increasing time and this may be due to a non-linear role of the membrane.

Besides studies on normal WBCs, Sung et al. (1979) also performed measurements on WBCs at different osmolarities of the suspending medium. These experiments show that if the osmolarity of the suspending fluid of a suspension of WBCs is increased, the volume of each cell is reduced and the viscosity, μ_3, and the two elastic moduli, K_1 and K_2, increase. On the other hand, if the osmolarity is decreased, the cells swell until the usual foldings of the membrane are smoothed out. In this condition, the cells behave in a more nearly elastic manner and the viscosity, μ_3, is reduced. These osmolarity effects on the material constants lend support to the theoretical model for the normal WBC and to the identification of the components responsible for its observed rheological behaviour.

4.3.4 Adhesion Forces

In the pipette and capillary stenosis experiments described above *in vitro* and in any observations *in vivo,* it is important to establish whether or not the WBCs adhere to the boundary walls. In the small venules, at least, it is clear from even casual observation of almost any microcirculation that there is a certain adhesion of the WBCs to the vessel walls. Thus, the WBCs may be at rest while RBCs flow past them or the WBCs may roll along on the wall, but move more slowly than the RBCs of the mainstream flow.

The considerations of whether or not, and if so, how, a WBC can adhere to a vessel wall or to an artificial surface is a matter of the biochemistry and molecular level forces involved between the outermost layers of the cell membrane and of the materials that comprise the surface layers of the wall. The nature of these bonds, the factors that influence them, and quantitative information on the strength of these bondings is highly desirable for a fundamental understanding of the adhesion process, but is largely unknown for any particular wall at present.

The process of the adhesion may be regarded from a macroscopic scale directly instead of considering the molecular details. In this view one would consider the energy of bonding of the WBC surface to the vessel wall or other solid boundary, ignoring the detailed biochemistry. The adhesion is characterised by a single adhesive or fracture energy, γ. This is the work that must be done to fracture a unit area of bonded surface (ergs/cm^2 = dyn/cm) and has the same units as surface tension. Such values are unknown at present, but it is possible to estimate the range in which the typical values must lie.

At the next level of abstraction, one considers the mechanics of a complete cell and the fluid mechanics of the flow. This allows estimates of the forces exerted on a cell by the blood flowing past it. Such estimates have been developed by Schmid-Schönbein et al. (1975) by a combination of large-scale experiments and *in vivo* measurements. Model experiments were made in which the force on a solid sphere (6.3 mm in diameter) was measured, using a suspension of flexible particles to simulate blood flowing past the sphere. From these large-scale measurements and considerations of dynamic similarity, it is possible to predict the force exerted on a WBC adhering to a vessel wall in terms of the haematocrit and velocity of the blood flow in the vessel. For the venules of the rabbit omentum it was estimated that a WBC sticking to the endothelial wall was subjected to a force of 4×10^{-5} dyn to 234×10^{-5} dyn. The exact value depended on the size of the vessel, velocity of blood flow and haematocrit as well as the size of the WBC. This force was further interpreted in terms of an average shear stress between the WBC and the vessel wall. The stress (τ_w) was found to range between 50 dyn/cm^2 and 1060 dyn/cm^2. This is not a direct estimate of the fracture energy, γ, but it does give the order of magnitude of the average stresses involved at the contact. These stresses are comparatively large and in the range that could cause endothelial cell damage if maintained for a period of the order of 1 h (Fry, 1969).

It should be noted, however, that in the glass capillary experiments discussed previously there is no evidence of adhesion being involved. The absence of any adhesion may be due to the fact that coagulation of the blood is prevented by EDTA, which is known to reduce the adhesion of WBCs to glass (see, eg., Garvin, 1961).

4.4 Rheology of WBCs in Pathological Conditions

On the basis of the behaviour of WBCs under normal circumstances a number of conditions can be identified where the influence of the WBCs on the microvascular blood flow may be of greater importance than is anticipated normally. Thus, leucocyte plugging (or leucocyte arrest) should be expected to occur more frequently if the number of WBCs is increased, if the WBCs become stiffer or if the driving pressure is decreased.

The two first conditions may prevail in leukemia. Lichtman and Weed (1972) have found in micropipette experiments that even normal, immature WBCs are more rigid than mature WBCs. Of leukemic cells, Lichtman (1973) finds myeloblasts to be particularly rigid and in fact impossible to suck entirely into a 5-μm orifice at a negative pressure of 250 mm H_2O. Similar data have been obtained for the same cells with the capillary stenosis technique (Bagge, 1976). Both types of experiments clearly indicate that the great stiffness of the myeloblasts is largely due to a large stiff nucleus.

Since the pressure drops in the capillary bed have been measured to be considerably lower (cf. Zweifach, 1974) there can be little doubt that cells of such rigidity must become trapped in microvascular beds. Bone capillaries, being rigidly enclosed in the Haversian channels, may be particularly prone to plugging by rigid WBCs, which may then explain some frequent symptoms of leukemia such as bone and joint pain (see also Lichtman, 1973).

Lodgement of circulating tumour cells and the development of metastases are poorly understood phenomena. Recent experiments in our laboratories, utilising *in vivo* perfusion of tumour cells in the mesoappendix of rats, indicate, however, that lodgement of the tumour cells can depend upon the rheological properties of the cells. Circulating tumour cells are spherical, as are the WBCs, but have diameters which are quite often much larger than the WBCs. The tumour cells deform in the visco-elastic manner typical for normal WBCs but are clearly stiffer and therefore become arrested more frequently and also, importantly, for much longer periods or even permanently in the capillaries. Experiments are at present under way to determine whether the trapped cells will eventually invade the tissue and form metastases.

In shock, the capillary perfusion pressure may be significantly reduced in many tissues and organs of the body. In skeletal muscle, which is a metabolically important tissue mass comprising 40% - 50% of the body wet weight, the lowered perfusion pressure in shock is accompanied by a marked heterogeneity of the capillary perfusion and widespread cellular dysfunction (cf. Amundson et al., 1979). Recent high-resolution vital microscopic observations of the microcirculation in the tenuissimus muscle in the cat during hemorrhagic shock revealed that a large proportion of the capillaries without flow contain one or several WBCs apparently blocking the lumen (Amundson et al., 1980; Bagge et al., 1980). Deformability tests in stenosed capillaries *in vitro* on WBCs collected from the cats before and after 2 h of shock show no quantitative changes in the mechanical properties of the cells (Bagge et al., 1980). These data support the idea that leucocyte plugging is mainly a pressure-dependent phenomenon.

Beside these direct observations of skeletal muscle in shock there is some additional evidence that leucocyte plugging also takes place in other organs. Thus Wilson (1972) found, in electron micrographs of lung tissue from dogs subjected to hemorrhagic shock, significantly higher numbers of WBCs, primarily granulocytes, lodged in the pulmonary capillary network than in normal dogs.

Despite the above information, it must be emphasized that there are not yet any quantitative data available to show the relative importance of leucocyte plugging in shock. To begin with, however, the important thing is to recognize that the WBCs are potential blockers of nutritive flow in low-flow states (cf. Brånemark, 1970).

Acknowledgements. This work is partially supported by grants from the Swedish Medical Research Council; B79-12X-00663-1413 and US NIH Grant HL-16851.

4.5 References

Adell R, Skalak R, Brånemark P-I, (1970) A preliminary study of rheology of granulocytes. Blut *21:* 91-105

Amundson B, Jennische E, Haljamäe H (1980) Correlative analysis of microcirculatory and cellular metabolic events in skeletal muscle during hemorrhagic shock. Acta Physiol Scand *108:* 147-158

Bagge U (1976) Granulocyte rheology. Blood Cells *2:* 481-490

Bagge U, Brånemark P-I (1977) White blood cell rheology. An intravital study in man. Adv Microcirc *7:* 1-17

Bagge U, Karlsson R (1980) Maintenance of white blood cell margination at the passage through small venular junctions. Personal communication

Bagge U, Johansson B R, Olofsson J (1977a) Deformation of white blood cells in capillaries. A combined intravital and electron microscopic study in the mesentery of rabbits. Adv Microcirc *7:* 18-28

Bagge U, Skalak R, Attefors R (1977b) Granulocyte rheology. Experimental studies in an in vitro micro-flow system. Adv Microcirc *7:* 29-48

Bagge U, Amundson B, Lauritzen C (1980) White blood cell deformability and plugging of skeletal muscle capillaries in hemorrhagic shock. Acta Physiol Scand *108:* 159-163

Bessis M (1973) Living blood cells and their ultrastructure. Springer-Verlag, Berlin-Heidelberg-New York

Brånemark P-I (1970) Schock. Stoffwechselveränderungen und Therapie: Pathophysiologie der Mikrozirkulation im Schock. Beziehung zu metabolischen Veränderungen. Internationales Symposium, Freiburg, 1969. F.K. Schattauer Verlag, Stuttgart, pp 33-43

Brånemark P-I (1971) Intravascular anatomy of blood cells in man. Karger, Basel-New York

Chien S (1975) Biophysical behaviour of red blood cells in suspensions. In: The red blood cell, vol II, 2nd ed. Academic Press, New York

Chien S, Luse S A, Bryant C A (1971) Hemolysis during filtration through micropores: A scanning electron microscopic and hemorheologic correlation. Microvasc Res *3:* 183-302

Clark E R, Clark E L, Rex R O (1936) Observation on polymorphonuclear leukocytes in the living animal. Am J Anat *57:* 123-173

Dintenfass L (1971) Blood microrheology-viscosity factors in blood flow, ischemia and thrombosis. Butterworths, London

Evans E A, Waugh R, Melnik L (1976) Elastic area compressibility modulus of red cell membrane. Biophys J *16:* 585-600

Fry D L (1969) Certain histological and chemical responses of the vascular interface of acutely induced mechanical stress in the aorta of the dog. Circ Res, *24:* 93-108

Garvin J E (1961) Factors affecting the adhesiveness of human leukocytes and platelets *in vitro.* J Exp Med *114:* 51-73

Johnson P C, Wayland H (1967) Regulation of blood flow in single capillaries. Am J Physiol *212:* 1405-1415

Krogh A (1922) The anatomy and physiology of capillaries. Hafner, New York, 1959 edition

Lichtman M A (1973) Rheology of leukocytes, leukocyte suspensions and blood in leukemia. J Clin Invest *52:* 350-358

Lichtman M A, Weed R I (1972) Alteration of the cell periphery during granulocyte maturation: Relationship to cell function. Blood *39:* 301-316

Murphy P (1976) The neutrophil. Plenum, New York

Nicoll P A, Webb R L (1946) Blood circulation in the subcutaneous tissue of the living bat's wing. Ann N Y Acad Sci *46:* 697-711

Palmer A A (1959) A study of blood flow in minute vessels of the pancreatic region of the rat with reference to intermittent corpuscular flow in individual capillaries. Q J Exp Phys *44:* 149-159
Rand R P, Burton A C (1964a) Mechanical properties of the red cell membrane. I. Membrane stiffness and intracellular pressure. Biophys J *4:* 115-135
Rand R P, Burton A C (1964b) Mechanical properties of the red cell membrane. II. Viscoelastic breakdown of the membrane. Biophys J *4:* 303-316
Robb H J, Jabs C (1968) Distortion and dynamics of cellular elements in the microcirculation. Angiology *19:* 602-611
Sandison J C (1932) Contraction of blood vessels and observations on the circulation in the transparent chamber in the rabbit's ear. Anat Rec *54:* 105-127
Schmid-Schönbein G W, Fung Y C, Zweifach B W (1975) Vascular endothelium-leukocyte interaction. Sticking shear force in venules. Circ Res *36:* 173-184
Schmid-Schönbein G W, Shih Y, Chien S, Usami S (1979) Morphometry of white blood cells. Abstract, 2nd World Congress for Microcirculation. Microvasc Res *17:* S43
Skalak R, Brånemark P-I (1969) Deformation of red blood cells in capillaries. Science *164:* 717-719
Skalak R, Tözeren H, Zarda R P, Chien S (1973) Strain energy function of red blood cell membranes. Biophys J *13:* 245-264
Sung K L P, Schmid-Schönbein G W, Tözeren H, Skalak R, Chien S (1979) Determination of rheological properties of white blood cells. Abstract, 2nd World Congress for Microcirculation. Microvasc Res *17:* S44
Tanner L Mc K, Scott R B (1976) A filtration model for study of leukocyte transit in the microcirculation. Am J Hematol *1:* 293-305
Tözeren H, Skalak R (1978) The steady flow of closely fitting incompressible elastic spheres in tubes. J Fluid Mech *87:* 1-16
Vejlens G (1938) The distribution of leukocytes in the vascular system. Acta Pathol Microbiol Scand, Suppl 33: 11-239
Wilson J W (1972) Leukocyte sequestration and morphologic augmentation in the pulmonary network following hemorrhagic shock and related forms of stress. Adv Microcirc *4:* 197-232
Zweifach B W (1974) Quantitative studies of microcirculatory structure and function. I. Analysis of pressure distribution in the terminal vascular bed in cat mesentery. Circ Res *34:* 843-857

5 The Measurement of Red Blood Cell Deformability

J.C. Barbenel

5.1 Introduction

At rest the normal red blood cell is a biconcave disc, but in flowing blood the cell may assume a variety of shapes, suggesting that it is highly deformable. The ability to undergo large deformations when subjected to stresses allows the red blood cells to pass through capillaries narrower than the resting erythrocyte diameter, and ensures that the blood remains a fluid at physiological haematocrits. The clinical importance of both these factors has led to a growing interest in the assessment or measurement of red-cell deformability. There are, however, a multiplicity of techniques in use, and the relationship between them and the parameters they measure is often unclear.

5.2 Measurement of Red Blood Cell Deformability

5.2.1 Anomalous Viscosity of Blood

Normal blood is non-Newtonian, the apparent viscosity being a function of shear rate (Fig. 5.1). At rates below c.$1 s^{-1}$ the red cells interact to produce rouleaux, and the presence of these structures of aggregated red cells produces a high viscosity. As the shear rate increases the structures are broken down and a suspension of individual red blood cells is produced. The change is associated with a progressive fall in apparent viscosity, and the suspension behaves as a Newtonian fluid. The apparent viscosity of the blood reaches an asymptotic value of c.5 mPa.s at high shear rates.

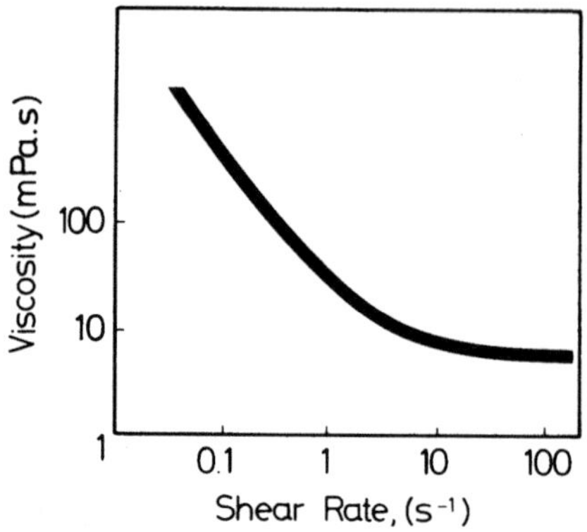

Fig.5.1. Variation of whole-blood viscosity with shear rate. The viscosity falls with increased shear rate reaching an asymptotic value

Suspensions of hardened red blood cells show a Newtonian viscosity which, at high shear rates, is much greater than that of normal cells (Chien et al., 1967). At haematocrits of 30% the viscosity of hardened cell suspensions is very similar to that of rigid spheres (Whitmore, 1968); at haematocrits of greater than 60% the suspension has the consistency of a solid. The difference in viscosity, at high shear rates, is due to the deformability of the red cells. It should be possible, in principle, to obtain information on the flexibility of red cells from the viscosity of the blood and plasma, at these high shear rates, and from the haematocrit. Unfortunately there is no generally accepted theoretical relationship connecting these variables.

The viscosity of a suspension of equal-sized, rigid, spherical particles was calculated by Einstein (1906), who showed that for small volume concentrations ($c \simeq 0.01$) the ratio, μ_r of the viscosity of the suspension to the viscosity of the suspending fluid, μ_o, was given by

$$\mu_r = 1 + 2.5c$$

Taylor (1932) extended the analysis to a dilute suspension of spherical liquid droplets having an internal viscosity, μ_i, calculating the ratio of viscosity to be

$$\mu_r = 1 + cT$$

$$\text{where } T = \frac{\mu_o + 2.5\mu_i}{\mu_o + \mu_i}$$

The analyses have been extended to include the effects of higher volume concentrations, and non-spherical particles. It was suggested by Dintenfass (1968) that the equivalent equation for blood of haematocrit H was

$$\mu_r = (1 - HT)^{-2.5}$$

The equation has been used to produce estimates of the red-cell viscosity, which appear to be realistic, although it is clear that the equation does have limited applicability (Dintenfass, 1975).

The limitations are clearly demonstrated by Sirs in Chapter 2, where the application of the formula leads to physically unrealistic red-cell internal viscosities at low haematocrits. The empirical extension of the theory proposed by Sirs, which introduces a numerical factor to account for the red-cell shape, yields realistic values of the red-cell viscosity. The difference in the numerical value of this shape factor in the two subjects reportd is, however, disturbing, as is the low value of 2.73, which is similar to the value of 5/2 expected for spheres. It must be concluded that the experimental evidence is, as yet, insufficient to assess the validity of the analysis.

5.2.2 Observation of Red Blood Cell Shape

The deformability of red cells has been assessed from the observation of the change in shape of cells subjected to external forces.

The response to shear stresses produced during flow has been examined *in vitro* in both couette and tube flow. Direct microscopic observations (Goldsmith, 1968; Schmid-Schönbein and Wells, 1969) have been made at low haematocrits. More indirect methods have utilised the diffraction pattern produced by passing a collimated monochromatic light beam, produced by a laser, through thin blood films (Bessis and Mohondas, 1975), or by the fixation of cells during flow (Sutera and Mehrjardi, 1975).

When the flow is slow, and the shear stresses small, the cells spin, but at higher shear stresses they travel with the major axis of the cell nearly aligned with the direction of flow. Visual observations show that the cell membrane appears to be in constant rotation about the cell contents (tank treading). Higher stresses result in the deformation of the cells into ellipsoidal shapes.

In order to produce high shear stresses, the viscosity of the suspending medium can be increased by the addition of dextran (Schmid-Schönbein and Wells, 1969) or Ficoll (Goldsmith, 1971). When compared with results obtained from cells suspended in plasma or saline it appears that cells suspended in high-viscosity media show greater deformation and less rotation at comparable shear rates (Chien, 1975). The importance of the viscosity of the suspending medium has been demonstrated by Fisher et al., (1978) who showed that it was more important than the shear rate in producing cell elongation. The concentration of red blood cells shows a similar effect, with an increase in cell concentration causing an increase in cell deformation (Goldsmith, 1971).

5.2.3 Centrifuge Techniques

The difference in flow behaviour shown by suspensions of normal and hardened red cells (section 2.1) is also reflected in differences in the packing behaviour of the red cells subjected to centrifugal force. Normal red cells can be packed at higher concentrations than are possible for hardened erythrocytes (Chien, 1975).

The centrifugal technique has been extensively used by Sirs (1970) to assess red-cell deformability. A constant centrifugal acceleration of 200 *g* was applied and the initial packing rate of the red-cell column, relative to the original column length, was used as an index of red-cell deformability.

The packing rate is controlled by several factors. Sirs (1968) suggested that the rate depended on the magnitude of the centrifugal acceleration, the haematocrit, and the density difference between cells and plasma, in addition to the red-cell flexibility. A later paper (Sirs, 1970) suggested that the effect of the density differences found in blood samples was negligible. There is disagreement about the influence of red-cell aggregations on the packing rate. Rampling and Sirs (1972) claimed that the presence of rouleaux had no effect at haematrocrits greater than 35%. Kiesewetter et al. in the review in this volume of methods of measuring the deformability of red cells believe that the packing velocity is mainly determined by the extent of red-cell aggregation (Chapter 1).

5.2.4 Microfiltration Methods

The majority of reported studies which have made routine clinical measurements of red-cell deformability have utilised some form of microfiltration technique. Blood (or suspensions of red cells) is filtered through membranes or microsieves which contain small pores. The technique was used by Prothero and Burton (1962), to investigate the resistance to flow in capillary channels. Blood was filtered through a Millipore filter in which the blood flow channels were irregular in both size and shape. Gregerson et al. (1967) introduced the use of polycarbonate microsieves to study red-cell deformability. There is a wide variety of microporous materials available, but Nuclepore* polycarbonate membranes, which have a relatively uniform pore diameter and density, are commonly used. Simple microfiltration systems, suitable for routine use, have been described, which incorporate such filters and standard filter holders (e.g. Reid et al., 1976).

There is no general agreement on the optimum diameter of the membrane pores. A lower limit was set by Chien et al., 1971, who showed that erythrocytes could pass undamaged through pores with a diameter of 3 μm or greater. A pore diameter greater than that of the red blood cell was suggested by Teitel (1977). The most commonly used value appears to be 5 μm, and it is clear that red blood cells deform in passing through such pores (Fig. 5.2).

The lack of general agreement on pore diameter is also apparent in other features of the method. The pressure gradient required to drive the blood through the microfilter may be obtained by the application of an elevated pressure (Lessin *et al.*, 1977) by a pressure reduction (Gregerson et al., 1967) or by the hydrostatic pressure generated by the column of blood undergoing filtration (Teitel, 1977). The variable measured as an assessment of the red blood cell deformability may be the time required to filter a known volume of blood (Reid et al., 1976), the filtration rate (Teitel, 1977) or more complex characterisations, e.g., the area beneath the applied pressure-time curve (Stuart and Kenny, Chapter 11, this volume).

The red-cell deformability determined by the filtration technique is an integrated value determined by all the cells passing through the filter. It is likely, therefore, that the results would be disproportionately influenced by the least deformable cells. Thus older red blood cells contain an increased mean haemoglobin concentration (Usami et al., 1971) and may be expected to have a higher internal viscosity and reduced deformability. The results of Kiesewetter et al. (Chapter 3) show, rather unexpectedly, that the range of passage times for individual normal red blood cells is small.

Microscopic examination of membranes used for microfiltration tests show the presence, within the micropores, of white blood cells (Fig. 5.3). These are considerably less deformable than red cells (Bagge et al., Chapter 4) and during the cell transit time block the pore to red blood cells.

*General Electric Co., Pleasantown, California, U.S.A.

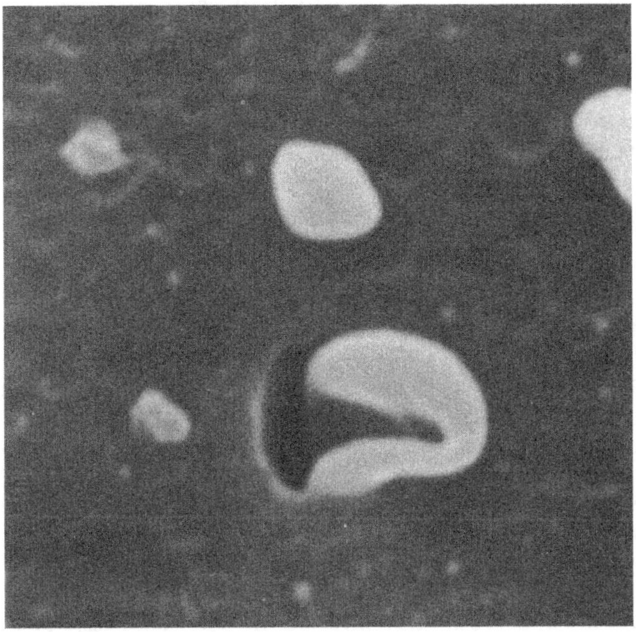

Fig.5.2. Scanning electron micrograph of Nuclepore filter (5 μm) with deformed red cell in situ ($\times$ 3000)

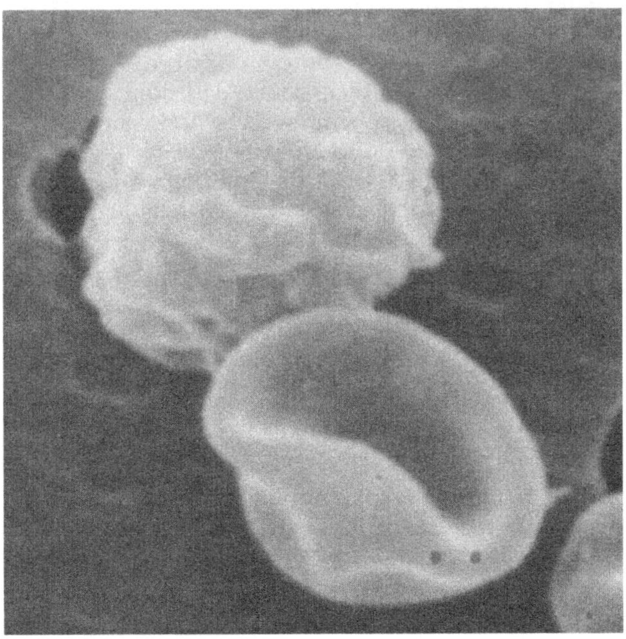

Fig.5.3. Scanning electron micrograph of 5 μm micropore blocked by a white cell ($\times$ 8000)

Microfiltration methods are relatively simple to apply and can give highly reproducible results. Drummond et al. (1980) found the coefficient of variation to be 3.7% when making seven replicated measurements, and Reid et al. (1976) found it to be 4.4% for triplicate measurements. The results of the tests must, however, be regarded as comparative assessments of red-cell deformability, and no method of analysis is available to obtain absolute values.

5.2.5 Measurement of Red Blood Cell Membrane Stiffness

The elastic modulus of the red-cell membrane has been obtained by a variety of experimental methods. Katchalsky et al. (1960) investigated the sphering of red cells in hypertonic solutions and Rand and Burton (1964) sucked portions of the cell membrane into micropipettes. The results of both experiments can be analysed if simplifying assumptions are made about the mechanics and geometry of the test (Fung and Tong, 1968). The analysis yields a parameter which is the product of the cell membrane stiffness and thickness, and the derived values are in the range of 10^{-6} to 10^{-4} N/m.

Tests in which red blood cells attached to a glass surface were elongated by shear stress produced by the flow of fluid over the cells, yielded estimates of the parameter which are very much smaller (Hoeber and Hochmuth, 1970). The difference appears to reflect the difference in test mode. The sphering and micropipette experiments produce both stretching of the cell membrane and an increase in surface area; the attached cell method produces elongations without increasing the surface area. The red-cell membrane appears to be resistant to an increase in area, and if this increase exceeds c. 3.5%, haemolysis may occur (Evans, 1973). The elastic modulus of the cell membrane is approximately four orders of magnitude greater during deformations producing area expansion compared to those at constant area (Sung and Chien, 1978).

The resistance to bending of the red-cell membrane has not been measured directly, the results of the experiments outlined above depending both on the resistance to extension and the resistance to bending. The flexural rigidity will, however, depend on the cube of the membrane thickness while the extensional stiffness will depend on the thickness only. For thin membranes this relationship makes it reasonable to assume that the membrane is highly flexible compared to its external stiffness.

Recent studies (e.g. Hochmuth et al., 1978) have shown that the red-cell membrane, like most biological materials, is not perfectly elastic, but shows significant visco-elastic behaviour.

5.3 Discussion and Conclusion

The nature of the response of red blood cells to applied forces depends on three primary factors. *The shape* of the resting cell provides a large ratio of surface area to volume. This allows the cells to deform to a wide variety of shapes without producing large local strains in the cell membrane, and without a major change in cell volume. *The mechanical properties* of the cell membrane in the deformed areas strongly influence the forces necessary to produce the deformations. The non-linear load-deformation properties of the membrane also control the shapes which may be assumed by the cell. *The cell content,* which is a solution or a suspension of haemoglobin has, in normal cells, a low but measurable viscosity (see Sirs, Chapter 2 of this volume).

Modifications in any of these primary determinants will produce changes in cell deformability. Thus the abnormal haemoglobin present in sickle-cell disease may lead to a large reduction in deformability (see Stuart and Kenny, Chapter 11). More commonly, however, associated changes occur in more than one of the factors. Alteration in the permeability of the cell membrane may lead to an increase in cell volume and to some degree of sphering. This change produces a decrease in the favourable surface-area-to-volume ratio, and increases the stiffness and resistance to bending of the cell by prestraining the non-linear cell membrane. In addition the haemoglobin concentration will fall, altering the viscosity of the cell contents.

The red blood cells are deformable in that they show differences from equivalent rigid structures, either as individual cells or in suspension. The way in which the differences appear is a function of the type of deformation produced by the test conditions. Thus the relative importance of the primary factors will depend on the test mode. The most clearly defined deformations are produced by the micropipette and cell tethering methods of investigating the cell wall. The well-defined geometry and boundary conditions make the tests amenable to mathematical analysis, which allows the quantitative comparison of cell membrane mechanics. Unfortunately the complexity of the methods makes them too difficult and time consuming for routine clinical application.

The remaining test methods, which have been applied for routine clinical studies, produce more complex and less clearly defined test conditions. It is, therefore, difficult to draw definitive conclusions but it would seem unlikely that all these methods measure similar aspects of cell deformability. Under the viscometric flow conditions used to measure the whole-blood viscosity, the deformability of the red cells appears as the continuous rotation of the cell membrane about the cell contents. It is probable that under these conditions the viscosity of the cell content is of major importance in determining the measured cell deformability. Indeed Dintenfass (1971) assumed that the cell could be considered as a fluid drop and the cell membrane properties ignored under such flow conditions. The "tank tread" motion of the red cell also strongly influences the results obtained by measuring the flow rate of blood passing through pores greater than the red-cell diameter (Kiesewetter et al., Chapter 3). The viscosity of the cell contents becomes relatively less important if the pores are smaller than the red cells and the results obtained using 5-μm-pore membranes are probably a measure of the resistance of the cell and cell membrane to deformation. It is not possible to provide a theoretical basis for the comparison of the results obtained in these different test modes.

The difficulty of comparison also applies to results obtained by what may be nominally the same technique. When using the microfiltration method it is necessary to use a single batch of filter membranes in order to ensure comparable pore size and density; the effect of the different driving pressures used by various workers is unclear, and so far uninvestigated.

In summary it is clear that there is a wide variety of methods available for the measurement of red-cell deformability, with the most commonly used and simplest being the microsieve technique. Unfortunately the ingenuity in producing new and modified test methods has not been apparent in the mathematical modelling of these tests. It is unfortunately not possible to obtain basic quantitative parameters of cell deformability from any of the test methods for routine clinical use. The tests provide valuable information on cell deformability and its alterations, but the results must be viewed as comparative assessments rather than measurements of deformability.

5.4 References

Bessis M, Mohondas M (1975) Diffractometric method for measurement of cellular deformability. Blood Cells *1:* 307-327

Chien S (1975) Biophysical behaviour of red cells in suspension. In: Surgenor D M (ed) Red blood cells II. Academic Press, New York, pp 1031-1133

Chien S, Usami S, Dallenback R J, Gregerson M I (1967) Blood viscosity. Science *157:* 825-831

Chien S, Luse S A, Bryant C A (1971) Haemolysis during filtration through micropores. A scanning electron microscopic and haemorheologic correlation. Microvasc Res *3:* 183-202

Drummond M M, Lowe G D O, Belch J J F, Barbenel J C, Forbes C D (1980) An assessment of red cell deformability using a simple filtration method. J Clin Pathol (in press)

Dintenfass L (1968) Internal viscosity of the red cell and a blood viscosity equation. Nature *219:* 956-958

Dintenfass L (1971) Blood microrheology. Butterworth, London

Dintenfass L (1975) Internal viscosity of the red cell. Biorheology *12:* 253-256

Einstein A (1906) Eine neue Bestimmung der Molekuldimensionen. Ann Physik Lpz *19:* 289

Evans E A (1973) New membrane concept applied to the analysis of fluid shear- and micropipette-deformed red blood cells. Biophys J *13:* 941-954

Fisher T M, Stöhr M, Schmid-Schönbein H (1978) Red blood cell (RBC) microrheology. In: Huang C R, Copley A L (eds) Biorheology, AlCheE Symposium, No. 182, *74:* 38-45

Fung Y C, Tong P (1968) Theory of sphering of red blood cells. Biophys J *8:* 175-198

Goldsmith H L (1968) The microrheology of red blood cell suspensions. J Gen Physiol *52:* 5s-27s

Goldsmith H L (1971) Deformation of human red cells in tube flow. Biorheology *7:* 235-242

Gregerson M I, Bryant C A, Hammerle W E, Usami S, Chien S (1967) Flow characteristics of human erythrocytes through polycarbonate sieve. Science *157:* 825-827

Hochmuth R M, Worthy P R, Smith S, Evans E A (1978) Viscosity of red cell membrane. In: Huang C R, Copley A L (eds) Biorheology, AlCHeE Symposium, No. 182, *74:* 1-3

Hoeber T W, Hochmuth R M (1970) Measurement of red blood cell modules of elasticity by *in vitro* and model cell experiments. Trans ASME, Ser D. *92:* 604-621

Katchalsky A, Kedam D, Klibanshy C, De Vreis A (1960) Rheological considerations of haemolysing red blood cells. In: Copley AL, Stainsby A (eds) Flow properties of blood and other biological systems. Pergamon Press, Oxford, pp 155-169

Lessin L S, Kurantsin-Mills J, Weems H B (1977) Deformability of normal and sickle erythrocytes in a pressure-flow filtration system. Blood Cells *3:* 241-262

Prothero J W, Burton A C (1962) The physics of blood flow in capillaries II. Biophys J *2:* 199-213

Rampling M W, Sirs J A (1972) The interaction of fibrinogen and dextrans with erythrocytes. J Physiol *223:* 199-212

Rand R P, Burton A C (1964) Mechanical properties of the red cell membrane I. Membrane stiffness and intracellular pressure. Biophys J *4:* 115-135

Reid H L, Barnes A J, Lock P J, Dormandy J A, Dormandy T L (1976) A simple method of measuring erythrocyte deformability. J Clin Pathol *29:* 855-858

Schmid-Schönbein H, Wells R (1969) Fluid drop-like transition of erythrocytes under shear. Science *165:* 288-291

Sirs J A (1968) The measurement of the haematocrit and flexibility of erythrocytes with a centrifuge. Biorheology *5:* 1-14

Sirs J A (1970) Automatic recording of the rate of packing of erythrocytes in blood by a centrifuge. Phys Med Biol *15:* 9-14

Sirs J A (1970) Automatic recording of the rate of packing of erythrocytes in blood by a centrifuge. Phys Med Biol *15:* 9-14

Sung K-L P, Chien S (1978) Viscous and elastic properties of human red cell membrane. In: Huang C R, Copley A L (eds) Biorheology, AlCHeE Symposium No. 182, *74:* 81-84

Sutera S P, Mehrjardi M H (1975) Deformation and fragmentation of human red blood cells in turbulent shear. Biophys J *15:* 1-10

Taylor G I (1932) The Viscosity of a fluid containing small drops of another fluid. Proc R Soc A *138:* 41-48

Teitel P (1977) Basic principles of the 'filtrability test' and analysis of erythrocyte flow behaviour. Blood Cells *3:* 55-70

Usami S, Chien S, Gregersen M I (1971) Viscometric behaviour of young and aged erythrocytes. In: Copley A L (ed) Haemorheology. Pergamon Press, Oxford, pp 266-270

Whitmore R L (1968) Rheology of the circulation. Pergamon Press, Oxford

Section B

General Aspects of Blood Rheology

6 Interaction of Vasomotion and Blood Rheology in Haemodynamics

H. Schmid-Schönbein

6.1 Introduction

The rheology of blood as an extremely non-Newtonian fluid has attracted the interest not only of physiologists but also of physicists, bioengineers and of course of clinicians. Unfortunately, the conclusions drawn from the *in vitro* measurements of apparent viscosity of blood in health and disease are often gross over-simplifications, ignoring the established facts about haemodynamics in general, let alone the important organ-specific differences in haemodynamics. It is therefore not surprising that the reaction of the general biomedical public is one of neglect or of almost emotional rejection — as evidenced in a recent editorial in an influential journal (Anonymous, 1977), in which the biological significance of the shear-rate dependence of apparent blood viscosity was totally denied.

At the other end of this credibility spectrum are markedly overenthusiastic *"simplificateurs terribles"* who attribute pathogenetic mechanisms (and thence disease processes) solely to an alleged "hyperviscosity" of blood, which they claim to have established by measuring apparent blood viscosity in large-bore viscometers. Unfortunately, classic viscosity concepts in the sense of continuum mechanics are not applicable to a haemodynamic analysis of the circulation, i.e.

for the flow analysis of blood perfusing a complex network of vessels of grossly different diameter under extremely variable driving pressures.

Anomalous flow behaviour is a feature primarily of mammalian, and especially human, blood. The rheology of blood containing nucleated red cells is far more "conventional" in that it primarily depends on the viscosity of plasma, the haematocrit value, and the temperature (Chien et al., 1971; Gaehtgens and Schmid-Schönbein, unpublished). In addition to these factors, in mammalian blood the driving pressure and the vessel radius have pronounced effects on the apparent viscosity. This "anomaly" of the macroscopic flow behaviour is a consequence of the flow-dependent change of the mechanical behaviour of the non-nucleated red cells. These can either passively participate in flow (by which mechanism they maximize the fluidity of blood) or they can be united into rouleaux and rouleaux networks. The latter give rise to "structural viscosity of blood", a rheological state that can potentially interfere with or abolish altogether the fluidity of blood. To make the situation even more complicated, the cells' ability to be deformed in flow depends not only on the prevailing driving pressures but also upon the micromechanical integrity of the red cells, a feature subject to changes in the metabolism of the cells themselves or of the parenchymal cells. Likewise, the tendency of aggregation can be grossly enhanced by local or general changes in the protein composition of the plasma, particularly the concentration of fibrinogen and other high-molecular-weight macromolecules. Last but not least, the locally effective haematrocrit level in the microcirculation is subject to vast variations. These are brought about by vasomotion (v.i.) and by the effect of white cells, which are much stiffer, on local blood velocity. In summary then, the composition of the blood and hence its fluidity can be vastly different in the microvessels (where all the important haemodynamic events take place) and macrovessels (from which the blood is taken for viscometry).

6.2 Present State of Blood Rheology

Most authors working in the field of blood microrheology agree to a remarkable extent about the underlying microrheological causes of the non-Newtonian flow behaviour of blood. As can be seen in a large number of recent review articles (Bicher, 1972; Braasch, 1971; Charm and Kurland, 1974: Chien, 1972; Dintenfass 1971; Larcan and Stoltz, 1970; Merrill, 1969; Schmid-Schönbein and Wells, 1971; Schmid-Schönbein, 1976; Wells, 1973) there is no more debate that red-cell deformation and red-cell aggregation are prime determinants of the shear-dependent apparent viscosity in large-bore rheometers *in vitro,* as well as phenomena associated with normal rapid and abnormally retarded flow in the microcirculation *in vivo.*

In an attempt to summarize the content of the above-mentioned work, the present author, an experimental haemorheologist with the bias of a physiologist with a medical background, has condensed our present ignorance as follows. Blood is an "anomalous fluid", the viscosity of which cannot be defined, as it varies with flow conditions. As blood is a dispersion of cells in plasma, the computed coefficient of apparent viscosity primarily depends on the true viscosity of the plasma and the effect of the dispersed cells on the flow of plasma. In addition forces associated with high-molecular-weight plasma proteins, e.g. fibrinogen and α_2- macroglobulin, aggregate the red cells into rouleaux and rouleaux networks. The transition from the viscidising effect of aggregates to the fluidising effect of cell adaptation in any given blood sample depends on the magnitude of the shearing forces. The red-cell aggregation due to rouleaux and rouleaux networks, and thence the structural viscosity of blood, is reversible by mere increase of shearing forces. Structural viscosity becomes a real property of blood whenever the flow forces are decreased or the shear resistance of the red-cell aggregates is increased. The deformability of red cells influences their effect on plasma flow, and also determines their survival (Weed, 1970). In order to appreciate the significance of *in vitro* measurements of haemorheologic variables, two additional points must be stressed. The value of plasma viscosity is a prime determinant of flow in the vast majority of the rapidly perfused small blood vessels. Although virtually irrelevant in such vessels, the haematocrit value is a prime determinant of blood viscosity in large-bore rheometers.

6.3 Factors Affecting Blood Rheology *In Vivo*

An evaluation of the haemodynamic relevance of rheological properties is far from clear. For obvious geometrical reasons one might predict a more pronounced role for rheological effects on microvascular than on macrovascular flow. However, in order to corroborate this intuitive hypothesis, we need to proceed in three stages. The first step is the identification of, and where possible the quantification of, the flow properties of blood elements, followed by an analysis of their flow behaviour when subjected to well-defined although artificial flow forces *in vitro*. Much has been done towards this end in the last decade. The second step is a description of the complex natural flow conditions in the microcirculation of various organs. This must be defined by dimensions, architectural arrangements of the different classes of vessels, and a quantification of the forces of flow. The third step involves predictions about the *in vivo* flow behaviour of the complex fluid, blood, in complex vascular beds, under varying flow conditions. This sequence of investigations will eventually lead to experiments *in vivo* and the formulation of testable hypotheses about the causes and consequences of intravascular flow or no-flow phenomena.

Quantitative information regarding the angio-architectural arrangement of different blood vessels in the organs of the body is not easily available. Anatomical information about the wall structure of different vessels in the micro- and macrocirculation is of little help in this respect. For a detailed hydrodynamic analysis, data of the following kind are required:

1) Specific measurement of diameter (minimum-maximum), taper, curvature, and branching of single vessels (arteries, arterioles, true capillaries, venules, and veins; and arterio-arterial, arterio-venous and veno-venous shunts).

2) A description of the distribution of individual vessel cross-sectional area, total cross-sectional area of any class of vessels, and effects of vasoconstriction and vasodilation on these parameters.

3) Accurate measurements of intravascular pressure and pulse pressure, pressure drop (ΔP), and pressure gradient ($\Delta P/L$, where L is vessel length), as well as transmural pressures, under normal conditions and a range of disturbed conditions.

Unfortunately, only very limited quantification of such factors is available. The few published reports only deal with limited types of vessels (e.g. small arterioles, venules and nutrient capillaries). Any textbook of anatomy will make it immediately evident that the angio-architectural patterns of such organs as skin and muscle, intestine and liver, lung, heart, kidney and brain are extremely heterogeneous because they are adapted specifically to the function of the organ. Compared to these variabilities, species differences among mammals are probably less significant.

In order to obtain at least an order of magnitude estimation of the distribution of vessel lengths and diameters, of vessel conductance, and of shear stresses in the macro- and microvessels, scattered data about these variables in the literature were combined into a hybrid model of the mammalian circulation. By computation, using the Hagen-Poiseuille Law, we arrived at the following conclusions (Schmid-Schönbein, 1976):

1) Under conditions of normal arterial pressure there is a progressive increase in shear stresses from the arteries (2.0 Pa) through the arterioles (8.0 Pa) to the nutritive capillaries (10.0 Pa), which have the highest shear stresses of all vessels.

2) Depending upon the total cross-sectional area, there is a conspicuous drop in shear stresses acting in the post-capillary venules (0.2-0.5 Pa), values increasing again as the blood flows from the small veins through the larger veins back to the heart (the lung circulation was not analysed).

3) In all blood vessels the shear stresses are so high that the blood can be assumed to exhibit very low viscosity (1.2-3.0 mPa.s) or conversely, high fluidity.

4) Any drop in the locally effective shear stresses is most likely to affect firstly, and to the greatest degree, the post-capillary venules. It can be predicted that here the blood might first lose its fluidity — a prediction borne out by numerous observations in intravital microscopy in experimental animals and in man. Here blood is fed from narrow single "bottle neck" capillaries into a system of interconnected, progressively widening vessels with limited supplies of vascular smooth muscle and a large total cross-sectional area.

The abundance of post-capillary venules, as found by Wiedeman (1963) in the subcutaneous tissue of the bat wing, is probably typical for the vasculature of the skin and for interstitial tissue and has also been described in the intestine and the liver. In muscle (including myocardium), such vessels are, by contrast, not prominent. Here the microvasculature is primarily constructed of comparatively long and narrow capillaries, often in a hairpin configuration. Keeping these differences in mind, it becomes evident that not only the absolute numbers but, more important, the ratio of total cross-sectional areas, and hence of hydraulic hindrances, mean flow rates, and shear stresses, between different classes of blood vessels are different from organ to organ. The work of Gross and Intaglietta (1973) illustrates this argument. They have compared four types of microcirculation: two muscular, an omental and a mesenteric. The distribution of resistance, calculated for different types of blood vessels, is quite variable. Recently, direct hydraulic mapping of cat mesentery and omentum was performed by Zweifach (1974) and Lipowski and Zweifach (1974). These studies have produced data that agree favourably with the computed values cited above.

Differences in resistance depend on the differences in the total cross-sectional areas of arterioles, capillaries and venules. These manifest themselves as a hydraulic by-product of an angiogenesis selected in phylogeny for certain purposes — specific exchange functions, filtration, reabsorption, thermoregulation, etc. The ratio of cross-sectional area of arterioles and venules, or arterioles and capillaries, is, of course, subject to local and general regulatory adjustments. Although primarily controlling tissue perfusion, these ratios cannot help but affect flow conditions. Under normal conditions low post-capillary resistance is beneficial to the perfusion of an organ. Low driving pressures, and thence low shear rates, suffice to drive back to the right atrium the blood that has served its purpose. However, in low flow states this benefit may turn into a disadvantage. The flow forces can then drop to values insufficient to maintain the fluidity of the blood. The shear stresses in the microcirculation can fall below normal values because of a number of factors: a fall in arterial pressure or a rise in venous pressure, steeper pre-capillary pressure gradients, vasoconstriction, or increased blood viscosity anywhere along the vascular tree. Shear rates in the venular part of the system drop critically whenever general vasoconstriction affects the arterioles more than it does the venules, when capillaries but not venules are shut off from the circulation, or when arteriolar vasoconstriction is associated with venular dilation (Schmid-Schönbein and Wells, 1971).

While under resting conditions the wall shear stresses in the pre- and post-capillary segments seem to be at least of the same order of magnitude, the response to a flow disturbance in the high-pressure system compared with the low-pressure system will be vastly different. In the high-pressure (the pre-capillary) system, the normal wall shear stresses are only a fraction of the maximum possible shear stresses. If an intravascular obstruction should occur in the high-pressure system, the full arteriovenous pressure gradient ($P_{ART} - P_V$) would immediately build up along the affected vascular segment and it would increase both the normal (pressure) forces and the shear (flow) forces. As a result, the obstruction would be flushed away, unless it is structurally capable of withstanding such high shear forces. In the low-pressure system an obstruction would have a far greater effect. When an obstruction is situated in the major proximal veins, the pressure gradient, and thus wall shear stresses, would be reduced. When situated in the smaller distal veins and venules, wall shear stresses could only be elevated to a small extent. When, however, the obstructed post-capillary segment has nearby collaterals that can shunt the flow around the obstructed segment, the wall shear stresses along the occluded segment can be varied by the flow dynamics in the shunting vessels. Whenever many of them are open, and when flow velocity in any single one is therefore low, wall shear stresses in the affected collaterals fall even more and may attain values insufficient to maintain the fluidity of the blood.

6.4 "Blood Fluidity" *In Vivo*

Attempts to gain information about blood viscosity *in vivo* date back to the time of Poiseuille, whose interest in this problem prompted the experiments that led to the discovery of the classical Poiseuille-Hagen law (1841). Many authors have attempted to assess "blood viscosity" in entire organs by exchange perfusion; blood and Newtonian reference fluids were compared and "viscosity values" were computed from the ratio of the flow rates and/or resistances. The validity of such measurements stands or falls with the confidence that both individual vessel diameters and the total number of perfused vessels remain absolutely invariant during the experiments. Although such confidence is not warranted, it must be accepted for the time being that blood viscosity in the normal vascular bed is low (Whittacker and Winton, 1933; Levy and Share, 1953; Hint, 1964; Djojosugito et al., 1970; Baeckstrom et al., 1971; Gaehtgens and Uekerman, 1973). The interpretation of the data of Whittacker and Winton (1933), who used cell-free plasma as a reference fluid, has to be modified due to a kinetic energy correction (Benis et al., 1973). Even when measured under conditions of reduced flow (Baeckstrom et al., 1971), the apparent viscosity as measured in isolated, dilated vascular beds was found to be lower than the apparent viscosity as measured in rotational viscometers, even when there was evidence of vascular blockade (v.i.).

All these data were obtained not only in the presence of normal arterial driving pressures but usually also under the conditions of maximal vasodilation. As a result, the local shear stresses in the microcirculation were probably higher than under normal *in vivo* conditions. One has to expect that the fluidity of blood is high under these conditions, simply because there are ideal conditions for the adaptation of the cells to flow.

There is much less information available on blood fluidity in individual blood vessels. Kurland et al. (1968) measured blood viscosity in a rat tail artery (218-380 μm) and found values similar to those in a viscometer. Meiselman et al. (1972), in a quasi *ex vivo* outflow technique, found a Fahråeus-Lindqvist effect. Recent data by Lipowski and Zweifach (1974) are most relevant in this respect. By measurement of flow, diameter, and upstream and downstream pressure, these authors computed an apparent viscosity of 1 mPa.s in a 10-μm capillary at 30 mm/s, but 20 mPa.s in the same capillary when the velocity was reduced to 0.2 mm/s. These data clearly underline the idea that a very pronounced Fahråeus-Lindqvist effect occurs in capillaries, but only in the presence of high shear forces. Using a microperfusion technique, LaCelle (1975) has also published evidence of low apparent viscosity in capillaries of the mouse cremaster muscle *in vivo,* and correlated these to similar viscosity in glass capillaries *in vitro.*

Our knowledge of the Fahråeus-Lindqvist effect in small blood vessels, as well as the hydrodynamic analysis of the vascular bed, makes all findings in vascular beds very plausible. Caused by their large number, and their narrow diameters, the hydrodynamic hindrance of the circulation resides in those paracapillary vessels in which the blood viscosity is equal to, or only slightly higher than, that of plasma. This statement is of course not new — it only repeats the explanation given by Whittacker and Winton (1933) for their classic experimental results.

6.5 Interaction of "Viscosity" and General Haemodynamics

As we have seen, it is justified to state that the blood viscosity value as measured in conventional viscometers is only of limited significance, as it applies only to the very few large-bore vessels in which the Fahråeus-Lindqvist effect does not operate. It does not apply to the vast majority of microvessels with diameters below about 300 μm, in which the mechanical energy generated by the heart is dissipated. Whether or not apparent blood viscosity in large-bore rheometers is higher or lower than normal by 10%-20% is therefore of very limited rheological relevance in itself. The relevance of elevated apparent viscosity greatly depends on the mechanism responsible for this change; thus it is indispensable to support capillary or rotational viscometry with additional direct measurements not only of haematocrit and plasma viscosity, but also of red-cell deformability and the tendency to red-cell aggregation.

The role of haematocrit abnormalities is especially difficult to predict. This variable, which so strongly affects the apparent viscosity in macrorheometers, may have a negligible effect on blood flow in rapidly perfused microvessels. On the other hand, elevated haematocrit levels may lead to a strongly increased yield shear stress of blood (v.i.) and hence might very effectively inhibit the recovery of blood flow following an accidental stop, as occurs frequently *in vivo,* especially under low flow conditions (Driessen et al., 1979).

When corrected for plasma viscosity, haematocrit and temperature, the data obtained in viscometers do, nevertheless, reflect the rheological potentials of suspensions of red blood cells under varying flow forces. These potentials are responsible on the one hand for the favourable flow behaviour of blood when it is flowing rapidly, but on the other hand, they can turn into the cause of prolonged stagnation of blood flow in microscopic blood vessels under conditions of circulatory insufficiency. Whenever relative apparent blood viscosity at high shear stresses (above 0.5 Pa) is normal, this indicates that in large vessels blood actually flows as an emulsion of perpetually deforming haemoglobin "droplets". When the same red cells approach the microvasculature, axial migration, reduction of effective haematocrit, and their assumption of the typical bullet shape minimizes the viscosity of blood (by maximizing the Fahråeus-Lindqvist effect) so that in the rapidly perfused true capillaries it is only slightly higher than, or equal to, plasma viscosity.

Considered in its relevance to the haemodynamics of the normal circulation, the viscometric measurement of high apparent blood viscosity due to red-cell aggregation in pre-static flow (shear stress less than 0.05 Pa) therefore reflects not an actual, but only a potential property of blood. This potential property may certainly become quite an effective property under conditions of pathological haemodynamics. Whenever the flow is chronically retarded — irrespective of the factor causing the flow retardation — the haemodynamic effects of apparent blood viscosity may become the critical and limiting factor governing flow in the affected vessels. Depending on the actual pressure gradient, vasomotor reserve, local haematocrit level, etc., the effect of elevated viscosity due to aggregation may then be *far more pronounced than predicted by rotational viscometry.* Due to locally acting low shear stresses, single vessels may be entirely clogged by cellular material (v.i.). Such a blockade was actually recorded by Djojosugito et al. (1970), who performed simultaneous measurements of the "viscosity" *in vivo* and the kinetics of transcapillary exchange (reduction of the permeability surface area product as computed from indicator diffusion studies).

The material responsible for blockage in narrow capillaries may be large leucocytes, or small platelet aggregates in arterioles, or it may be red-cell aggregates in any blood vessel with reduced blood flow. Leucocytes and platelet aggregates may act as functional microemboli, because their mechanical properties are sufficient to withstand the forces acting in the normal microcirculation. On the other hand, the mechanical properties of the red-cell aggregates only allow them to resist the forces of the disturbed, hypoperfused microvasculature. For all their reversibility and for all their frailty that allows normal flow forces to clear the affected vessel, the red-cell aggregates can lead to prolonged vascular obstruction in a major fraction of microvessels under conditions of sustained hypoperfusion, especially if accompanied by local or generalized changes in haematocrit or aggregating tendency. Whether a flow retardation is caused by rheological or by haemodynamic factors, the consequences mainly derive from the highly variable flow behaviour of both red cells and their aggregates. Even if there is not aggregation, axial migration no longer occurs (Devendran and Schmid-Schönbein, 1975; Goldsmith and Beitel, 1970; Merrill, 1969), the beneficial effects of the lubricating plasma layer and the reduced effective haematocrit tend to be lost, and the red cells might establish contact with the capillary wall (Lighthill, 1968; Fitzgerald, 1969). If the individual red cells are unable to display their normal flow properties through mere lack of adequate shear forces, the concentrated suspension of red cells exerts a highly effective resistance against the mutual motion of fluid lamellae (and thence flow). Blood then assumes the properties of a conventional suspension.

When, in addition, the red cells are aggregated, the blood assumes the properties of a "reticulated" suspension. At the unusually high volume fraction near 50%, all network-like, structured suspensions are flow resistant and/or exhibit elastic properties. In blood, a two-phase suspension, cell aggregates as such exert haemodynamic effects only under the following conditions:

1) They reach across planes of shear.

2) They withstand the forces acting between these planes of shear.

3) They are kept from evading these forces (as in the case of axial migration or settling).

4) They impede the motion of fluid lamellae past each other.

In rotational viscometers (Couette viscometric flow) the motion away from the shear forces (shear-induced phase separation) is governed by parameters different from those in perfused natural or glass capillaries, where axial migration and sedimentation lead to shearing mainly in the lubricating plasma layer. When viscosity is low, phase separation obscures the fact that the cell aggregates comprising the non-sheared portion have lost their fluidity ("plug flow").

In vascular networks, red-cell aggregation, accompanied by haemoconcentration, is frequently observed. From these phenomena observed *in vivo* and the measured properties of concentrated, aggregated red-cell suspensions *in vitro,* it must be inferred that a state of irreversible stagnation may easily be reached due to changes in rheological behaviour. When associated with a reversible loss of blood fluidity, it may trigger a sequence of effects with positive feedback. Flow retardation, formation of red-cell aggregates and increase in viscosity may lead to a further decrease in velocity, unless compensated by an increase in cardiac output, arterial pressure or diameter of the affected vessel or vessels immediately upstream or downstream.

6.6 Vasomotor Control of the Haematocrit Value

It has long been observed that the haematocrit in the microvessels is by no means constant, as it varies not only with time but also with the location of the vessel in the terminal network. As a rule, the haematocrit in the microvessels is lower than that in the central arteries and veins, a phenomenon explained in part by the faster flow rate of the red cells in the axial core of the microscopic blood vessels relative to the slowly moving plasmatic marginal layer. As first elaborated by Fåhraeus, this is one of the important consequences of the tendency to axial migration found in red-cell suspensions. There may be other mechanisms, such as red-cell screening, red-cell shunting, uneven distribution of red cells at intersections and disturbance effects of leucocytes, that affect the distribution of cells and plasma, and hence the haematocrit value in the terminal network (for a review see Klitzman and Duling, 1979).

Under conditions of hyperaemia, there is elevation not only of the overall flow rate but also of the local haematocrit (Mchedlishvili, 1969; Johnson et al., 1971; Klitzman and Duling, 1979). Under such conditions, the peripheral haematocrit can be as high as the central one. Under pathological conditions of slow flow, in which the vascular permeability is abnormally high, the haematocrit can be very high, even higher than the central haematocrit. In extreme cases the haematocrit rises to 90% or more with solid compaction of cells in stasis. This phenomenon is caused by intravital settling (Knisely, 1965; Heidtmann et al., 1979), as well as by haemoconcentration due to abnormal vascular permeability. It must be stressed, however, that the stasis phenomenon and the subsequent haemoconcentration is greatly augmented by a local or general retardation; in other words, it is associated with low shear stress levels and hence a low apparent fluidity of the blood.

6.7 Relative Role of Vasomotor Control and Blood Rheology

6.7.1 Feedback Loops in Blood Flow

In vivo, reduction of arterial pressure alone does not necessarily lead to critically reduced flow rate, nor to elevated viscosity. Hypoperfusion is immediately followed by the sequence of events known as "autoregulation", i.e. a myogenic or a metabolically controlled vasodilatation tending to re-establish blood flow

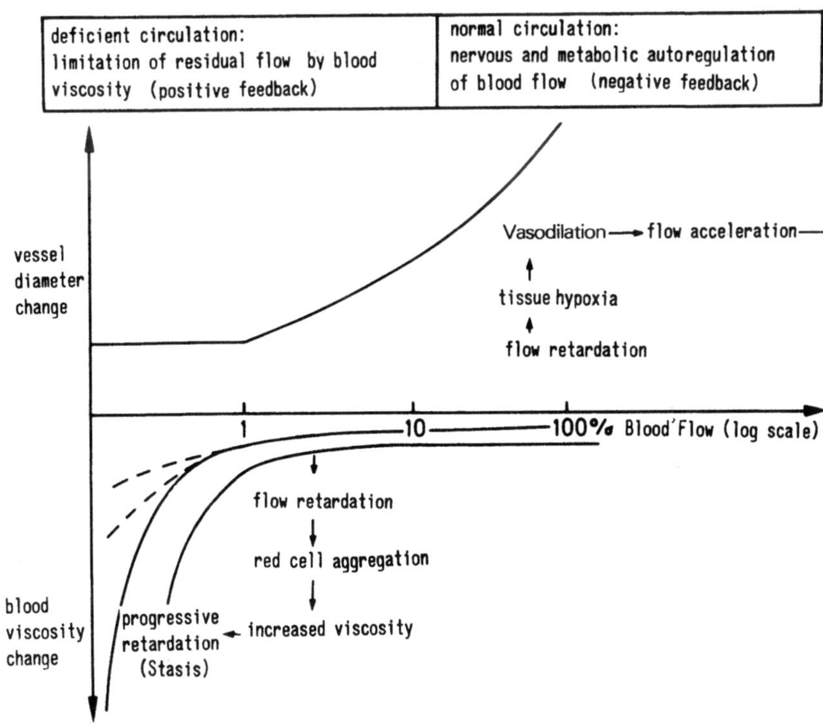

Fig.6.1. Schematic representation of effects of "blood viscosity" changes and of diameter changes on blood flow. (For explanation see text)

(negative feedback). Whether or not the circulation deteriorates by the described chain of reactions therefore depends on the relative gain of the positive feedback loop by increased viscosity and that of the compensating negative feedback loop by autoregulation. This interdependence is schematically depicted in Fig. 6.1. Very obviously, under all physiological conditions flow velocity is high, viscosity low, and only slightly shear dependent (low gain of the viscosity loop). At the same time, the vasomotor reserve is high, because the arteriolar vessels are under tone and can greatly improve flow velocity by very limited vasodilation — a factor not only restoring flow but also improving the fluidity of the blood. Therefore, the physiological regulation of the peripheral blood supply is without doubt dominated by vasomotor factors (high gain of vasomotor loop), and blood viscosity factors play a negligible role.

6.7.2 Control of Vascular Resistance

For ontological reasons beyond the scope of the present report, the pre-capillary arterioles are richly endowed with vascular smooth muscle, whereas the true capillaries have none, and the post-capillary venules have very little. Based on the ratio of wall thickness to vessel radius, the smallest arterioles are the most "powerful" blood vessels. Thus, the terminal arterioles are not necessarily the vessels with highest resistance, but certainly the vessels with the best control of the resistance. Based on the measured values of the pressures in small arteries (300 μm, where pressure is about equal to aortic pressure, i.e., 90 mm Hg), at the capillary end of arterioles (equal to the arteriolar end of capillaries, i.e. pressure 35-40 mm Hg), the highest pressure drop (ΔP), and hence the highest resistance, has been attributed to the arterioles. Because, however, the resistance should be calculated on the basis of the pressure gradient ($\Delta P/l$), the generally accepted concept of "high resistance in arterioles" should be revised and extended to the concept of arteriolar control of resistance.

Indeed, the arterioles are unique in their ability to control flow. They not only modulate the resistance along any given arteriole, but also the number of exchange capillaries open to the circulation. If the terminal ends of arterioles (5 μm) are strongly constricted, the lumen may shrink to the point where flow is interrupted (sometimes with the help of an inherently rigid leucocyte). The terminal arterioles may thereby assume the action of a functional sphincter. The concept of "pre-capillary sphincters" has been the subject of much controversy. In the present context, the functional sphincter behaviour, rather than any specific task of a specialised structure, is discussed. Because of this action, not only is the blood flow in all parts of the circulation effectively controlled, but also the pressures, pressure gradients, and thus the shear stresses and their distribution. Arteriolar dilatation reduces the pressure drop in the arterioles, but it increases the pressure, and thence pressure drop, in all downstream vessels. Conversely, arteriolar constriction increases the pressure in the arteries and the pressure drop along arterioles, but it decreases the pressure, pressure drop, and the shear stresses in all downstream vessels.

Despite their relative deficiency of smooth muscle, the resistance to flow in the post-capillary vessels may increase under certain conditions. It must be emphasised, however, that the evidence of an increased "post-capillary resistance" is derived indirectly, i.e. from measurements of the trans-capillary exchange dynamices (e.g., Oberg, 1964). In this context, terms such as "post-capillary sphincters" have been used. It must be emphasised that there is no convincing anatomical evidence for the presence of a vascular region that might serve as such a sphincter, nor is there any physiological evidence that the post-capillary venules might be under isolated control of metabolic, hormonal or nervous factors. We must, therefore stress that post-capillary resistance is not yet attributable to vascular or other (e.g., rheological) factors. However our personal bias favours the rheological hypothesis.

6.7.3 Blood Flow in Pathological Conditions

In grossly pathological conditions of circulatory deficiency, the situation in hypoperfusion may be totally different from hypoperfusion in physiological conditions. The reduced flow velocity has several different consequences. Hypoperfusion leads to chronic hypoxia, vasodilator metabolites are produced by the parenchymal cells, and they paralyse the vascular smooth muscle. This may proceed up to the point where the vasomotor reserve is exhausted. The reduced linear flow velocity in any single blood vessel, on the other hand, not only increases blood viscosity but now shifts the viscosity profile to its steep portion (high gain of the positive feedback viscosity loop; Fig 6.1.). In other words, under conditions of chronic or prolonged circulatory deficiency, the gain of the vasomotor negative feedback loop goes to zero, while at the same time the gain of the viscosity positive feedback loop goes to its maximum. Consequently, the deficient circulation is limited by the actual level of apparent viscosity in the respective blood vessels. Haematocrit level and aggregation tendency affect the critical shear rates below which aggregation occurs, but even more so the gain of the viscosity loop (position and steepness of the viscosity profile).

One additional principal difference between vasomotor control and viscous limitation of perfusion follows from the same argument. A vasomotor shutdown of blood vessels, especially small arterioles and their peripheral ends, the functional pre-capillary sphincters (Folkow, 1964; Johnson, 1967) is highly effective in producing a rapid but short-lasting stoppage of flow. However, metabolic autoregulation will paralyse the constriction, and soon restore flow. Any kind of maintained flow retardation and/or stasis must therefore be caused by an intravascular obstacle. The physiological, and even more so the pathologically enhanced, cell aggregates are candidates likely to be held responsible for such a state of maintained intravascular stagnation. Thus rheological changes are not likely to initiate, but are likely to sustain, arrest of flow.

Previously, red-cell aggregates were assumed to simply plug the arteriolar inflow into the capillaries. This assumption, advocated primarily by anatomists observing the "sludge phenomenon" (e.g. Knisely, 1965), has never been corroborated by direct measurement, and is in conflict with established physical properties of red-cell aggregates. Red-cell aggregates, in contrast to platelet aggregates, have been shown to be dispersed by shear stresses as low as 0.2 Pa in normal human blood, 0.05 Pa in most laboratory animal species, and at most 2.0 Pa in severe human disease (Schmid-Schönbein, 1976). In the light of this limited shear resistance of the red-cell aggregates, which have been described as very "frail" or "loose" by observers of the microcirculation *in vivo* (e.g. Brånemark, 1971), it is very unlikely that even the most severe form of aggregates should be able to produce a primary arteriolar obstruction. If such an aggregate should lodge in an arteriole, the full arterio-venous pressure gradient would then act upon it, easily producing shear stresses in the order of 10 - 100 Pa: values which are most likely sufficient to flush away the obstacle. Therefore, a more complex rheological theory had to be developed in order to explain the unequivocally established fact that red-cell aggregates can indeed lead to permanent occlusion of microvessels both in man and in experimental animals in low flow states.

It has long been established (Bloch, 1972; Branemark, 1971; Ditzel, 1959; Gelin and Zederfeldt, 1961; Goldstone et al., 1971; Knisely, 1965; Schmid-Schönbein et al., 1967; Thorsen and Hint, 1950; Thuranskii, 1957) that both *in vivo* and *in vitro* red-cell aggregation depends on the equilibrium between the shearing forces of the blood flow and the adhesive forces acting between red cells. In particular, reduction of flow forces (Schmid-Schönbein et al., 1971; Gaehtgens et al., 1966) as they occur *in vivo* following arterial clamping, or in the course of haemodynamic reflexes involving pre-capillary vasoconstriction (orthostasis, cold pressor tests) is accompanied by phenomena of intravascular aggregation in the venules of the bulbar conjunctiva. Also, an increase in the adhesive properties of red-cell aggregates produces the phenomenon of intravascular aggregation in most vessels, even at normal perfusion pressure. Slight pressure reduction enhances the effect, and an increasing number of vessels become occluded by erythrocyte masses, which either move extremely slowly or are completely stagnant — only to resume motion upon re-establishment of normal perfusion pressure.

The microscopic analysis of the microcirculation reveals that the red cells very quickly change their shape and also that red-cell aggregates are formed and dispersed within a fraction of a second. In other words, we can assume that the blood *in vivo* is capable of changing its physical properties as quickly as *in vitro*. However, normal red-cell aggregates in human blood are extremely shear labile. This frailty notwithstanding, they are frequently observed in the microcirculation, preferentially in the venules, but often in the arterioles also (Knisely, 1965). Prolonged stasis or even flow reversal is not an uncommon finding in the microcirculation. The venules, as a predilectory site for intravascular red-cell aggregation with concomitant retardation, have several haemodynamic peculiarities that might predispose them towards low shear ranges. The mere fact that in most organs the total cross-sectional area of all venules is considerably higher than that of all other classes of vessels is presumably most important. While under normal perfusion conditions this fact is advantageous, under pathological conditions it carries the risk that venular shear stresses are subcritical, while the shear stresses in all other blood vessels are above the critical value necessary to keep red-cell aggregates dispersed. The walls of most of the venules appear to contain very little, if any, vascular smooth muscle, and exhibit a high permeability. On the other hand, the arterioles immediately upstream are strongly invested with vascular smooth muscle and appear to have very low permeability ("gradient of vascular permeability"). A general vasoconstriction, in which venules are spared due to their lack of effector muscle cells, might also lead to a preferential fall of venular shear stresses and an increase in viscosity in the venules. As a consequence, a shift in the equilibrium between pre-capillary and post-capillary resistance occurs, a factor discussed by Chien (1969). In those venules that do possess vascular smooth muscle, the preferential loss of venular tone, as occurs early in haemorrhagic shock, when arteriolar tone is still high has a similar effect (Goldstone et al., 1971; Hutchins et al., 1973). If loss of tone is associated with enhanced permeability, local haemoconcentration occurs with a threefold effect:

1) Increase in haematocrit.
2) Increase in plasma-protein concentration.
3) Increase in the shear resistance of red-cell aggregates.

Blood that slowly enters such a venular segment with quite normal rheology might undergo secondary changes which turn it into a highly concentrated, highly adhesive mass of red cells which strongly resists flow. Furthermore, since the static network of cell aggregates has elastic properties (and thus stops red-cell flow) it does leave free spaces that allow plasma to flow in an orthograde direction, thereby producing local haemoconcentration.

A comparison of the data on shear resistance of red-cell aggregates *in vitro*, and the data on shear forces normally acting *in vivo* (Lipowski, 1975) makes it highly unlikely that in the normal microcirculation red-cell aggregates should produce arteriolar plugging by microembolisation, as previously suggested by Knisely (1965) or Gelin and Zederfeldt (1961). Since whole blood yields to forces as low as 0.1 Pa, and since, moreover, even the most pronounced red-cell aggregates are incapable of withstanding shear forces higher than 2.0 Pa, one has to accept the fact that it is *a priori* unlikely that cell aggregates should ever be able to resist the kind of shear forces in such vessels (as computed or measured), especially when these are positioned in series to the "main arteriovenous pathway". Even a venule positioned in series is unlikely to be blocked by even the most adhesive red-cell aggregates observed so far, because following such a blockade, an increasing portion of the arteriovenous pressure gradient will build up upstream of the occluded venule and tend to flush away the cell aggregates — unless there is a bypass parallel to the affected vessel (see section 6.3).

6.7.4 The Concept of Collateral Blood Viscidation

In the microcirculation of most tissues, however, vessels can actually be bypassed, as the vast majority of them are positioned parallel to the main arteriovenous pathway. Here, a definition of the term "main arteriovenous pathway" is in order. We define this as the one channel connecting artery and vein that has the smallest hydraulic resistance as an in-series vessel. Very obviously, such a vessel can be, but need not always be, anatomically defined. Moreover, it need not always be the same single vessel, but vessels may take turns in this function, adapting to spontaneous, hormonal, metabolic or nervous change in vascular tone. Obviously long, narrow nutritive capillaries are unlikely candidates for this function, whereas any type of short arteriovenous pathway, especially when dilated, is predisposed to operate as the main arteriovenous channel.

The old controversy about the absence or presence of anatomically pre-formed shunt vessels, as well as the endless discussion about the organisation of microvascular beds with "thoroughfare" and/or "preferential channels" (Chambers and Zweifach, 1944) has been based too much on anatomical arguments (Illig, 1961; Staubesand, 1974). The simple consequences of Hagen-Poiseuille's law make it unnecessary to search for anatomical channels to explain shunting. Any arteriovenous channel, irrespective of its diameter, that is endowed with smooth muscle, and is normally kept under tone, may become the main arteriovenous pathway. It may always serve this function, or it may assume this function whenever the vascular smooth muscle is relaxed. The haemodynamic and metabolic situation of circulatory deficiency not only creates the bypass for a vessel occluded by an aggregate, but may likewise give rise to the rheological conditions (flow with progressively decreasing shear rate) that initiate the progressive formation of red-cell aggregates within capillary networks. Venules admittedly are the most likely sites for low shear stress flow in parallel vessels, but other microvessels may be subject to a similar shift in forces. The highly flow-dependent "viscosity" of the blood, the general fall in shear stresses during hypoperfusion in combination with dilatation, and the preferential reduction of shear stress in parallel vessels must be held responsible for the frequent observation of heterogeneous, non-uniform perfusion of the microcirculation seen in various disease states, i.e., flow stagnation in one vessel, but normal or even increased blood flow in its immediate neighbours.

In order to explain such flow heterogeneities on the basis of the flow dependence of apparent blood viscosity, we have proposed the concept of "collateral blood viscidation", by which a rheological occlusion in certain vessels of a network, despite the maintenance of flow in others, is explained. Recent work by Kiesewetter et al. (1979) has clearly shown that in capillary networks with highly different vascular lengths, stagnation occurs in the longer vessels (and hence the vessels with the lower shear stresses for any given pressure difference between the branch-point and the conflux). The fluid dynamic principle is depicted schematically in Fig. 6.2. Flow in one branch (A) leads to a pressure gradient ($P_1 - P_2$) which also acts along the branches (B) or (B′). However small

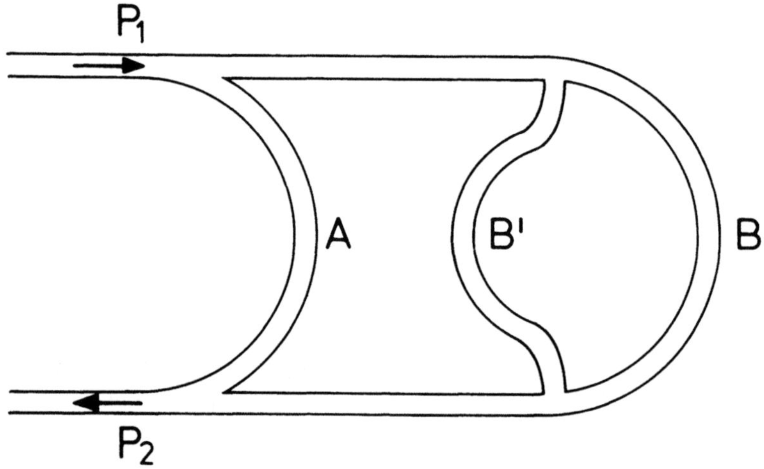

$$\tau_{w_B} = \frac{(P_1 - P_2) \cdot d_B}{4\, l_B} \qquad P_1 - P_2 = f\left(\frac{1}{\bar{V}_A} / l_A / \eta / \frac{1}{d_A^4}\right)$$

$$\text{Stasis: } \tau_{w_B} < \tau_y$$

Fig.6.2. Schematic representation of concept of "collateral blood viscidation" in branching and collaterals of microcirculation. Flow over branch A causes pressure drop (P_1-P_2), which is cause of flow over B and B′ in true fluids. After generalised reduction of flow forces (e.g. in low flow states) P_1-P_2 decreases (as function of velocity in A, length of A as compared to B, viscosity of blood in A and especially after dilation in A). Wall shear stress in B or B′ (τ_{wB}) directly proportional to P_1-P_2. Spontaneously arrested blood will not resume flow if τ_w is below yield stress of red-cell aggregates: collateral blood viscidation results this pressure gradient, it must lead to flow. If, however, the perfusing fluid possesses a yield shear stress, flow in B may come to a stop as long as the ratio $\Delta P/1$ is below the yield value. Since in turn the pressure gradient along vessel A is inversely proportional to the mean flow velocity and the fourth power of the vessel diameter, it is easy to understand why preferential channels, i.e. vessels which can dilate after losing muscular tone, especially when slowly perfused, have such a small viscous energy dissipation that wall shear stresses in B and B′ also fall below the "yield shear stress" of the blood elements, giving rise to permanent stagnation. Moreover, flow across B and B′ may fall to zero if an aggregate straddles the bifurcation or if the wall shear stresses along B and B′ are below the analogue yield shear stress of the perfusate. In actual fact, both mesenteric and muscular microcirculations (Eriksson and Myrhage, 1972) display the full spectrum of flow anomalies from sluggish flow of aggregated blood to full stasis with local haemoconcentration (due to sedimentation and possibly enhanced filtration).

The hydrodynamics in channel A will also affect the events in B and B′. A resumption of the driving pressure or the flow velocity to the initial high values will increase the pressure gradient and produce a more than linear increase in flow in B and B′, provided it is sufficient to disaggregate the red cells and to restore the emulsion behaviour of the blood. Conversely, a further drop of flow in A (which may be caused by factors acting either upstream or downstream) will potentiate the flow deterioration in B and B′. Most of all, a low flow in A, associated with vasodilatation in A, will lead to a precipitous drop in driving pressure, because according to the Hagen-Poiseuille law the pressure drop is inversely proportional to the fourth power of the radius of channel A.

The work of Kiesewetter et al. (1979) has shown that a definite yield shear stress of human blood can be measured in a simple tube network constructed according to Fig 6.2. As was expected, the yield shear stress greatly depends on the microrheological properties of the blood investigated (tendency to aggregation, plasma viscosity, deformability of red cells) but primarily on the haematocrit level. Below 35% haematrocrit, we could never observe any yield shear stress of blood, irrespective of the microrheological properties of the blood. Above 45% haematocrit, each increase of the haematocrit level leads to a steep increase in the measured yield shear stress, a rise from 45% to 55% leading to a 10-fold increase in the minimum pressure necessary to re-establish flow in the long branch of the network. It should be mentioned in closing that the very same blood samples examined in a conventional rotational viscometer did not show a trace of yield shear stress. As discussed at length elsewhere (Schmid-Schönbein and Wells, 1971), this is the consequence of a phase separation effect between red-cell aggregates and a lubricating plasma film.

6.8 Summary and Conclusions

1) Because of the unique micromechanics of the red blood cells, mammalian blood can exist and function as a concentrated dispersion of unparalleled fluidity under the normal perfusion conditions existing in the microvasculature.

2) Upon a reduction of flow forces blood assumes quite conventional flow properties of suspensions or reticulated suspensions. These have a low fluidity and are notoriously thixotropic. Consequently, the loss of fluidity will depend on

both the remaining flow forces and the composition of the blood, in particular the concentration of macromolecules, haematocrit value and tendency to aggregation.

3) As the red-cell aggregates have a very low resistance to hydrodynamic dispersion by the arterial flow forces that are compatible with the survival of a subject, they are unlikely to be able to serve as a plug for such vessels. Red-cell aggregates are, however, easily capable of interfering with flow through collaterals bypassing any main arteriovenous pathway. The well-established fact of intravascular aggregation and retardation is explained by the assumption of a collateral blood viscidation. Rheological changes are not likely to initiate, but are capable of sustaining, arrest of flow. Arrested circulation allows coagulatory processes to take place, leading to a truly irreversible state.

4) In normal human subjects variations in blood fluidity are unlikely to regulate or control haemodynamics under normal perfusion conditions. They are, however, very likely to limit perfusion in chronic low flow states. Rheological occlusion, based on the presence of a finite yield shear stress of blood, has recently been demonstrated using an *in vitro* model of the capillary network.

5) It can thus be concluded that the predictive value of *in vitro* viscometry is extremely limited because, depending on the local flow conditions in the tissues, viscometry in macroscopic tubes and rotational viscometers can both *underestimate* and *overestimate* the viscous hindrance presented to the motion of blood in the microvessels in disease.

6. References

Anonymous (1977) Hyperviscosity in disease (Editorial). Lancet *ii:* 961-962

Baeckstrom P, Folkow B, Kendrick E, Lofving B, Oberg B (1971) Effects of vasoconstriction on blood viscosity in vivo. Acta Physiol Scand *81:* 376-384

Benis A M, Chien S, Usami S, Jan K-M (1973) Inertial pressure losses in perfused hindlimb: a reinterpretation of the results of Whittacker and Winton. J Appl Physiol *34:* 383-389

Bicher H I (1972) Blood cell aggregation in thrombotic processes. C .C. Thomas, Springfield, Illinois

Bloch E H (1972) Sludged blood, human disease and chemotherapy. In: Oxygen transport to tissue. Pharmacology, mathematical studies and neonatology. Bruley D F, Bicher H I (eds) Plenum, New York pp 641-645

Braasch D (1971) Red cell deformability and capillary blood flow. Physiol Rev *51:* 679-701

Brånemark P I (1971) Intravascular anatomy of blood cells in man. Karger, Basel

Chambers R, Zweifach B W (1944) Topography and function of the mesenteric circulation. Am J Anat *75,* 173-182

Charm S E, Kurland G S (1974) Blood flow and microcirculation. J. Wiley and Sons, New York-London-Sydney-Toronto

Chien S (1969) Blood rheology and its relation to flow resistance and transcapillary exchange, with special reference to shock. Adv Microcirc *2:* 89-103

Chien S (1972) Present state of blood rheology. In: Messmer K, Schmid-Schönbein H (eds) Hemodilution, theoretical basis and clinical application. Karger, Basel, New York pp 1-40

Chien S, Usami S, Dellenback R J, Bryant C A (1971) Comparative haemorheology — haematological implications of species differences in blood viscosity. Biorheology *8:* 35-57

Devendran T, Schmid-Schönbein H (1975) Axial concentration in narrow tube flow for various RBC suspensions as function of wall shear stress. Pflueger's Arch *355:* R-19

Dintenfass L (1971) Blood microrheology-viscosity factors in blood flow, ischaemia and thrombosis. Butterworths, London

Ditzel J (1959) Relationship of blood protein composition to intra-vascular erythrocyte aggregation (sludged blood). Acta Med Scand *164* (Suppl. 343): 1-63

Djojosugito A M, Folkow B, Oberg B, White S (1970) A comparison of blood viscosity measured in vitro and in a vascular bed. Acta Physiol Scand. *78*, 70-84

Driessen G K, Heidtmann H, Schmid-Schönbein H (1979) Effect of haematocrit on red cell velocity in the capillaries of rat mesentery during hemodilution and hemoconcentration. Pfluegers Arch *380:* 1-6

Eriksson E, Myrhage R (1972) Microvascular dimensions and blood flow in skeletal muscle. Acta Physiol Scand *86:* 211-222

Fitzgerald J M (1969) Mechanics of red cell motion through very narrow capillaries. Proc Roy Soc London B *174:* 193-227

Folkow B (1964) Autoregulation in muscle and skin. Circulat Res, XIV and XV, Suppl. I: 19-24

Gabelnick H, Litt M (1973) Rheology of biological systems. Charles C. Thomas, Springfield, Illinois

Gaehtgens P, Schmid-Schönbein H, Dickmans H A, Hirsch H (1966) Über das Auftreten von Erythrocyten-Aggregaten in Abgängigkeit vom Schergrad in vivo. Pfluegers Arch *291,* R-36

Gaehtgens P, Uekermann U (1973) The apparent viscosity of blood in different vascular compartments of the auto-perfused canine foreleg, and its variation with haematocrit. Bibl Anat *11:* 76-82

Gelin L E, Zederfeldt B (1961) Experimental evidence of the significance of disturbances in the flow properties of blood. Acta Chirurg Scand *122:* 336-342

Goldsmith H L, Beitel L (1970) Axial migration of red cells in tube flow. Fed Proc *29:* 319-321

Goldstone J, Hutchins P M, Schmid-Schönbein H, Urschel C, Sonnenblick E, Wells R (1971) Correlation of microvascular and rheological factors in haemorrhagic shock. In: Ditzel J Lewis D (eds) Proc. 6th Europ. Soc. Microcirculat; Karger, Basel, pp 150-155

Gross J F, Intaglietta M (1973) Effects of morphology and structural properties on microvascular haemodynamics. Bibl Anat *11:* 532-539

Heidtmann H, Driessen G, Haest C W M, Kamp D, Schmid-Schönbein H (1979) The influence of rheological factors on the recovery of the microcirculation following arterial hypotension. Microvasc Res *18:* 449-457

Hint H C (1964) The flow properties of erythrocyte suspensions in isolated rabbit ear; the effect of erythrocyte aggregation, haematocrit and perfusion pressure. Bibl Anat *4:* 112-118

Hutchins P M, Goldstone J, Wells R E (1973) Effects of haemorrhagic shock on the microvasculature of skeletal muscle. Microvasc Res *5:* 131-140

Illig L (1961) Die terminale Strombahn. Capillarbett und Mikrozirkulation. In: Pathologie und Klinik. Springer, Berlin p 159

Johnson P C (1967) Autoregulation of blood flow. Gastroenterology *52:* 435-441

Johnson P C, Blaschke J, Burton K S, Dial J H (1971) Influence of flow variations on capillary haematocrit in mesentery. Am J Physiol *221:* 105-112

Kiesewetter H, Schmid-Schönbein H, Radtke H, Stolwerk G (1979) In vitro demonstration of collateral blood viscidation: Flow measurement in a model of vascular networks. Microvasc Res *17:* S-72

Klitzman B, Duling B R (1979) Causes of low microvascular haematocrit in hamster cremaster capillaries. Microvasc Res *17:* S-70

Knisely M H (1965) Intravascular erythrocyte aggregation (blood sludge). In Hamilton W F, Dow P (eds) Handbook of physiology, Sect 2 Vol III. Amer Physiol Soc, Washington D C, pp 2249-2292

Kurland G S, Charm S E, Brown S, Tousignant P (1968) Comparison of blood flow in a living vessel and in glass tubes. In: Hemorheology. Copley A L (ed) Pergamon Press, Oxford, pp 609-615

LaCelle P L (1975) Pathologic erythrocytes in the capillary microcirculation. Blood Cells *1:* 269-284

Larcan A, Stoltz J-F (1970) Microcirculation et hémorheologie. Masson, Paris

Levy M, Share R L (1953) The influence of erythrocyte concentration upon the pressure flow relationship of the dog's hind limb. Circ Res *1:* 247-255

Lighthill M J (1968) Pressure-forcing of tightly fitting pellets along fluid filled elastic tubes. J Fluid Mech *34:* 113-143

Lipowski H H (1975) In-vivo study of the rheology of blood in the microcirculation. Ph.D. Dissertation, Univ. of California, San Diego

Lipowski H H, Zweifach B W (1974) Network analysis of microcirculation of cat mesentery. Microvasc Res *7:* 73-83

Mchedlishvili G I (1969) The conjectural role of the Fahråeus-Lindqvist rheological phenomenon in some microcirculatory events. Bibl Anat *10:* 66-73

Meiselman H J, Frasher W G, Jr, Wayland H (1972) In vivo rheology of dog blood after infusions of low molecular weight dextran or saline. Microvasc Res *4:* 399-412
Merrill E W (1969) Rheology of blood. Physiol Rev *49:* 863-888
Oberg B (1964) Effects of cardiovascular reflexes on net capillary fluid transfer. Acta Physiol Scand *62:* 1-98
Schmid-Schönbein H (1976) Microrheology of erythrocytes, blood viscosity, and the distribution of blood flow in the microcirculation. Int Rev Physiol *9:* 1-62
Schmid-Schönbein H, Wells R E (1971) Rheological properties of human erythrocytes and their influence upon the "anomalous" viscosity of blood. Ergeb Physiol Biol Chem Exp Pharmacol *63:* 147-219
Schmid-Schönbein H, Gaehtgens P, Hirsch H (1967) Über eine neue Methode zur Untersuchung der rheologischen Eigenschaften von Erythrozythen-Aggregaten. Pfluegers Arch *297:* 107-114
Staubesand J (1974) Arterio-venöse Anastomosen. In: Heberer G, Rau G, Schoop W (eds) Angiologie. Grundlagen, Klink and Praxis, 2 Auflg. Thieme Verlag, Stuttgart p 127
Thorsen G, Hint H (1950) Aggregation, sedimentation and intravascular sludging of erythrocytes. Acta Chirurg Scand, Suppl *154:* 1-50
Thuranskii K (1957) Der Blutkreislauf der Netzhaut. Ungarische Akademie der Wissenschaften Budapest
Weed R I (1970) The importance of erythrocyte deformability. Am J Med *49:* 147-150
Wells R E (1973) The rheology of blood. In: Zwiefach B W, Grant L (eds) The inflammatory process, 2nd ed. Academic Press, New York-London, p 221
Whittacker S R F, Winton F R (1933) The apparent viscosity of blood flowing in the isolated hindlimb of the dog, and its variation with corpuscular concentration. J Physiol (Lond) *78:* 339-369
Wiedeman M (1963) Dimensions of blood vessels from distributing artery to collecting vein. Circulat Res *12:* 375-381
Zweifach B W (1974) Quantitative studies of microcirculatory structure and function, I. Circulat Res *34:* 843-857

7 Measurement of Whole-blood Viscosity

J. A. Dormandy

7.1 Introduction

Very few "advances" in medicine are really new and the measurement of whole-blood viscosity is certainly not an exception. Clinical interest in abnormalities of blood viscosity was probably greater at the beginning of the century than now; in Albutt's standard medical textbook, published in 1915, there was a whole chapter devoted to diseases due to abnormal blood viscosity (Albutt, 1915). There was, however, considerable confusion regarding the actual measurement of blood viscosity, with different laboratories using different equipment obtaining results which could in no way be compared. The cause of this confusion was explained by Hess's discovery in 1915 that blood was non-Newtonian, and to define its rheological characteristics, viscosity measurements would have to be carried out at a range of constant definable shear rates (Hess, 1915). The clinical interest in haemorheology therefore ceased until such measurements became possible with the advent of rotational viscometers in the 1960s.

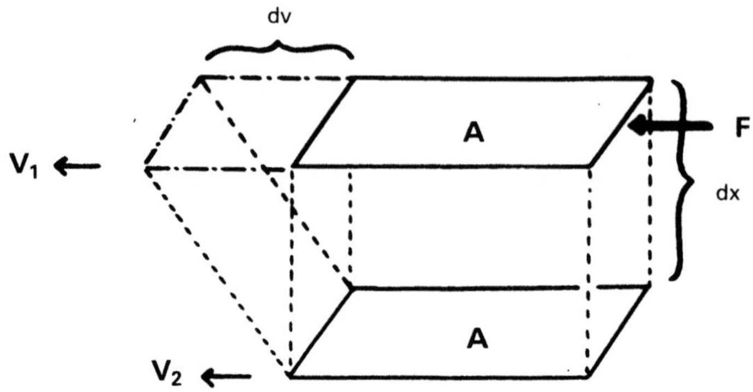

Shear Rate = Velocity Gradient = $\frac{dv}{dx}$

Shear Stress = Force Per Area = $\frac{F}{A}$

Viscosity = $\frac{\text{Shear Stress}}{\text{Shear Rate}}$

Fig.7.1. Definitions of viscosity, shear rate and shear stress

Following a brief definition of the basic concepts, the general techniques of viscometry will be discussed and the currently available commercial viscometers will be compared.

The viscosity of a fluid is due to the internal friction between adjacent layers. In streamlined flow, where adjacent layers move parallel to each other, the velocity difference is a measure of the shearing within the flowing fluid and this velocity gradient is termed the shear rate. The velocity of flow and the shearing within the fluid is produced by a force which is called the shear stress. The velocity of the fluid is then defined as the ratio of the shear stress to the shear rate it produces.

Figure 7.1 illustrates these definitions: A represents two layers of fluid flowing in relation to each other at velocities v_1 and v_2, pushed by a force F. The shear rate is the velocity gradient, dv/dx, where v is the difference between v_l and v_2 and x is the distance between the layers of fluid being considered. The units of shear rate will be distance per time divided by distance, which equals the reciprocal of time, usually expressed as inverse seconds (s^{-1}). The shear stress is expressed as force per unit area, F divided by A, and will be in Pascals (Pa). The viscosity of the fluid, being the shear stress over the shear rate it produces, will be expressed in Pascal seconds (Pa.s). The relative viscosity is the whole-blood viscosity divided by that of its plasma.

Newtonian fluids at a constant temperature have a single constant viscosity as the ratio of shear stress to shear rate is constant and a certain change in the shear stress applied to the fluid will produce an exactly proportional change in the resulting shear rate. This situation is shown diagrammatically by the continuous lines in Fig. 7.2. The slope of the straight line in Fig. 7.2a will represent the viscosity of the fluid. Blood, however, is non-Newtonian because as the shear

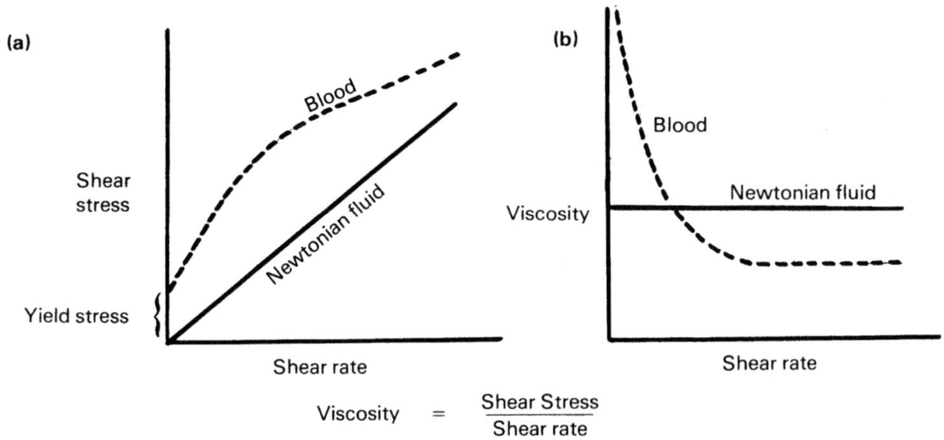

Fig.7.2. Non-Newtonian viscosity of blood. (a) Shear stress vs. shear rate (b) Viscosity vs. shear rate

stress decreases, the resulting shear rate decreases out of proportion, as shown by the interrupted lines in Fig. 7.2. The ratio of shear stress to shear rate, that is the viscosity, thus increases at lower shear rates. This non-Newtonian behaviour means that it requires a *relatively* larger force to move blood slowly than to move it fast. As can be seen from Fig. 7.2b, at higher shear rates blood does become Newtonian and normal blood reaches a constant viscosity value at shear rates above 100 s^{-1}.

Another feature of the non-Newtonian behaviour of blood is that it possesses a yield stress. This is the minimum shear stress necessary to produce any shear rate or flow. The yield stress has to be exceeded to begin flow in stationary blood and similarly flow will cease if the shear stress falls below the yield stress. This is also illustrated diagrammatically in Fig. 7.2.

Thixotropy is sometimes mentioned in relation to blood viscosity and is applied to the time dependence of viscosity; that is, a spontaneous alteration in viscosity as a consequence of the fluid being exposed to a shearing force. It has never been properly characterised for human blood largely because of methodological problems.

7.2 General Techniques in Viscometry

7.2.1 Sampling and Anti-coagulants

Venous blood is used routinely, and it is important to avoid stasis. (If it is only possible to insert the needle if the vein is artificially distended, then the occlusion should be released for a minute before a sample is withdrawn.) To avoid excessive trauma to the blood during withdrawal, a wide-bore needle (21 gauge) and minimum suction on the plunger should be used. The specimen is then placed in an anti-coagulant container and gently mixed. If allowance is made for slight differences in haematrocrit, none of the commonly used anti-coagulants other than citrate or oxalate, which may cause shrinkage of the red cell, have a significant effect on the blood viscosity (Merrill, 1969; Galluzzi et al., 1964; Meiselman et al., 1973). This has been confirmed by measurements carried out immediately without anti-coagulation. The most commonly used anti-coagulants in the literature are solid lithium heparin and ethylene diamino tetra acetic acid (EDTA).

The packed cell volume must be measured on the same sample of blood as is used for measuring viscosity. A duplicate microhaematocrit technique is perfectly adequate.

7.2.2 *Storage and Delay Before Measurement*

One of the biggest logistic problems in viscosity is the very marked effect of storage on the physical properties of whole blood. The results of measurements of whole-blood viscosity begin to change in an unpredictable fashion within a few hours of withdrawal of the blood (Barres, 1969). If the sample has been kept at room temperature all measurements should be carried out within 4-8 h (Zingg et al., 1973; Reid, 1977). At 4°C the sample may maintain its essential physical characteristics for approximately 12 h (Rosenblum and Warren, 1973). These changes will be greater if measurements are made at low rather than high shear rates.

7.2.3 *Temperature of the Viscometer*

Measurements of viscosity are usually carried out at 37°C and all viscometers should be equipped with systems for thermostatically controlling the temperature of the sample. Where readings have to be taken rapidly after delivery of the sample into the machine, as in the Contraves viscometer, equilibration of the sample to 37°C may be a problem. In this case, the blood, pipettes and the sample holder, if this is removable, should be pre-heated to 37°C in an incubator.

The viscosity of normal plasma increases by 2%-3% for every degree fall in temperature from 37°C to 15°C (Harkness, 1971). The parallel increase in whole-blood viscosity is of the same order so that in normal subjects the relative viscosity of blood is unaltered between 15°C and 37°C (Barbee, 1973; Chien et al., 1971; Goyle and Dormandy, 1976). It may be clinically relevant to carry out measurements of blood viscosity at temperatures below 37°C, for instance in investigating patients with Raynaud's phenomenon or the effect of hypothermic surgery. In some patients grossly abnormal patterns of viscosity have been described at temperatures below 37°C (Goyle and Dormandy, 1976).

7.2.4 *Calibration of Viscometer*

All viscometers should be re-calibrated every day, preferably at the beginning and end of each series of measurements to ensure that there has been no drift. Although theoretically the results of measurements can be calculated directly in absolute units in some viscometers, in practice it is wiser to calculate all readings on the basis of a daily calibration control. There are a number of calibration fluids available, usually oils, which all have the inevitable drawback of being Newtonian. Accurately standardised non-Newtonian fluids with blood-like behaviour do not exist.

7.2.5 *Determination of Yield Stress*

The yield stress of normal blood is very low and direct measurements in viscometers are of doubtful validity because of the many inevitable artefacts at low shear rates. The variable shear stress Deer viscometer should be able to measure yield stress directly, while in other conventional rotational viscometers theoretically the residual torque, after stopping movement of the sample, should also be a direct measure of the yield stress (Charm and Kurland, 1974). In practice the simplest and probably the most accurate determination of yield stress is by extrapolation from viscosity measurements at high shear rates assuming a linear relation between the square root of the shear rate and the square root of the shear stress, as originally suggested by Casson (Cokelet et al., 1963; Merrill et al., 1969).

7.2.6 *Correction for Haematocrit*

The haematocrit is by far the most critical determinant of the whole-blood viscosity *in vitro* and most workers have used some method for attempting to eliminate this effect for the following reasons: firstly, because the haematocrit in different parts of the circulation can vary at any one time as well as from hour to hour depending on the subject's activity. Secondly, the effect of the haematocrit on the measured viscosity can be so overwhelming that the influence of other determinants cannot be analysed until the haematocrit effect has been removed. Lastly, and possibly most importantly, there is mounting evidence that although haematocrit is an important determinant of viscosity in larger vessels and as measured in viscometers, the *in vivo* viscosity of blood in the majority of the circulation is independent of haematocrit (Skalak et al., 1972; Schmid-Schönbein, 1976). In order to compare the effective viscosity in the microcirculation of two examples of blood, it is therefore necessary to eliminate differences due to haematocrit, which only influence the viscosity of bulk blood.

The most unifying concept of blood viscosity and its dependence on shear rate as well as haematocrit is the concept of the effective cell volume put forward by Chien (1970). This suggests that at low shear rates the effective cell volume and therefore the viscosity is increased because red-cell aggregates "immobilise" a larger volume of surrounding plasma than individual red cells. For similar reasons the effect of haematocrit will be greater at low than high shear rates. (The effect of red-cell deformability on blood viscosity can also partly be explained on this basis, as the effective cell volume of a deformable red cell will be less than that of a rigid particle of the same volume.) A number of different mathematical relations between haematocrit and viscosity have been postulated, ranging from a double arithmetic linear to a double logarithmic linear relationship. It now seems likely that the simplest relationship, true within the physiological range, is most closely described by the following formula:

$$\log \eta = \log \eta o + kC$$

where C is the packed cell volume expressed as a percentage, η is the viscosity of that packed cell volume, $\log \eta o$ is the intercept when $C = 0$ and the slope of the line is k. The value of k increases as the shear rate gets less, in other words the effect of haematocrit becomes more critical at lower shear rates. This relationship is approximately true within the physiological range of haematocrit and it has been verified experimentally by several workers, with coefficients of correlation between 0.85 and 0.98 (Begg and Hearns, 1966; Weaver et al., 1969; Gregerson et al., 1965).

The most obvious way to determine what the viscosity of a sample would have been at a standard haematocrit is to artificially reconstitute the sample at that haematocrit. This involves centrifuging the sample and then removing some of the plasma or red cells to reproduce the chosen standard haematocrit. But there are two drawbacks to this technique. Firstly, it requires double the volume of blood so as to allow measurements at both the original (or native) haematocrit and at the reconstituted standard haematocrit. Secondly, inaccuracies may be introduced by the process of reconstitution achieving exactly the correct new haematocrit, and partly by altering the concentration of both the formed elements and the plasma constituents.

Alternatively, a correction for the haematocrit effect can be applied by calculating the regression equation for the measured viscosity and the original haematocrit. This equation can then simply be applied to correct the measured viscosity to what it would have been at the standard haematocrit (usually 45%).

7.3 Viscometers

7.3.1 Wells-Brookfield Viscometer

This was the first commercial rotational viscometer, originally described in 1961 (Wells et al.) and capable of measuring whole-blood viscosity at a range of specified constant shear rates. In all rotational viscometers the sample is sheared between two surfaces, of different complexity, moving in relation to each other. Usually one surface is static and the other can be rotated at varying speeds equivalent to different single shear rates. With the correct geometry, the whole sample can be exposed to this single shear rate. The stress exerted on one or the other surface by the sheared blood is then recorded. (The Deer viscometer, to be described later, is an exception to this principle.) The Wells-Brookfield viscometer has a cone-on-plate configuration and Fig. 7.3 illustrates the principle on which it works.

A 1-ml blood sample is placed in a detachable sample cup which is surrounded by a water jacket to control the temperature. When replaced in the viscometer a shallow cone comes into contact with the surface of the sample. The cone is rotated on the surface of the sample by a variable-speed motor to which it is connected by a berylium copper spring suspension. The torque in this spring gives a measure of the drag applied to the cone by the blood and is recorded on a rotating scale at the top of the viscometer. Different speeds of rotation will expose the sample to different shear rates (from 230 to 1.15 s^{-1}) and the shear stress can be calculated directly from the torque on the spring suspension. The cone-on-plate configuration ensures that the whole sample is exposed to the same shear rate. The velocity difference between the two surfaces will increase towards the periphery, but so will the distance between them, thereby ensuring that the velocity gradient or shear rate remains constant throughout the sample. Measurements with this machine are most accurate at maximum torques, that is, at high shear rates. The coefficient of variation of individual readings on different aliquots of the same sample varies from 2% at high shear rates (230 s^{-1}) to over 15% at the lower shear rates. The minimum shear rate at which reliable

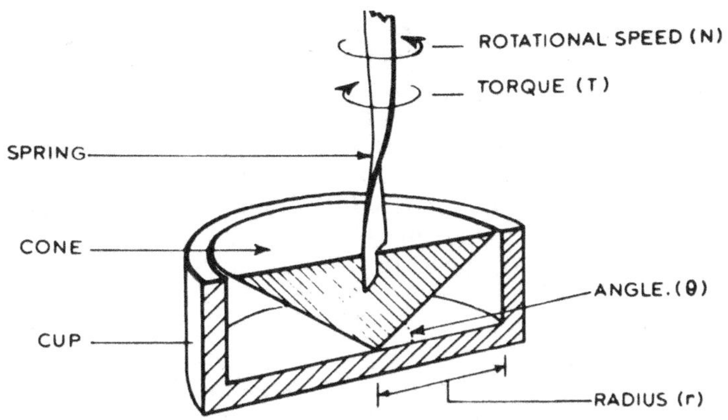

SHEAR RATE	$\dfrac{\frac{2\eta N}{60}}{\sin\theta}$	3.83N	s^{-1}
SHEAR STRESS	$\dfrac{T}{\eta r3}$	$\dfrac{T}{28.8}$	dynes cm^2
VISCOSITY	$\dfrac{\text{Shear Stress}}{\text{Shear Rate}}$	$\dfrac{T}{110\,N}$	poise

Fig.7.3. Principle of Wells-Brookfield viscometer

readings can be obtained is probably 23 s^{-1}, where the coefficient of variation is around 8%.

7.3.2 Contraves Viscometer

The Contraves viscometer is a more sensitive instrument than the Wells-Brookfield because one component of the sample chamber is used to apply the shear rate, while the stress is measured by the other component. (Compare with the Wells-Brookfield, where the cone suspended on the spring is used to apply the shear and to measure the stress, the cup being totally passive.) Figure 7.4 illustrates the principle of its operation (Oiknine et al., 1976; Spinelli and Meier, 1974).

A 0.7-ml sample is introduced into a stainless steel cup, surrounded by a water jacket which controls the temperature. A small cylindrical stainless steel bob, held by a suspension wire, is lowered into the sample cup. The bob and cup are designed so that the distance between them, containing the sample, is small compared with the cross-sectional radii of the bob and cup. Under these circumstances, the shear rate is nearly constant within the sample between the cup and the bob. The cup is rotated at a known speed by a motor, while the bob remains stationary. The output of the motor unit can be adjusted so that the rotational speeds correspond to shear rates of 90 s^{-1} down to approximately 0.01 s^{-1}. The torque applied to the bob through the sample fluid produces a turning effect on the bob, which is prevented from twisting by an equal and opposing torque, applied by an electromagnet. An automatic feedback mechanism from a mirror attached to the suspension wire regulates the strength of the electromagnetic field. The shear stress in the test fluid is directly related to the measured torque.

A time-dependent record of the electrical output, that is, the current necessary to prevent the bob from turning, has to be made at low shear rates. There are considerable problems involved with interpreting these time-dependent tracings at low shear rates, but the most commonly used technique is to extrapolate the initial linear decline on torque back to the time when the bob was lowered into the sample, as illustrated in Fig. 7.5 (Oiknine et al., 1976; Schmid-Schönbein, 1976).

7.3.3 Other Viscometers

These viscometers are more rarely used at the moment, either because they are unnecessarily elaborate or because they have only recently become commercially available.

The Weissenberg Rheogoniometer is the most sophisticated apparatus available and has long been used for measuring the viscosity of materials other than blood, such as various pastes, suspensions, emulsions and creams. Various platens can be fitted and either cone-on-plate or concentric cylindrical configurations are available for use with blood. It is very versatile, for instance it can apply an oscillatory or sinusoidal movement to the sample. It can therefore be adapted to assess the viscoelastic properties of red cells (Chien et al., 1975) as well as measuring whole-blood viscosity. Its main disadvantages are its cost, the relatively large volumes of samples required and its greater complexity; it is therefore rarely used in clinial haemorheology.

The Deer rheometer is a recent viscometer which is fundamentally different from those previously described. It neither rotates one component of the sample container while measuring the torque in the other, as in the Contraves or the Weissenberg machines, nor does it measure the torque in the transmission of the rotating element as in the Wells-Brookfield. Instead it applies a constant torque to a cylinder suspended in the sample and measures the resultant rate of rotation of the cylinder. While all the other rotational viscometers measure the shear stress at a fixed predetermined shear rate, the Deer viscometer measures the shear rate produced by a fixed shear stress. This effect is achieved by a special electric motor which can be driven at a predetermined stress. This system has the theoretical advantage of being more relevant to the circulation, where one is usually interested in the flow that will result when a certain force is applied to the blood, rather than the force necessary to produce a certain flow. In practical terms, it means that yield stress and viscoelastic properties of blood can be measured more directly.

Haake is a company which has had considerable experience with industrial viscometers, but they have only recently introduced the Biovisco, which is specially designed for blood. It is basically of the same design as the Contraves, with some theoretical advantages, but at the moment there is no published work with this machine. It could prove to be the ideal machine for sensitive routine measurements.

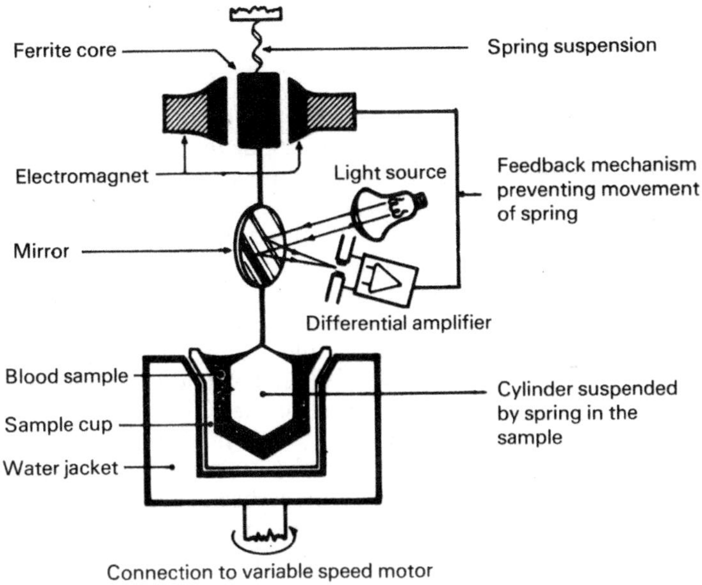

Fig.7.4. Principle of Low Shear Contraves viscometer

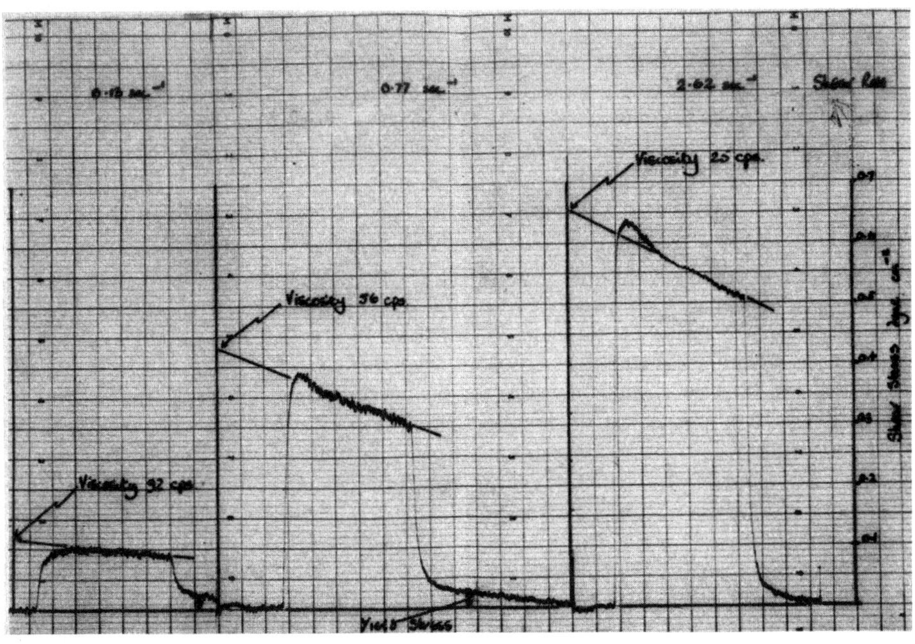

Fig.7.5. Time-dependent tracings of shear stress in Low Shear Contraves viscometer, at three different shear rates

7.4 Conclusions

7.4.1 Physiological Relevance of in Vitro Viscosity Measurements

It is appropriate to mention briefly the evidence suggesting that the *in vitro* measurements just described have a physiological relevance. This evidence falls into three groups. In a number of isolated perfusion experiments in animals the viscosity of blood could be related to the known viscosity of an artificial fluid or plasma by measuring the flow rate at fixed perfusion pressures. These values for *in vivo* viscosity could be related to simultaneous measurements of the viscosity of the same samples in viscometers. In most of these experiments a very accurate correlation was found. In many cases, however, the increase in viscosity at low flow rates measured *in vivo* was less than would be predicted from *in vitro* measurements, suggesting that the non-Newtonian behaviour of blood is less apparent in the living circulation than in viscometric measurements (Skovborg et al., 1968; Rasmussen, 1972; Baeckström et al., 1971). The second group of experiments relate measured viscosity to actual *in vivo* blood flow. For instance, in some dog experiments, changes in viscosity have been related to changes in total cardiac output, which were largely due to alterations in stroke volume rather than heart rate (Dormandy, 1970). In humans, changes in blood viscosity measured *in vitro* have been related to changes in particular circulations such as the leg (Dormandy, 1971) or the brain (Thomas et al., 1977).

Finally, there is now a mass of evidence that a high measured blood viscosity is accompanied by clinical states of circulatory insufficiency (Somer, 1975; Editorial, 1971; Burge et al., 1975; Wells, 1970; Dormandy et al., 1973a,b; Barnes et al., 1977.) Also, therapeutic measures shown to lower blood viscosity have been seen to improve patients with diseases of blood flow and have been supported by objective measurements of flow (Laks et al., 1974; Yates et al., 1979; Dodds et al., 1979; Dormandy et al., 1977).

7.4.2 Comparison of Available Viscometers

There have been few published comparisons of the results obtained with different types of viscometers but while the results using plasma are usually similar, the whole-blood viscosity is often 10% to 20% higher in a concentric cylindrical compared with a cone-on-plate rotational viscometer (Charm and Kurland, 1969; Merrill, 1969). The choice of viscometer should be determined by the purposes for which the measurements are needed. For routine measurement of plasma viscosity, the Harkness capillary tube viscometer is the best buy. For measurement of whole-blood viscosity, one of the rotational machines is required. The Wells-Brookfield is about a tenth of the cost of the others, but is only reliable at high shear rates and is therefore unsuitable for research purposes. The other four are in the same price bracket. The Deer and Haake machines are relatively new and insufficient experience makes a definitive judgement impossible; they both have features which may make them ultimately more useful. For clinical use the Contraves is undoubtedly the most practical of the established widely used machines.

7.5 Summary

There are now generally accepted techniques for measuring whole-blood viscosity using commercially available viscometers, which give reproducible results. There seems little doubt left that these results have physiological validity and that viscosity has a place in clinical medicine in the investigation and treatment of circulatory disorders.

7.6 References

Albutt C (1915) Diseases of the arteries including angina pectoris. Macmillan & Co, London

Baeckström P, Folkow B, Kendrick E, Löfving B, Oberg B (1971) Effects of vasoconstriction on blood viscosity in vivo. Acta Physiol Scand *81:* 376-384

Barbee J H (1973) The effect of temperature on the relative viscosity of human blood. Biorheology *10:* 1-5

Barnes A J, Locke P, Scudder P R, Dormandy T L, Dormandy J A, Slack J (1977) Is hyperviscosity a treatable component of diabetic microcirculatory disease? Lancet *ii:* 789-791

Barres J P (1969) Blood rheology. General review. Bibl Haemat *33:* 277-283

Begg T B, Hearns J B (1966) Components in blood viscosity. The relative contribution of haematocrit, plasma fibrinogen and other proteins. Clin Sci *31:* 87-93

Burge P S, Johnson W S, Prankerd T A J (1975) Morbidity and mortality in pseudopolycythaemia. Lancet *i:* 1266-1268

Charm S E, Kurland G S (1969) A comparison of pipette, cone on plate and capillary tube viscometry for blood. Bibl Anat *10:* 85-92

Charm S E, Kurland G S (1974) Blood flow and microcirculation. John Wiley & Sons, New York, London

Chien S (1970) Shear dependence of effective cell volume as a determinant of blood viscosity. Science 168: 977-979

Chien S, Usami S, Dellenback R J, Bryant C A (1971) Comparative haemorheology — Haematological implications of species differences in blood viscosity. Biorheology *8:* 35-57

Chien S, King R G, Skalak R, Usami S, Copley A L (1975) Viscoelastic properties of human blood and red cell suspension. Biorheology *12:* 341-346

Cokelet G R, Merrill E W, Gilliland E R, Shin H (1963) The rheology of human blood measurement near and at zero shear rate. Trans Soc Rheol *7:* 303-304

Dodds A J, O'Reilly M J G, Yates C J P, Cotton L T, Flute P T, Dormandy J A (1979) Haemorheological response to plasma exchange in Raynaud's syndrome. Br Med J *2:* 1186-1187

Dormandy J A (1970) Clinical significance of blood viscosity. Ann Roy Coll Surg Engl *47:* 211-228

Dormandy J A (1971) Influence of blood viscosity on blood flow and the effect of Low Molecular Weight Dextran. Br Med J *2:* 716-719

Dormandy J A, Hoare E, Colley J, Arrowsmith D E. Dormandy T L (1973a) Clinical haemodynamic, rheological and biochemical findings in 126 patients with intermittent claudication. Br Med J 4: 576-581

Dormandy J A, Hoare E, Khattab A M, Arrowsmith D E, Dormandy T L (1973b) Prognostic significance of rheological and biochemical findings in patients with intermittent claudication. Br Med J 4: 581-583

Dormandy J A, Goyle K B, Reid H L (1977) Treatment of severe intermittent claudication by controlled defibrination. Lancet *ii:* 625-626

Editorial (1971) Hyperviscosity syndrome. Br med J 2: 184-185

Galluzzi N J, Delashmutt R K, Connolly V J (1964) Failure of anticoagulants to influence the viscosity of whole blood. J Lab Clin Med *64:* 773-777

Goyle K B, Dormandy J A (1976) Abnormal blood viscosity in Raynaud's phenomenon. Lancet *1:* 1317-1318

Gregerson M I, Chien S, Peric B, Taylor H (1965) Investigations of blood viscosity at low rates of shear. Bibl Anat *7:* 383-384

Harkness J (1971) The viscosity of human blood plasma; its measurement in health and disease. Biorheology *8:* 171-193

Hess W (1915) Gehorcht das Blut dem allgemeinen Strömungsgesetz der Flüssigkeiten? Arch Ges Physiol *162:* 187-197

Laks H, Pilon R N, Kloevekorn W P, Anderson W, MacCallum J R, O'Connor N E (1974) Acute haemodilution: its effect on haemodynamics and oxygen transport in anaesthetized man. Ann Surg *180:* 103-109

Meiselman J J, Frasher W G, Wayland M (1973) In vivo haemorheology employing outflow viscometric techniques. Biorheology *10:* 361-373

Merrill E W (1969) Rheology of blood. Physiol Rev *49:* 863-867

Merrill E W, Cheng C S, Pelletier G A (1969) Yield stress and endogenous fibrinogen. J Appl Physiol *26:* 1-3

Oiknine C, Mandret G, Goldman M (1976) Determination du comportement rhéologique du sang a l'aide d'un viscosimètre a cylindres coaxiaux. Biorheology, *13:* 127-132

Rasmussen S N (1972) Influence of plasma hypertonicity on blood viscosity in vitro and in an isolated vasculator bed. Acta Physiol Scand *84:* 472-481

Reid H (1977) Erythrocyte deformability: its measurement in normal man and in patients with peripheral circulatory disease. Ph.D. Thesis, Univ. of London

Rosenblum W I, Warren E W (1973) Elevation of blood viscosity produced by shearing in a rotational viscometer and its inhibition by refrigeration. Biorheology *10:* 43-49

Schmid-Schönbein H (1976) Microrheology of erythrocytes, blood viscosity and the distribution of blood flow in the microcirculation. Int Rev Physiol *9:* 1-62

Skalak R, Chien P H, Chien S (1972) Effect of haematocrit and rouleaux on apparent viscosity in capillaries. Biorheology *9:* 67-82

Skovborg F, Neilsen A V, Schlichtkrull J (1968) Blood viscosity and vascular flow rate. Scand Clin Lab Invest *21:* 83-88

Somer T (1975) Rheological basis of the hyperviscosity syndrome of plasma cell dyscrasias. Bibl Anat *13:* 105-106

Spinelli F R, Meier C D (1974) Measurement of blood viscosity. Biorheology *11:* 301-306

Thomas D J, Marshall J, Ross-Russell R W, Wetherley-Mein G, Du Boulay G H, Pearson R, Symon L, Zilkha E (1977) Effect of haematocrit on cerebral blood-flow in man. Lancet *ii:* 941-942

Weaver J P A, Evans A, Walder D N (1969) The effect of increased fibrinogen content on the viscosity of blood. Clin Sci *36:* 1-10

Wells R E, Denton R, Merrill E (1961) Measurement of viscosity of biologic fluids by cone plate viscometer. J Lab Clin Med *57:* 646-656

Wells R (1970) Syndromes of hyperviscosity. N Eng J Med *283:* 183-186

Yates C J P, Berent A, Andrews V, Dormandy J A (1979) Increase in leg flow by normovolaemic haemodilution in intermittent claudication. Lancet *ii:* 166-168

Zingg W, Sulev J C, Morgan C D (1973) Study of possible sources of error in clinical blood viscosity determinations with the Wells-Brookfield viscometer. Biorheology *10:* 509-515

8 Measurement of Plasma Viscosity

J. Harkness

8.1 Introduction

Plasma is the liquid part of human blood; in the plasma are suspended the cellular elements — erythrocytes, leucocytes and platelets. By volume, plasma usually forms somewhat more than half of the blood.

Plasma is a solution of large molecules of the plasma proteins in an aqueous solution of salts and of minute amounts of many substances such as hormones, metabolic products, etc. Serum is the liquid extruded when plasma or whole blood clots by the conversion of soluble fibrinogen to insoluble and contracting fibrin.

Interest in the viscosity of whole blood dates from the middle of the last century and the quantitative studies of the physician Poiseuille. When, by the

Table 8.1 Contribution of plasma constituents to the increase of the mean normal plasma viscosity (1.64 mPa.s at 25°C) over the viscosity of water (0.8937 mPa.s at 25°C)

	Mean concentration (g/litre)	% of Total	%[a] Effect on viscosity	Characteristics of protein molecules			
				Molecular weight	Molecular dimensions (Å)		
					Length (L)	Diameter (D)	L/D ratio
Protein	78		99				
Albumin	45	58	36	69000	150	38	3.95
Globulin	30	38	41	160000	235	44	5.34
Fibrinogen	3	3.8	22	341000	700	38	18.4
Non-proteins			1-2				

[a]Lawrence (1961)

beginning of this century, clinicians showed interest in the rheology of blood, they worked on whole blood and serum. The viscosity of serum was only one of its physical properties which were recorded in a variety of diseases. Similar studies were made on plasma when preservation of non-clotted specimens by the addition of anti-coagulants became popular in the 1920s. It must be stressed, however, that there was never any suggestion that the serum viscosity or plasma viscosity could be used as a clinical index in itself.

The concept of measuring the plasma viscosity as a clinical pathology procedure stems from the researches of Whittington, who studied the physics of the erythrocyte sedimentation rate test (BSR; ESR) (Whittington, 1942), and who, in collaboration with Miller, showed a correlation between the clinically assessed severity of pulmonary tuberculosis and the degree of increase in the plasma viscosity (Whittington and Miller, 1942). We have continued to investigate the relationships between the clinical states in a wide variety of diseases and the changes in the plasma constituents and the resulting plasma viscosities.

In the late 1940s other workers measured the plasma viscosity, mainly in rheumatic diseases, and confirmed its value as an index of the severity of disease processes. At the same time comparison was made with the ESR results measured in parallel and there was universal agreement that the plasma-viscosity test was superior in all aspects; indeed, this critical assessment of the ESR test now revealed it to be an insensitive and unreliable index. At the same time there were difficulties in the laboratories over instrumentation, choice of anti-coagulant, choice of standard temperature, and the prolonged technician time required by the viscometers then in use. By 1950 the plasma-viscosity test had virtually lapsed.

Interest revived and general acceptance followed after the design by the author of a semi-automated viscometer (Harkness, 1963; Whittington and Harkness, 1965) whereby the viscosity could be measured with precision using small volumes with temperature control and using one stipulated anti-coagulant.

Early studies of the viscosity of the ultrafiltrates of serum and plasma had shown just how little is contributed to the viscosity of the parent liquid by the non-protein constituents. Even the extreme changes of uraemic and diabetic comas produce no significant effect on the serum or plasma viscosity. The viscosity of the plasma depends almost entirely upon its protein fractions. The effect on the viscosity by the different proteins increases with molecular size and as the molecular shape departs from the spherical, as indicated by the length/diameter ratio. Thus the large elongated molecule of fibrinogen exerts a viscosity effect in the plasma equal to half of that of the globulin, although its concentration is only one-tenth (Table 8.1); similarly, globulins have a greater effect on the viscosity than the albumin although the globulin concentration is lower.

The question as to whether plasma is a Newtonian liquid has not yet been resolved. Nevertheless, for practical clinical purposes, it can be treated as if it were truly Newtonian so that the rate of shear of the viscometer is not important. This is in contrast to the cell-plasma mixture of the original blood, which is markedly non-Newtonian and whose apparent viscosity value depends upon the rate of shear at which the determination was carried out.

8.2 Technique of Plasma Viscosity Measurement

8.2.1 Preparation of the Patient

In section 8.3 below, it will be shown how little physiological stimuli affect the plasma viscosity. Provided such gross stimuli as ingestion of pints of fluid or exercise to exhaustion can be avoided, no precautions need be demanded.

8.2.2 Collection of Venous Blood Samples

Using a dry sterile plastic syringe and a large-bore needle, a venous blood sample is collected from the antecubital fossa with minimal stasis.

8.2.3 Preparation of Plasma Specimen

8.2.3.1 Natural plasma: Immediate transfer of the blood to a plastic centrifuge tube for a 3-min high-speed separation will allow a viscosity measurement before coagulation.

8.2.3.2 Anti-coagulants: When compared with natural plasma, all anti-coagulants in general use today have a measurable effect upon the plasma viscosity; some cause an increase, others a decrease. EDTA (sequestrene) at a concentration of 2 mg per ml of blood has least effect (increase by 0.6%).

8.2.3.3 Centrifugation: Centrifugation at 3000 *g* for 3-5 min produces adequate separation. Prolonging the spinning to 30 min in order to remove the platelets yields the same viscosity result. It is essential that the centrifuge tube be stoppered to prevent the loss of fluid.

8.2.3.4 Delay in measurement: In blood specimens, protein-free fluid passes slowly from the plasma into the erythrocytes and thereby the plasma protein concentration and the plasma viscosity increase.

Plasma separated at any time of the day of collection may be tested on that day. If separated and transferred to a small sterile stoppered plastic container it can remain on the bench at room temperature for up to a week unaltered; it must not be stored in a refrigerator as the very large protein-lipid complexes are broken down by the cold into smaller, less viscous molecules.

Provided that the erythrocytes have initially sedimented rapidly, that the specimen remains undisturbed on the bench, and that the specimen has not been remixed prior to centrifuging, on the following day a satisfactory result can be obtained. If the cells have initially settled slowly, or if the specimen has been remixed after settlement, the plasma viscosity will have been increased by 0.03-0.08 mPa.s. Postal specimens of whole blood will suffer in this way: separated plasma is much less susceptible to damage and has been transmitted successfully by post. Specimens of blood which are 2 days old or more should not be examined.

8.2.4 Viscometers

The type and model of the viscometer to be selected will be governed by the requirements of clinical pathology laboratories in their attempts to meet the demands by their clinician colleagues for large numbers of tests. It is possible to perform several hundred tests per day (about 50 per hour) and these are best handled by a single technician and a single instrument. The viscometer should be robust and simple to use. The volume of the specimen should be small — e.g. 0.2-0.5 ml. The greatest source of error lies in tiny blood clots or specks of fibrin blocking orifices and capillaries. These must be observable and capable of simple and rapid removal. To meet the assured detection of the 0.05 mPa.s difference between two samples (which is an indication of a significant change) requires that the overall experimental error of the plasma viscosity measurement does not exceed 1%. Such a degree of precision is more likely to be reached by capillary than rotational cylinder or cone-and-plate instruments.

A suitable model of a capillary viscometer capable of measuring both plasma and whole blood has been previously described (Harkness and Whittington, 1971). This has been modified so that an average normal blood will have a Newtonian rate of shear at the wall equal to 360 $^{-1}$, which is within the range of values for the major blood vessels in the human (Whitmore, 1968). For a plasma viscosity equal to 1.63 mPa.s at 25°C, the Newtonian rate of shear at the wall equals 900 s^{-1}. A variety of working temperatures is available.

8.2.5 Temperature

For most plasmas — whether from healthy or diseased subjects — the sensitivity to temperature change is approximately 2%-3% per degree Celsius between say 15° and 40°C. Table 8.2 presents the range of normal plasma-viscosity results determined at three temperatures.

When the plasma viscosity is being measured as a clinical pathology index of disease activity, the author uses and recommends 25°C rather than the more obvious 37°C. Again this choice is governed by the need to keep the experimental error below 1%. Opportunities to introduce errors increase with the difference between the ambient room temperature and the working temperature — the measurement at 25°C is within the capacity of all grades of medical laboratory technicians, whereas working at 37°C requires the attentions of an expert.

Table 8.2 The plasma viscosity in normal subjects

Temperature °C	Sex	Number of tests	Plasma viscosity (mPaS) Range	Mean	Standard deviation
20	—	—	1.70-1.915	—	—
25[a]	Male	96	1.50-1.72	1.64	0.049
	Female	63	1.50-1.72	1.63	0.056
	Combined	159	1.50-1.72	1.64	0.052
37	—	—	1.16-1.35	—	—

[a]Harkness (1971)

8.3 Plasma Viscosity

8.3.1 Variations in Health

In a healthy adult population the plasma-viscosity values fall within a restricted range (see Table 8.2). This corresponds to similarly restricted ranges for the plasma total protein and the plasma protein fractions. This implies a strict dynamic equilibrium between the formation of new proteins at a variety of sites in the body to compensate for continual protein destruction.

Longitudinal studies have shown that a healthy individual keeps his or her own plasma-viscosity value remarkably constant, even for decades. A temporary change by 0.03-0.05 mPa.s may occur in response to physiological stimuli (Gibson, 1949; George, 1967). Because of the width of the normal range occasionally it is found that a subject whose plasma-viscosity value usually lies at the lowest margin of this range has had a significant increase (i.e. in excess of 0.05 mPa.s) while still remaining within the range.

8.3.2 Sex

The ranges, means, and standard deviations for normal plasma-viscosity values are the same for both sexes (cf. ESR).

8.3.3 Age

The figures given in Table 8.2 apply from the age of 3 years up to 65 years.

The author knows of no plasma-viscosity work on children under 3 years of age, but from the reported figures for their plasma proteins, their plasma-viscosity levels would be expected to be low. At birth, a baby's plasma contains less protein per millilitre than its mother's. The difference is specially marked in the IgA globulin fraction, and is less so in the IgG; however, the IgG falls further during the first few months. Thereafter the protein concentrations gradually increase, to reach the adult levels about the age of 3-5 years.

After middle age there is a slight but continual increase in the plasma fibrinogen. There is no agreed figure to indicate by how much the normal range should be extended upwards in the over-65s to allow for this fibrinogen increase.

8.3.4 Diet

The ingestion of ordinary amounts of food and drink in ordinary meals does not affect the viscosity of the plasma. The lactescent plasma resulting from a high-fat meal has the same viscosity as the clear pre-meal specimen. Sato (1936) found that the ingestion of 1 litre water or physiological saline produced a haemodilution and a reduction of the plasma viscosity by 0.05-0.08 mPa.s.

8.3.5 Exercise

In the short term, violent exercise can produce dehydration and increased plasma viscosity. In the long term, Charm et al. (1979) have recently reported that the mean plasma viscosity of a group of "joggers" was lower than the mean of a matched group of "non-joggers". One hour's continual swimming and the exercise involved in a laboratory technician's working day failed to make any consistent or significant change in the plasma viscosity.

8.3.6 Menstruation

No systematic study of this has yet been made but menstruation does not upset the steady levels of serial testing in female patients. (See however Chapter 16.3.2-Ed.).

A study is required in subjects who suffer from pre-menstrual tension where water retention is held to be a causal factor.

8.3.7 Pregnancy

During the course of a normal pregnancy there is a continual increase in the plasma fibrinogen and a progressive fall in the globulin and albumin. These protein changes have counteracting effects upon the plasma viscosity with the result that the plasma viscosity remains within the usual normal range for the first and second trimester. (See however Chapter 16.5.2—Ed.) In most normal pregnancies the plasma viscosity level is within the normal range during the third trimester also (c.f. ESR).

8.4 Plasma Viscosity in Disease

The plasma viscosity in disease has been reported on by at least 20 groups who studied serial tests in individuals and statistical analyses of large groups of patients. [See Review Article, Harkness (1971), for extensive bibliography.] The following summary is derived from these reports.

For the plasma-viscosity test, diseases may be divided into two groups. In the first there is little or no change in the plasma protein pattern so that the plasma viscosity also shows little or no increase — for example, in the degenerative conditions such as osteoarthritis. In the second group there is definite change in the plasma protein pattern and a corresponding increase in the plasma viscosity — for example, in acute and chronic infections, severe tissue damage, collagen diseases, metastatic malignancy, etc. The incidence of this second group is shown by the claim by Lawrence (1961) for a 59% incidence of abnormal plasma viscosities in patients on admission to a medical ward (c.f. 43% with abnormal ESRs.).

In acute inflammation (e.g. pneumonia) or acute tissue destruction (e.g. cardiac infarction) the plasma protein change begins within a few hours. Within 24 h the fibrinogen may have increased by 1 g/litre; and such is the effect of its large elongated molecule upon viscosity, that the plasma viscosity will have become abnormally high. Claims for the sensitivity of the plasma-viscosity test are based upon the speed of such changes. Direct estimation of the fibrinogen is a much slower and more complex technique than its indirect estimation by viscometry.

Another protein which increases in the first few days of an acute illness is the α_2-macroglobulin whose large molecule (molecular weight = 820,000) has a high viscosity influence. The α_1 globulin may also be increased early, but only temporarily. By the end of the first week the γ globulins will have increased and the albumin will have fallen.

In chronic diseases (e.g. pyogenic infection, syphilis, tuberculosis, rheumatoid arthritis, systemic lupus erythematosis, temporal arteritis, and metastatic malignancy) there is usually an increase in the plasma viscosity, reflecting a raised globulin, a lowered albumin and a normal or raised fibrinogen. The degree of the increase in the viscosity and the upset of the protein pattern is related to the clinically assessed severity of the disease process. Changes in the elevated plasma-viscosity values during serial testing reflect either a change in the main disease or the action of some intervening complication.

"Failures" of the plasma-viscosity test do occur. The incidence of false positives (i.e. a raised plasma viscosity without a clinical explanation) becomes lower as the clinical examinations become more thorough. For example, among hospitalized mentally defective patients their are some who make no complaint and who have no disease detectable by their clinician but nevertheless have an abnormal plasma viscosity and an abnormal plasma protein pattern; the proportion is even higher in patients with Down's syndrome; similar patients living in the general community do not show either of these abnormalities. Eastham et al. (1965) explained the difference as the hospital patients' reaction to their exposure to cross-infection, which is frequent in institutions.

False negatives (i.e. normal plasma viscosities in the presence of obvious disease) result from the severe depression of plasma protein formation such as occurs in the terminal stages of malignancy. Some patients will show no change in their plasma proteins and their plasma viscosity when suffering from diseases in which most patients do have plasma changes. This difference in response merits an intensive search for an explanation rather than a mere dismissal as a failure of the tests.

The clinician can use the plasma-viscosity test in several ways. In diagnosis, an abnormal plasma viscosity is very strong evidence of the presence of a disease known to cause such an elevation; similarly, a normal plasma viscosity is against its presence. In prognosis, a guide to the severity of the disease comes from the amount by which the plasma viscosity is increased. Changes in the plasma viscosity arising during serial testing mirror the underlying disease processes. In treatment, the plasma viscosity reflects the patient's response; cure or quiescence may be indicated by the attainment and maintenance of normal plasma-viscosity levels.

8.5 Comparison of the Plasma Viscosity and the ESR

The early papers, referred to above, gave much space to demonstrating the superiority of the plasma-viscosity test over the ESR. This evidence need not be repeated. As more laboratories adopt the plasma-viscosity test as the technique of choice, and discard the ESR, the conflict can now be regarded as settled.

There is one situation only in which the ESR retains a significant importance. In temporal arteritis and its kindred conditions, stopping steroid treatment too soon, before the disease is totally quiescent, may incur the dreadful complication of blindness. The ESR reaches normal levels earlier than the plasma viscosity in the majority of patients and it is then essential for treatment to be continued at least until the plasma viscosity also becomes and remains normal. There are a few instances where it is the ESR which persists at abnormal levels and stopping treatment at this stage is perilous. The ESR should be retained solely as a monitor for normal plasma-viscosity values during steroid therapy of temporal arteritis.

8.6 Hyperviscosity Syndromes

The symptoms of the patients who suffer from this syndrome arise from the marked increase in the viscosity of the plasma and these are relieved rapidly when treatment, such as plasmapheresis, produces a reduction in the viscosity. There are two main diseases. Multiple myeloma has a plasma viscosity increase due to monoclonal protein fractions with a size and shape similar to the proteins normally present in the plasma; the IgG is increased in about 80% of patients, and the IgA in about 20%. Waldenström's macroglobulinaemia has an increased IgM globulin fraction which has a very large molecule formed from the union of five normal-sized globulin molecules and which sometimes has the characteristics of a cryoprotein.

Viscometric studies of the plasma are very applicable to these two conditions. A chance discovery of a very high plasma viscosity may be the first clue to the diagnosis in a patient with a vague illness; a macroglobulin in the plasma may be identified by the reduction in its viscosity when treatment of a plasma specimen with dilute mercaptoethanol causes the breakdown into five smaller, less viscous globulin molecules; a cryoglobulin can be identified in a much shorter time and with greater confidence than with the standard stand-in-the-refrigerator-overnight technique by measuring the viscosity at two temperatures, when the difference may far exceed the normal 2%-3% per degree Celsius (Harkness and Whittington, 1970); and the response to treatment can be monitored by serial testing. When attempting to estimate the strain on the heart due to the high viscosity of the blood, results obtained at 25 °C cannot be used and it is absolutely necessary for the viscosity measurements to be made at actual body temperature. Instances have been found where the extremely high plasma viscosity at 25 °C was due to cryoglobulins which increased the viscosity by 400% on the temperature being reduced by 12 °C (instead of the usual 30%); such a plasma and a severe anaemia could combine to produce a blood whose viscosity at 37°C was near the upper limit of the normal range.

8.7 Conclusions

It is a simple technical procedure to measure the viscosity of plasma with speed and accuracy.

The plasma viscosity is an index of the plasma proteins.

The plasma proteins and the plasma viscosity change in some diseases. The degree of the plasma-viscosity change bears a close relationship to the patient's general reaction in response to these diseases.

A knowledge of the patient's plasma viscosity will aid the clinician in diagnosis, prognosis and treatment.

8.8 References

Charm S E, Paz H, Kurland G S (1979) Reduced plasma viscosity among joggers compared with non-joggers. Biorheology *16:* 185-191

Eastham R D, Jancar J, Duncan E H (1965) Plasma viscosity in mental deficiency and Down's syndrome. Br J Psychiatry *111:* 999-1002

George W H S (1967) Relative plasma viscosity with reference to patients with Diabetes Mellitus. Proc Assoc Clin Biochem *4:* 169-174

Gibson H J (1949) Observations on the relative viscosity of blood plasma in comparison with other empirical blood tests on the rheumatic diseases. Proc Roy Soc Med *42:* 734-744

Harkness J (1963) A new instrument for the measurement of plasma viscosity. Lancet *ii:* 280-281

Harkness J (1971) The viscosity of human blood plasma; its measurement in health and disease. Biorheology *8:* 171-193

Harkness J, Whittington R B (1970) Blood plasma viscosity: an approximate temperature — invariant arising from generalized concepts. Biorheology *6:* 169-187

Harkness J, Whittington R B (1971) The viscosity of human blood plasma: its change in disease and on the exhibition of drugs. Rheol Acta *10:* 55-60

Lawrence J S (1961) Assessment of the activity of disease. H K Lewis, London

Sato J (1936) On micro-method of viscosity measurement using Mess Determann's viscosimeter. Nagoya J Med Sci *10:* 187-190

Whitmore R L (1968) Rheology of the circulation. Pergamon Press, Oxford

Whittington R B (1942) Blood sedimentation; study in haemo-mechanics Proc Roy Soc London *131:* 183-190

Whittington R B, Harkness J (1965) A new capillary-type viscometer. In: Copley A L (ed) "Symposium on Biorheology, Proc of Fourth International Congress on Rheology", Part 4. Interscience, John Wiley, New York, pp 267-269

Whittington R B, Miller A K (1942) Haemomechanics and sedimentation rates. Tubercle *23:* 195-204

Section C

Blood Rheology, Blood Flow and Vascular Occlusion

9 Relationship of Blood Rheology to Blood Flow

D. Charlesworth

9.1 Introduction

The relationship between volume flow rate and the pressure gradient in a rigid pipe is expressed in Poiseuille's law; in mathematical terms this is:

$$Q = \mathrm{k}\,\frac{(P_2 - P_2)d_i^4}{L}$$

and a variant of this is:

$$\bar{V} = \frac{(P_1 - P_2)r_i^{\,2}}{8L\eta}$$

where Q = volume rate of flow
V = mean velocity, $Q/\pi r_i^2$
k is a constant dependent upon the viscosity of the fluid
$(P_1 - P_2)$ is the drop in pressure along a tube, length L.
L = length of tube
d_i = internal diameter of tube
r_i = internal radius of tube
η = viscosity of fluid

This relationship holds good only in certain specific circumstances i.e. steady flow of a Newtonian fluid in a pipe with rigid walls. This clearly is not the case when one considers the flow of blood which is pulsatile and passes through arteries with visco-elastic walls. However if we accept for a moment that Poiseuille's law is a reasonable approximation in these circumstances it is clear that a reduction in the viscosity of blood would increase mean velocity in the arteries, and this would be a logical way of improving an ailing circulation.

The arterial system can be considered to consist of two parts, the major arteries through which blood is transported to the periphery and the arterioles and capillaries which together form the peripheral resistance. The role of the main arteries is to transport blood with as little loss of energy as possible. This function is compromised by arteriosclerosis. Before we consider what effect a reduction in the viscosity of blood would have on the circulation of a patient with arteriosclerosis, we should consider first the effect this might have on the transport of blood through the large arteries.

Mean pressure in the arch of the aorta and its major branches is maintained at a fairly steady level by a complex series of physiological reflexes, and the rate at which blood flows to the periphery is controlled by the peripheral resistance. It is usual to say that blood flows along a pressure gradient but more precisely it flows along an energy gradient (Burton, 1965). Mean flow is away from the heart, but the direction of flow is reversed in the major arteries for part of the cardiac cycle.

In arteriosclerosis the major arteries become less compliant, stenosed, and eventually may be blocked completely. Blood is carried round a block by small "collaterals" which are in effect long stenoses and represent a major discontinuity in the artery. Minor degrees of stenosis produce little or no measurable effect on mean flow but they do alter the amplitude of the pressure and flow waves in addition to reducing the energy of the pulse. Initially potential energy is lost but when this is exhausted, kinetic energy will be lost and in consequence the velocity of flow reduced (Beurger and Hwang, 1974; Cave and Charlesworth, 1978). Energy is lost by reflections at discontinuities in the artery and in disturbed flow, i.e. as noise and heat in turbulent flow distal to a stenosis. The net result is that, in patients with arteriosclerosis, when blood eventually reaches the arterioles there has been a considerable loss of pulsatile energy. What we should consider is whether this loss will be augmented by a reduction in the viscosity of blood.

When flow is laminar and the velocity of flow fairly low, rate of flow is directly proportional to the pressure gradient. But at higher velocities flow may become turbulent, in which case the pressure gradient is proportional to the square of the flow rate. The point at which transition from laminar to turbulent flow takes place can be predicted from the Reynolds number (*Re*). This is the ratio of inertial to viscous forces.

$$Re = \frac{\bar{V}d}{\nu}$$

where $\bar{V}$ = mean flow
d = diameter of tube
and ν = kinematic viscosity = $\frac{\text{viscosity}}{\text{specific gravity}}$

As the Reynolds number is increased small disturbances in flow are less well damped and a critical Reynolds number is reached at which point flow becomes turbulent. The critical Reynolds number is about 2000-2300, and factors which lead to turbulence are abrupt changes in the calibre of the artery or roughness at the wall. Arteriosclerosis provides just such circumstances and a reduction in viscosity will increase any tendency to turbulent flow. When flow is pulsatile the extent to which the velocity profile differs from that in laminar flow can be expressed as a dimensionless parameter, the Womersley number, α:

$$\alpha = \frac{d}{2}\sqrt{\frac{\omega}{\nu}} \quad \text{where } \omega = \text{angular velocity}$$

If the Reynolds number is plotted against α for flow in the major arteries of several species including man, the points fall close to the boundary which marks the transition from laminar to turbulent flow (Nerem and Seed, 1972). The inference is that in normal people it would take only minor disturbances to create turbulence.

Since viscous forces affect these relationships we can make several predictions about what is likely to happen if the viscosity of blood is reduced in patients with arteriosclerosis. The results would be:

An increase in the likelihood of flow being turbulent and hence an increase in energy losses.

Viscous forces damp out reflections from the peripheral vascular bed and from bifurcations, stenoses and sections of the arteries which have become incompliant. A reduction in the viscosity of blood may result in an increase in the amount of energy lost due to reflections.

9.2 Methods

We carried out several experiments, the results of which have a bearing on these problems. We carried out experiments on a model (Fig. 9.1) in which flow was

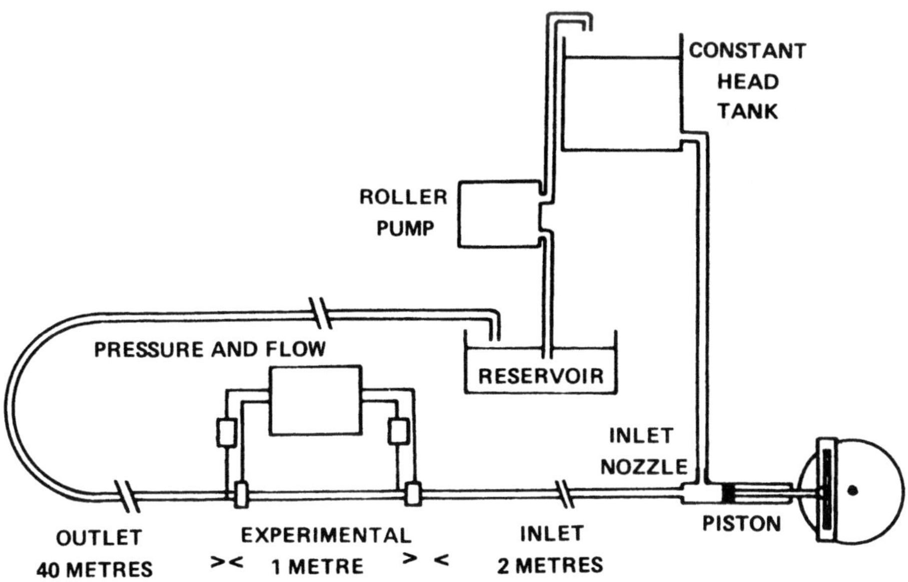

Fig.9.1. The hydraulic model

pulsatile and the conditions of flow (pressure amplitude, rate of flow, *Re* and α) mimic closely those in the femoral artery in man. Axis-symmetrical stenoses of known length and diameters were introduced into the model and their effect on energy losses calculated. Results showed that loss of pulsatile energy was proportional to the bore and length of a stenosis and that losses were greater for less viscous fluids.

9.3 Results

The total resistance of the tube plus stenosis varied for different degrees of constriction and for fluids of different viscosities. There was an increase in the resistance of all stenoses when fluids of high viscosity were used (Fig. 9.2).

The effects of an arterial stenosis can be quantitated in terms of energy losses. A combination of haemodynamic variables, *Re* and α, can be derived to give a dimensionless index of unsteady flow, the Strouhal number.

$$\text{Strouhal No.} = \frac{\alpha^2}{\pi Re}$$

Using this concept it is possible to predict both the area of a stenosis and the energy losses across it (Reul et al., 1972). Energy losses calculated from our experiments were compared with the Strouhal number (Fig. 9.3). Although our results differed from those of Reul and his associates there is a relationship between the two and losses were greater for less viscous fluids.

We also calculated peripheral resistance and input impedance from measurements (taken at operation) of pressure and flow in the popliteal arteries (Humphreys et al., 1976). Measurements were repeated after an infusion of 1 litre Dextran 40 and viscosity was measured from samples of blood taken at the time of each measurement. Results showed that after an infusion of 1 litre Dextran 40 the packed cell volume decreased in all patients (mean decrease 14.9% ± 1.4%). All the parameters of blood viscosity changes, i.e. viscosity at two low rates of shear (0.675 s^{-1} and 0.078 s^{-1}) and the asymptotic viscosity ($p < 0.01$; Students' *t* test). When the viscosities were converted to a standard haemotocrit of 45%, using regression equations calculated from a large group of normal volunteers, there were no changes from the pre-infusion values. In every patient mean flow increased by an average of 33%. Mean pressure increased slightly but varied from person to person. The peripheral resistance decreased in all patients (mean change 16.2 ± 3.5%; $p < 0.005$). There was no correlation between the fall in viscosity and the change in peripheral resistance, but there was a good correlation between change in resistance and the initial resistance ($r = 0.75$). These results suggest that the fall in peripheral resistance was due to the increase in blood volume and that small changes in the viscosity of blood brought about by drugs are unlikely to improve total blood flow.

We also reduced the viscosity of the blood in a group of patients with pre-gangrene (Humphreys et al., 1977). Patients were given injections of Malayan pit viper venom (Ancrod/Arvin) which lowered the level of plasma fibrinogen. Reductions in fibrinogen levels were highly significant ($p < 0.001$) as were the decreases in all the indices of blood viscosity (viscosity at low shear, asymptotic viscosity and yield stress). The greatest change was in the corrected yield stress of

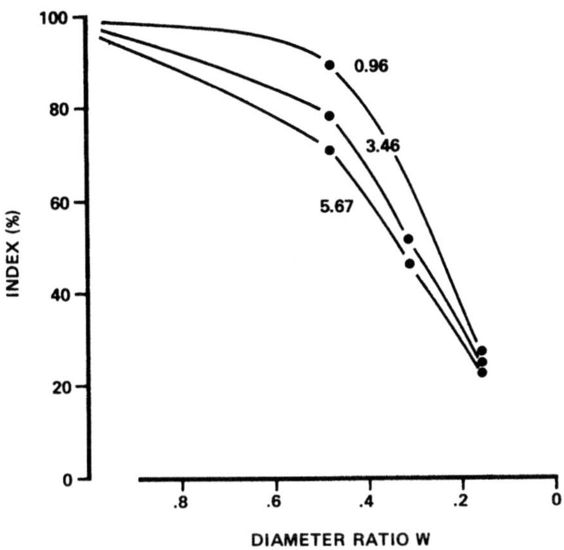

Fig.9.2. Graph of change in resistance (expressed as the ratio of resistance of tube to resistance of stenosis) on the vertical axis, and the degree of stenosis (expressed as a diameter ratio *W*) in the tube. The lines represent fluids of different viscosities: saline (0.97 cP), fluid A (3.46 cP), and fluid B (5.67 cP). 1 centipoise (cP) = 1mPa.s

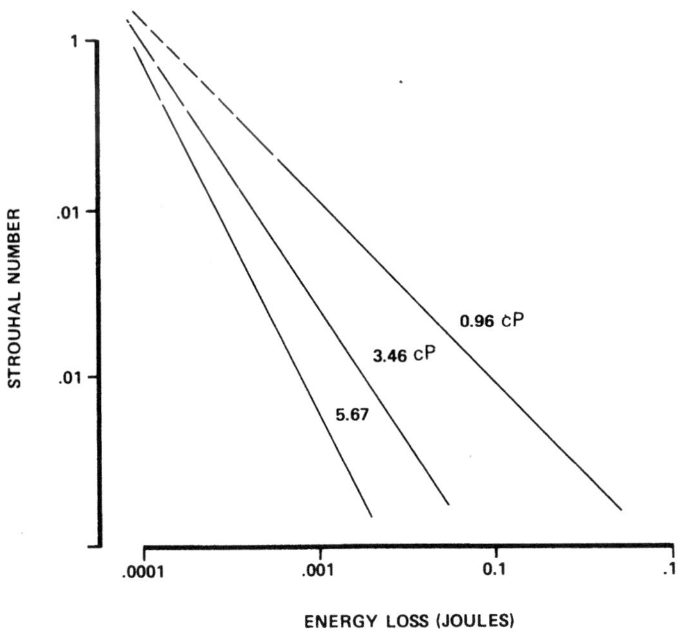

Fig.9.3. A graph of the Strouhal number against calculated energy loss, in Joules. The three lines represent three fluids of different viscosities (saline, fluid A, and fluid B, as Fig.9.2). It can be seen that at any Strouhal number energy losses are greater for less viscous fluids. 1 centipoise (cP) = 1 mPa.s

30.2% ($p < 0.001$). Despite these changes there was no change in the energy of the flow pulse at the knee or ankle and no increase in mean pressure in the arteries at the ankle. Clinically the changes had little or no effect on the patients' symptoms or the course of their illness.

9.4 Discussion

These results could have been predicted if one considers the effects which decrease in viscosity has on the expenditure of pulsatile energy. Energy losses depend upon the visco-elastic properties of the vessels, reflections, disturbed flow and frictional losses at the wall. A reduction in the viscosity of blood will increase these losses and when power is dissipated in this way it will be reflected by a decrease in flow. It seems likely that in patients with arteriosclerosis affecting the major arteries, the additional loss of pulsatile energy brought about by a decrease in the viscosity of blood is likely to outweigh any advantage which might accrue from a reduction in peripheral resistance.

9.5 References

Beurger R, Hwang N H W (1974) Critical arterial stenosis. A theoretical and experimental solution. Ann Surg *180:* 39-50

Burton A C (1965) Physiology and biophysics of the circulation. Medical Year Book Pub, Chicago, Chapter 6

Cave F D, Charlesworth D (1978) Unpublished data in Ph.D. Thesis, F D Cave, University of Sussex

Humphreys W V, Walker A, Cave F D, Charlesworth D (1976) The effect of an infusion of low molecular weight Dextran on peripheral resistance in patients with arteriosclerosis. Br J Surg *63:* 691-693

Humphreys W V, Walker A, Charlesworth D (1977) A report on the use of Arvin in patients with pregangrene of the lower limb. Br J Surg *64:* 31-33

Nerem R M, Seed W A (1972) An *in-vivo* study of aortic flow disturbances. Cardiovasc Res *6:* 1-14

Reul H, Schoenmackers J, Starke W (1972) Loss of pressure, energy and performance at simulated stenoses in pulsatile quasiphysiological flow. Med Biol Eng *10:* 711-718

10 Haematocrit, Blood Viscosity, Cerebral Blood Flow, and Vascular Occlusion

T.C. Pearson, P.R.D. Humphrey, D.J. Thomas and
G. Wetherley-Mein

10.1 Introduction

Initially the evidence for a positive relationship between venous haematocrit (VH) and vascular occlusive events will be examined. In the next section the effect of haematocrit change on the rheological characteristics of blood will be presented. This will be followed by presenting evidence for the influence of blood viscosity on cerebral blood flow. Finally, the association between VH, blood viscosity, cerebral blood flow and vascular occlusive events will be discussed.

10.2 Venous Haematocrit and Vascular Occlusive Events

The relationship between VH and vascular occlusive events is considered in five main groups of patients. First, in primary polycythaemia at presentation;

secondly in primary polycythaemia in the treatment phase; thirdly, in patients with abnormally elevated haematocrits but without true polycythaemia; fourthly, in individuals with haematocrit values in the normal range; and finally, in patients with peripheral limb ischaemia.

10.2.1 Vascular Occlusive Events in Primary Polycythaemia at Presentation

An elevated VH is seen in the majority of patients with primary proliferative polycythaemia (polycythaemia rubra vera) at presentation. Although there are some discrepancies in the classification of the presenting clinical features of primary proliferative polycythaemia (PPP), the incidence of vascular complications has been found to be high in all published series (Norman and Allen, 1937; Lawrence *et al.*, 1953; Barabas *et al.*, 1973). Pearson (1977) found that 57%, 33 out of 61 patients, presented with vascular complications, 14% of which were venous. In idiopathic erythrocytosis (IE), patients with definite polycythaemia but without the other features of PPP or secondary polycythaemia, a similar high incidence of vascular complications occurred at presentation. Nineteen (63%) out of 30 of these patients presented in this way (Pearson and Wetherley-Mein, 1979). Venous occlusive events accounted for 26% of these vascular complications.

Two relevant features of both PPP and IE are the elevated VH and the vessel wall changes due to the age of these patients at presentation. It is probably these factors that give rise to the high incidence of vascular occlusion in both conditions. In PPP, unlike IE, quantitative and qualitative platelet changes are present. These platelet changes are probably the factor producing microvascular lesions which cause the symptom of painful burning of the fingers and toes and may produce pre-gangrenous changes of the toes. This type of presentation was seen in 11 out of 61 patients with PPP but in none of the patients with IE. The fact that similar microvascular lesions are seen in primary thrombocythaemia without polycythaemia (Preston et al., 1974; Singh and Wetherley-Mein, 1977) also supports the view that these symptoms and lesions in PPP are due to platelet abnormalities.

The site and sex distribution of the arterial complications in PPP show differences from those of arteriosclerosis. Barabas et al. (1973) compared their findings in PPP with those of Taylor and Calo (1962) in arteriosclerosis. They found that in PPP men and women were almost equally affected, while in arteriosclerosis males predominated, and that stroke was more common than "coronary artery disease" in PPP, while the reverse was true in arteriosclerosis. This disparate incidence of stroke and coronary thrombosis in PPP has also been established by other observers (Videbaek, 1950; Chievitz and Thiede, 1962; Pearson, 1977). Chievitz and Thiede, for example, found that 32% of 250 patients with PPP had a cerebral thrombosis during the course of their disease. Death due to cerebral thrombosis was five times that in a comparable, age-matched, non-polycythaemic population. In contrast, the incidence of coronary thrombosis they found in PPP was the same as that of the non-polycythaemic population. These findings suggest that some parts of the circulation are particularly affected by the blood changes in PPP.

10.2.2 Vascular Occlusive Events in the Treatment Phase of Primary Polycythaemia

In a retrospective study, Pearson and Wetherley-Mein (1978) found that there was a positive correlation between the incidence of vascular occlusive episodes and VH during the treatment of patients with PPP. Vascular occlusive episodes were shown to rise from 0.2 episodes/10 patient-years in the VH range 0.40-0.44, to 7.5 episodes/10 patient-years in the VH range of 0.60 and above. The high incidence in the high VH range is not particularly surprising and resembles the incidence in untreated patients. More important was the observation that the incidence of episodes (0.92/10 patient-years) was greater when the VH was in the high normal range than in the low normal range (0.20/10 patient-years). This reflects the findings in the normal population to be discussed later.

During the follow-up period, venous and arterial occlusive events occurred with equal frequency (Pearson and Wetherley-Mein, 1978). However, at presentation arterial events predominated. This difference is probably an artefact due, first, to the fact that during follow-up patients are closely supervised and, secondly, to the not uncommon failure to consider a diagnosis of PPP when patients present with venous occlusive events.

In the same study, it was found that there was an increased incidence of occlusive episodes at higher platelet counts, particularly above 400×10^9/litre, but this was not statistically significant. Dawson and Ogston (1970) also found a higher incidence of thrombosis in the follow-up period of patients with PPP and platelet counts in excess of 270×10^9/litre than they did in those with platelet counts below this level. In other studies (Szur et al., 1959; Chievitz and Thiede, 1962) no correlation between a high platelet count and a tendency to thrombosis was demonstrated. Thus the evidence that an elevated platelet count is a causal factor in large-vessel occlusion is contradictory and probably not statistically significant. It is clear that the VH level is the major factor and that the level of the platelet count is less important.

During treatment of idiopathic erythrocytosis a similar retrospective study has been performed (Pearson, 1977). A rather smaller number of patient-years was examined, 56 compared with 332 patient-years in the PPP study. With this limitation on the results of the study, it was found that the incidence of vascular occlusive events was six times more common in patients who were maintained with a VH in excess of 0.50 compared with those with lower values. Once again in this study, as in the PPP study, cerebrovascular complications exceeded coronary vessel complications by a ratio of approximately 5:1.

10.2.3 Vascular Occlusive Events in Patients with Abnormally Elevated Haematocrits Without True Polycythaemia

A number of terms have been used to describe patients who have an abnormally elevated haematocrit, some with a reduced plasma volume but none with a definite increase in their red-cell mass. These terms include Gaisbock's syndrome, stress polycythaemia (Lawrence and Berlin, 1952), relative polycythaemia (Davies et al., 1973) pseudopolycythaemia (Kaung and Peterson, 1962) and spurious polycythaemia (Brown et al., 1971). Although there is lack of uniformity of a descriptive title, there is little doubt that these patients have a high incidence of vascular occlusive disease (Kaung and Peterson, 1962; Russell and Conley, 1964; Hall, 1965; Burge et al., 1975; Weinreb and Shih, 1975).

Burge et al. (1975), in a follow-up study of 23 patients, found that the observed mortality rate, mostly due to occlusive vascular lesions, was approximately six times greater than the expected rate for age- and time-matched controls. In addition, three further patients had non-fatal arterial occlusions. In a larger study of 62 patients, Weinreb and Shih (1975) found that the incidence of thrombo-embolic complications for the entire group was 30% during the period of study. They concluded, however, that the non-haematological parameters, particularly hypertension, were the major factors of significance in the substantial morbidity in their patients.

10.2.4 Vascular Occlusive Events Related to Haemoglobin or Venous Haematocrit Level in the Normal Population

In the normal population the relationship between haemoglobin or VH level and vascular occlusive events has received considerable attention. A large epidemiological investigation has been conducted in Framingham, Massachusetts, and the 22-year follow-up has been reported. This study has shown that in the 50-59-year-old group the incidence of cerebral infarction is approximately twice as high when haemoglobin values are above 15 g/dl in males and 14 g/dl in females as in individuals with lower haemoglobin values (Dawber et al., 1977). It was also found that both hypertension and cigarette smoking correlated with high haemoglobin values and the incidence of cerebral infarction (Kannel et al., 1972). When allowance was made for these factors the haemoglobin level/cerebral thrombosis relationship was less strong. Tohgi et al. (1978) reported on the relationship between the incidence of cerebral infarction, often of subclinical type, as found at post-mortem examination and the preceding VH value in a study of 432 individuals with a mean age of 77 years. They found that the incidence of cerebral infarction was greater at the higher haematocrit value. For example, the incidence was 43.6% with haematocrits between 0.46 and 0.50 and 18.3% with values between 0.36 and 0.40. The risk of cerebral infarction with high normal VH values was augmented by severe atherosclerosis and age.

Many authors have studied the relationship between VH and coronary artery disease with conflicting conclusions. Conley et al. (1964) found that the VH of patients with myocardial infarction was no greater than for normal individuals. In contrast, DePasquale and Burch (1963) and Mayer (1965) found significantly higher VH values in patients with myocardial infarction than in control individuals. Similarly, Nicolaides et al. (1977) found that patients with angina had higher haematocrits than controls. In a prospective study, Abu-Zeid and Chapman (1976) did not find that VH and haemoglobin levels contributed significantly to the development of ischaemic heart disease. These authors mentioned that other risk factors, in particular cigarette smoking and cholesterol level, had been positively correlated with haemoglobin level (Elwood et al., 1970). In addition, Abu-Zeid and Chapman (1976) thought that ischaemic heart disease might be a causative factor in the development of higher haemoglobin values and this could explain the higher haematocrit values found in some retrospective studies. In the prospective Framingham study reported by Dawber et al. (1977) there was a positive relationship between high haemoglobin level and the risk of myocardial infarction. In females in the age range 50-59 years, the incidence of myocardial infarction was twice as high with haemoglobin values in excess of 14 g/dl as with values below this figure, although in males the association was rather weaker. A similar relationship in females at all ages, but particularly up to the age of 65 years, was found by Elwood et al. (1974) in a prospective study. In their study, the incidence of death due to circulatory disease was twice as high with haematocrits above 0.42 as with haematocrits below this level. The distribution of deaths due to myocardial infarction alone was similar and death certified to be due to myocardial infarction in females under the age of 65 years did not occur below a haematocrit of 0.38.

10.2.5 Peripheral Limb Ischaemia and Venous Haematocrit

It has been shown that patients with lower-limb ischaemia have higher haemoglobin values than age-matched control individuals (Bouhoutsos et al., 1974). These authors also assessed the frequency of post-operative complications in patients undergoing reconstructive arterial surgery. Post-operative haemorrhage or thrombosis, or cardiac arrest at operation occurred in 2.3% of patients with pre-operative haemoglobin values below 16 g/dl, in 14.5% with values in the range of 16-17.4 g/dl, and in 62% with values above 17.5 g/dl. These results indicate an association between pre-operative haemoglobin value and complication rate. In addition, the incidence of complications was significantly greater with pre-operative platelet counts in excess of 380 × 10^9/litre (Bouhoutsos et al., 1974). Bailey et al. (1979) assessed the success or failure of limited amputation in the feet of diabetic patients. The most significant finding of these authors was that pre-operative haemoglobin levels were significantly lower in patients whose amputations healed than in those with a failure of healing. Bailey et al. (1979) found that in all 18 patients with a pre-operative haemoglobin of less than 12 g/dl, amputation was successful, whilst all 30 amputations in those with a pre-operative level above 13 g/dl required more proximal surgery or had failed to heal after 1 month. The authors felt that higher blood flow and improved oxygen delivery to the limb at lower viscosity values associated with the lower haemoglobin and haematocrit values was responsible for the successful outcome in this situation.

10.2.6 Summary

There is unequivocal evidence that there is an association between the high venous haematocrit (VH), as seen at presentation in primary proliferative polycythaemia and idiopathic erythrocytosis, and vascular occlusive risk. Similarly, in these conditions, the higher the VH in the follow-up period, the greater the incidence of vascular occlusion.

Patients with increased VH without true polycythaemia have an increased incidence of vascular occlusion. In the normal population there is some evidence of an association between high normal haemoglobin values or high normal VH and the incidence of cerebral infarction. Evidence linking myocardial infarction and haemoglobin level or VH is rather less certain. Lastly, it has been suggested that in diabetic and non-diabetic patients with peripheral limb ischaemia the complication rate of limb surgery is greater the higher the haemoglobin value pre-operatively.

10.3 Haematocrit and *In Vitro* Rheology

All the observations indicating a relationship between VH and vascular occlusive disease obviously make it necessary to consider the relationship between VH, blood viscosity and blood flow. The haematocrit is the single most important factor influencing blood viscosity at a constant temperature and shear rate. In the physiological range the log viscosity/haematocrit relationship gives a straight line. The haematocrit has a progressively greater influence on blood viscosity as the shear rate at which measurements are made is reduced. Thus, at a high shear rate of 230 s^{-1}, viscosity increases by approximately 30% when the haematocrit is raised from 0.40 to 0.50, whereas at a low shear rate of 0.77 s^{-1} viscosity increases by approximately 70% with a similar rise in haematocrit. Thus the concept, often quoted, that blood viscosity only rises rapidly with increases in haematocrit above 0.50, is not strictly valid.

There are attractive forces between red cells which permit their interlocking into three-dimensional networks upon cessation of flow. The force required to overcome these attractive forces and to restart flow is the yield stress (Merrill et al., 1963). Using the Casson plot method for determining yield stress (Merrill et al., 1965), Pearson and Ring (1976) determined the yield stress for samples from patients with treated primary polycythaemia. The haematocrits ranged from 0.397 to 0.510. The best fit was found to be between log yield stress and haematocrit ($r = 0.50$, $p < 0.005$). Examination of the correlation line showed that the yield stress almost doubles when the haematocrit is increased from 0.40 to 0.50 (7.5 to 14.6 mPa). When the slope of this correlation line was used to "correct" the individual yield stress values to a haematocrit of 0.46, it was found that there was a significant positive correlation between the plasma viscosities and the "corrected" yield stress of the samples ($r = 0.69$, $p = < 0.001$). Thus increases in haematocrit and plasma viscosity have a marked effect on the yield stress of a blood sample and this could have important implications if stasis occurs *in vivo*.

10.4 Venous Haematocrit and Cerebral Blood Flow

There have been a number of reports of low cerebral blood flow (CBF) in patients with PPP (Kety, 1949; Nelson and Fazekas, 1956; Gottstein, 1965; O'Brien, 1971). In a larger study, Thomas et al. (1977) confirmed the reduction in CBF with a mean value of only 37.9 ml/100 g/min in 16 patients with PPP compared with 69.1 ml/ 100 g/min in controls. Reducing the VH by venesection from 0.536 to 0.455 increased the CBF to 62.7 ml/100 g/min. Venesection produced a substantial fall in whole-blood viscosity. The fall in oxygen carriage per unit volume of blood expected from the reduction in haematocrit was considered insufficient for the magnitude of the rise in CBF and it was postulated that part of the improvement in CBF following venesection was due to the fall in whole-blood viscosity.

More recently Humphrey et al. (1979a) have studied a group of patients with equivocal or considerable elevation in the haematocrit due to low plasma volume rather than elevated red-cell mass. These patients also had a low CBF. Venesection reduced the mean haematocrit from 0.521 to 0.432 and increased the mean CBF from 47.1 to 65.0 ml/100 g/min. There was a significant fall in whole-blood viscosity associated with the rise in CBF.

It would thus appear that a high VH, whatever the cause, whether it be a high red-cell mass or low plasma volume, is associated with a low CBF. Following venesection there is a fall in the VH with a significant rise in the CBF. It is postulated that the fall in whole-blood viscosity is one of the factors responsible for this rise in CBF. It also seems likely that haematocrits around the upper limit of normal, in which the red-cell mass is below that at which a diagnosis of definite polycythaemia is made, may in some patients be associated with a low CBF.

In anaemia, the CBF is high. This was first described by Scheinberg (1949) in a group of seven patients with iron-deficiency anaemia. Heyman et al. (1952) described a further eight patients with anaemia due to a variety of causes in which the CBF was high. Gottstein (1978) has recently confirmed the high CBF in patients with a low haemoglobin.

There thus seems little doubt that in patients with a high haematocrit the CBF is low, and in those with a low haematocrit the CBF is high. Two possible factors are likely to account for this. The first is the fall in oxygen carriage per unit of blood as the haematocrit declines (Paulson et al., 1979). The second is the fall in whole-blood viscosity which occurs as the haematocrit is reduced. Using Shapiro's nomogram (Shapiro, 1973), it is possible to estimate the fall in oxygen-carrying capacity per unit volume of blood associated with a fall in haemoglobin. In both the study of Thomas et al. (1977) and Humphrey et al. (1979a) the fall in oxygen carriage was not sufficient to account for the observed rise in CBF. It was felt that part of the rise in CBF as the VH fell was secondary to the fall in whole-blood viscosity.

In an attempt to obtain more information on the roles of these two factors, Humphrey et al. (1979b) measured the CBF, VH and blood viscosity in a group of paraproteinaemic patients. The results in these patients were compared with those for patients with an equivalent degree of anaemia but without paraproteinaemia. The mean CBF in the paraproteinaemic patients with a mean haematocrit of 0.342 was 60.8 ml/100 g/min. The group of anaemic patients without any protein abnormality, who had a similar mean haematocrit of 0.325, had a significantly higher mean CBF of 80.8 ml/100 g/min. The whole-blood and plasma viscosity were significantly higher in the paraproteinaemic patients. Since the oxygen carriage in the two groups was likely to be similar, the lower CBF in the paraproteinaemic patients was thought to result from the higher blood and plasma viscosity. When CBF was plotted against VH in these paraproteinaemic patients, a weak correlation coefficient of -0.487 ($p<0.05$) was found. When CBF was plotted against blood viscosity, however, the correlation coefficients were -0.781 ($p<0.001$) at a low shear rate ($0.67\ s^{-1}$) and -0.684 ($p<0.001$) at a high shear rate ($91\ s^{-1}$). This closer correlation between CBF and blood viscosity than between CBF and VH again tends to confirm that viscosity is one of the factors determining CBF. There is, however, convincing evidence that oxygen carriage also affects CBF and it seems therefore likely that blood viscosity and oxygen delivery to the brain are independent factors which both affect CBF.

10.5 The Relationship Between Venous Haematocrit, Blood Viscosity, Blood Flow, and Vascular Occlusion

A number of facts are established by the evidence which has already been presented. First, there is a clear relationship between the level of the venous haematocrit and the incidence of occlusive vascular lesions, particularly in the cerebral circulation. Secondly there is an unequivocal inverse relationship between VH and cerebral blood flow. Finally there are well-defined *in vitro* relationships between haematocrit and whole-blood viscosity. Because of these relationships it is tempting to assume a causal sequence in which a rise in venous haematocrit produces an increased blood viscosity and a decreased cerebral blood flow with vascular occlusion as a possible sequence. Such an assumption is not justified. For example, as Schmid-Schönbein (1977) stressed, observed *in vitro* viscosity values on a venous blood sample give limited information on *in vivo* behaviour. Furthermore, a number of other factors such as arterial wall changes, platelet distribution and coagulant activity under different flow conditions and the effect of anatomical variations in the blood vessels need to be considered.

Platelets play an essential part in thrombus formation, particularly in the arterial circulation. Elevated haematocrit values might have circulatory implications unrelated to viscosity changes. There is an axial migration of red cells leaving a relatively red-cell-free "plasmatic zone" at the walls in some parts of the arterial tree. Displacement of other blood elements, particularly platelets, occurs into this "plasmatic zone" (Palmer, 1966). The relatively cell-free zone becomes narrower as the haematocrit increases (Bugliarello et al., 1965). If red-cell aggregation occurs to any extent in the axial stream, further displacement of platelets towards the periphery will occur (Palmer, 1966). This contact between vessel wall and platelets, particularly if the vessel wall is abnormal in any way, might be the stimulus which initiates thrombosis (Dintenfass, 1971) independently of any first-order viscosity effect.

In veins, there is evidence that blood stasis and a decrease in blood flow is related to a high incidence of thrombus formation (McLachlin et al., 1960; Clark and Cotton, 1968). Diminished blood flow possibly allows the accumulation of coagulant substances normally cleared or diluted (Sevitt, 1973). Thrombus is often seen extending from valve pockets and there is phlebographic evidence that the points of maximum stasis along relatively stagnant channels are the valve pockets and venous saccules (McLachlin et al., 1960). It would seem reasonable, therefore, to implicate the haematocrit effect on blood viscosity and slow circulation as a factor in the developmet of venous thrombosis.

While these factors may well be involved in thrombus formation, particularly in atherosclerotic vessels, there is still good reason to implicate the haematocrit effect on blood viscosity and flow in the circulation as a factor in the development of occlusive vascular lesions. The clinical and rheological observations already discussed give clear support to the view that virtually all patients with abnormally high VH should be venesected to reduce the risk of occlusive vascular lesions. The view that patients with high normal or even normal VH values, particularly those who for other reasons are at risk of occlusive vascular lesions, should have prophylactic venesection to maintain the VH at low normal or even lower than normal levels, is still a matter for debate. When there are vessel wall abnormalities, it has been suggested that the optimal VH is lower than the normally accepted value (Burch and DePasquale, 1963; Yates et al., 1979). Lundsgaard-Hansen (1979) has opposed this view on the grounds that at VH values lower than normal a reduction in tissue oxygen delivery occurs which outweighs the rheological advantages. This is an area of considerable practical importance and the time now appears appropriate to initiate proper trials of venesection which would resolve the problem.

10.6 Conclusions

Elevated venous haematocrit (VH) is associated with a high incidence of vascular occlusion, particularly in the cerebral circulation. There is some evidence of an increased risk of vascular occlusion at high normal compared with low normal VH values.

The haematocrit is the single most important factor in determining whole-blood viscosity. Evidence has been presented to show that blood viscosity is probably a factor in determining cerebral blood flow.

The precise causative link between elevated VH and vascular occlusion has not been established. Changes in blood viscosity and flow could be expected to be implicated. In the majority of clinical situations, patients with elevated VH levels should be treated to lower them. When the VH falls within the normal range, it is suggested that it is now appropriate to initiate clinical trials of venesection in those patients with high normal VH values at risk of vascular occlusion.

Acknowledgements. We wish to thank our collaborators in the cerebral blood flow studies (Professor G H du Boulay, Professor J Marshall, Dr R Ross Russell, Professor L Symon and Dr E Zilkha). DJT and PRDH received financial support from the Medical Research Council. TCP received financial support from the Endowment Fund of St Thomas's Hospital and from the Merchant Taylors' Company.

10.7 References

Abu-Zeid H A H, Chapman, J M (1976) The relation between haemoglobin level and the risk for ischaemic heart disease: A prospective study. J Chron Dis *29:* 395-403

Bailey M J, Yates C J P, Johnston C L W, Somerville P G, Dormandy J A (1979) Preoperative haemoglobin as predictor of outcome of diabetic amputations. Lancet *ii:* 168-170

Barabas A P, Offen D N, Meinhard E A (1973) The arterial complications of polycythaemia vera. Br J Surg *60:* 183-187

Bouhoutsos J, Morris T, Chavatzas D, Martin P (1974) The influence of haemoglobin and platelet levels on the results of arterial surgery. Br J Surg *61:* 984-986

Brown S M, Gilbert H S, Krauss S, Wasserman L R (1971) Spurious (relative) polycythaemia. A nonexistent disease. Am J Med *50:* 200-207

Bugliarello G, Kapur C, Hsiao G (1965) The profile viscosity and other characteristics of blood flow in a non-uniform shear field. Proc. 4th int. Cong. Rheology. Interscience, New York, p 351

Burch G E, De Pasquale N P (1963) Phlebotomy. Arch Intern Med *111:* 687-695

Burge P S, Johnson W S, Prankerd T A J (1975) Morbidity and mortality in pseudopolycythaemia. Lancet *i:* 1266-1269

Chievitz E, Thiede T (1962) Complications and causes of death in polycythaemia. Acta Med Scand *172:* 513-523

Clark C, Cotton L T (1968) Blood flow in deep veins of leg. Recording technique and evaluation of methods to increase flow during operation. Br J Surg *55:* 211-214

Conley C L, Russell R P, Thomas C B, Tumulty P A (1964) Haematocrit values in coronary artery disease. Arch Intern Med *113:* 170-176

Davies S W, Glynne-Jones E, Lewis E P (1973) False polycythaemia. J Clin Pathol *26:* 987

Dawber T R, Wolf P A, Colton T, Nickerson R J (1977) Risk factors: Comparison of the biological data in myocardial and brain infarctions. In: Zülch K J, Kaufmann W, Hossmann K A, Hossmann V (eds) Brain and heart infarct. Springer-Verlag, Berlin-Heidelberg-New York, pp 226-252

Dawson A A, Ogston D (1970) The influence of the platelet count on the incidence of thrombotic and haemorrhagic complications in polycythaemia vera. Postgrad Med J *46:* 76-78

DePasquale N P, Burch G E (1963) Hematocrit in women with myocardial infarction. JAMA *183:* 142-143

Dintenfass L (1971) Blood microrheology — viscosity factors in blood flow, ischaemia and thrombosis. London, Butterworth, p 267
Elwood P C, Sweetnam P, Mahler R, Moore F, Welsby E (1970) Association between circulating haemoglobin level, serum cholesterol, and blood pressure. Lancet *i:* 589-590
Elwood P C, Benjamin I T, Waters W E, Sweetnam P M (1974) Mortality and anaemia in women. Lancet *i:* 891-894
Gottstein U (1965) Physiologie und Pathophysiologie des Hirnkreislaufs. Med Welt *15:* 715-726
Gottstein U (1978) Cerebral blood flow, $CMRO_2$ and CMR glucose in patients with hypo and hyper chronic anaemia and polycythaemia. 9th Int. Salzburg Conf. on Cerebrovascular Disease (to be published)
Hall C A (1965) Gaisbock's disease: Redefinition of an old syndrome. Arch Intern Med *116:* 4-9
Heyman A, Patterson J L, Duke T W (1952) Cerebral circulation and metabolism in sickle cell and other chronic anaemias, with observations of the effects of oxygen inhalation. J Clin Invest *31:* 824-828
Humphrey P R D, du Boulay G H, Marshall J, Pearson T C, Ross Russell R W, Symon L, Wetherley-Mein G, Zilkha E (1979a) Cerebral blood flow and viscosity in relative polycythaemia. Lancet *ii:* 873-877
Humphrey P R D, du Boulay G H, Marshall J, Pearson T C, Ross Russell R W, Slater N G P, Wetherley-Mein G, Zilkha E (1979b) Cerebral blood flow, haematocrit and viscosity in patients with paraproteinaemia: the importance of viscosity. (In preparation)
Kannel W B, Gordan T, Wolf P A, McNamara P (1972) Hemoglobin and the risk of cerebral infarction: the Framingham study. Stroke *3:* 409-420
Kaung D T, Peterson R E (1962) Relative polycythemia or pseudopolycythemia. Arch Intern Med *110:* 456-460
Kety S S (1949) The physiology of the human cerebral circulation. Anesthesiology *10:* 610-614
Lawrence J H, Berlin N I (1952) Relative polycythemia — the polycythemia of stress. Yale J Biol Med *24:* 498-505
Lawrence J H, Berlin N I, Huff R L (1953) The nature and treatment of polycythemia. Medicine (Baltimore) *32:* 323-388
Lundsgaard-Hansen P (1979) Haemodilution — New clothes for an anemic emperor. Vox Sang *36:* 321-336
McLachlin A D, McLachlin J A, Jory T A, Rawling E G (1960) Venous stasis in the lower extremities. Ann Surg *152:* 678-685
Mayer G A (1965) Hematocrit and coronary heart disease. Can med Ass J *93:* 1151-1153
Merrill E W, Gilliland E R, Cokelet G, Shin H, Britten A, Wells R E (1963) Rheology of human blood, near and at zero flow. Effects of temperature and haematocrit level. Biophys J *3:* 199-213
Merrill E W, Margetts W G, Cokelet G R, Gilliland E R (1965) The Casson equation and rheology of blood near zero shear. Proc int Congr Rheology New York, Wiley and Sons, 135-143
Nelson D, Fazekas J F (1956) Cerebral blood flow in polycythemia vera. Arch Intern Med *98:* 328-331
Nicolaides A N, Horbourne T, Bowers R, Kidner P H, Besterman E M (1977) Blood viscosity, red-cell flexibility, haematocrit, and plasma fibrinogen in patients with angina. Lancet *ii:* 943-945
Norman I L, Allen E V (1937) The vascular complications of polycythemia. Am Heart J *13:* 257-273
O'Brien M D (1971) Some neurological aspects of dementia. Geriontologica Clinica *13:* 339-349
Palmer A A (1966) Platelet and leucocyte skimming. 4th Europe Conf Microcirculation, Cambridge. Bibl Anat *9:* 300-303
Paulson O B, Henriksen L, Smith R J (1979) The effect of haemodilution on cerebral blood flow and blood gases in patients with polycythaemia. 9th Intern. CBF Symposium, Tokyo, pp 588-589
Pearson T C (1977) Clinical and laboratory studies in the polycythaemias. M.D. Thesis (London University)
Pearson T C, Ring C P (1976) Unpublished observations
Pearson T C, Wetherley-Mein G (1978) Vascular occlusive episodes and venous haematocrit in primary proliferative polycythaemia. Lancet *ii:* 1219-1222
Pearson T C, Weatherley-Mein G (1979) The course and complications of idiopathic erythrocytosis. J Clin Lab Haematol *1:* 189-196
Preston F E, Emmanuel I G, Winfield D A, Malia R G (1974) Essential thrombocythaemia and peripheral gangrene. Br Med J *3:* 548-552
Russell R P, Conley C L (1964) Benign polycythemia: Gaisbock's syndrome. Arch Intern Med *114:* 734-740
Scheinberg P (1949) The effects of postural changes, stellate ganglion block and anaemia on the cerebral circulation. J Clin Invest *28:* 808-809

Schmid-Schönbein H (1979) Rheological studies of the blood under normal and pathological conditions. In: Zülch K J, Kaufmann W, Hossmann K A, Hossmann V (eds) Brain and heart infarct. Berlin-Heidelberg-New York, Springer-Verlag, pp 96-106

Sevitt S (1973) Recent advances in thrombosis. Edinburgh and London, Churchill Livingstone, p 33

Shapiro B A (1973) The clinical application of blood gases. Year Book Publishers, Chicago, p 37

Singh A K, Wetherley-Mein G (1977) Microvascular occlusive lesions in primary thrombocythaemia. Br J Haematol *36:* 553-564

Szur L, Lewis S M, Goolden A W G (1959) Polycythaemia vera and its treatment with radioactive phosphorus. Q J Med *29:* 397-424

Taylor G W, Calo A R (1962) Atherosclerosis of arteries of the lower limb. Br Med J *1:* 507-510

Thomas D J, du Boulay G H, Marshall J, Pearson T C, Ross Russell R W, Symon L, Wetherley-Mein G, Zilkha E (1977) Cerebral blood flow in polycythaemia. Lancet *ii:* 161-163

Tohgi H, Yamanouchi H, Murakami M, Kameyama M (1978) Importance of the haematocrit as a risk factor in cerebral infarction. Stroke *9:* 369-374

Videbaek A (1950) Polycythaemia vera; course and prognosis. Acta Med Scand *138:* 179-187

Weinreb N J, Shih C-F (1975) Spurious polycythemia. Semin Hematol *12:* 397-407

Yates C J P, Andrews V, Berent A, Dormandy J A (1979) Increase in leg blood-flow by normovolaemic haemodilution in intermittent claudication. Lancet *ii:* 166-168

11 Sickle-cell Disease and Vascular Occlusion

J. Stuart and M.W. Kenny

11.1 Introduction

Patients with a clinically severe form of sickle-cell disease (homozygous sickle-cell anaemia, sickle-cell/β-thalassaemia, or sickle-cell/Hb C disease) are prone to recurrent episodes of limb or trunk pain known as the painful, vaso-occlusive crisis. The major factor causing vascular occlusion, usually in the terminal arterioles and capillaries (Diggs, 1965), is the interlocking of rigid, sickled erythrocytes. This process is commonly regarded as the end-stage of a cycle of sickling, stasis, low pH and local hypoxia (Harris et al., 1956) but platelets and fibrin are also involved in the vascular obstruction (Rosenblate et al., 1970). It is not clear whether sickling, hypoxia, platelet and coagulation activation, or some other factor is the primary event which initiates this cycle and ultimately leads to stasis.

Various systemic factors such as infection, pyrexia or chilling have been associated with the onset of crisis (Konotey-Ahulu, 1965) but more than 50% of crises may occur in the absence of any apparent precipitating factor (Diederich et al., 1976). When such a factor can be identified, the precise mechanism which initiates irreversible sickling and vascular occlusion is still unclear. It is therefore appropriate to study the rheological and thrombotic changes of sickle-cell crisis in order to determine which are primary events of aetioloical significance and which are secondary phenomena.

11.2 Laboratory Studies

11.2.1 Whole-blood Viscosity

When blood from asymptomatic sickle homozygotes is deoxygenated *in vitro* its viscosity, particularly when measured at a high shear rate, increases progressively as the oxygen saturation falls below 85% and rigid, sickled cells are formed (Usami et al., 1975; Chien, 1977). A similar process occurs as a result of deoxygenation *in vivo* and causes hyperviscosity and vascular stasis (Harris et al., 1956). While this may be the end-stage of most vaso-occlusive crises, severe deoxygenation is rarely the initiating event.

Oxygenated blood taken from asymptomatic homozygotes has a higher viscosity than normal blood, provided adjustment is made for the haematocrit difference (Chien et al., 1976; Chien 1977). This hyperviscosity is found over a wide range of shear rates (Chien et al., 1976; Chien, 1977) and is predominantly due to the irreversibly sickled cells (ISC) which remain in the circulation during the asymptomatic steady state (Chien et al., 1970). When the ISC are removed by centrifugation, however, the viscosity of the remaining non-ISC is still higher than oxygenated normal blood (Chien et al., 1970). These observations were made on erythrocytes suspended in Ringer's solution, to exclude the red-cell aggregating effect of plasma fibrinogen and globulin, and it was therefore presumed that the non-ISC also has reduced deformability, but to a lesser degree than the ISC.

11.2.2 Metabolic Changes in Sickled Cells

The nature of the metabolic defect which causes an increase in internal viscosity and membrane rigidity of the sickled cell is now better understood. The ISC has a reduced level of ATP and potassium and an increased content of calcium (Glader and Nathan, 1978; Eaton et al., 1979). Deoxygenation and sickling increase membrane permeability to calcium so that the sickled cell contains 3-10 times the calcium content of normal erythrocytes (Eaton et al., 1973). Although these metabolic changes are most obvious in the ISC, an increase in calcium is also found in the non-ISC (Eaton et al., 1973; Palek, 1977).

The water content and osmolality of the ISC are decreased (Glader et al., 1978) and there is an associated increase in mean cell haemoglobin concentration (MCHC), internal viscosity and membrane lipid (Bertles and Milner, 1968; Brownlee et al., 1974; Glader and Nathan, 1978; Serjeant et al., 1978). In addition to the effect of the higher MCHC in raising internal viscosity, sickle-cell haemoglobin behaves as a non-Newtonian fluid on deoxygenation and has a higher viscosity at low shear rates (Chien, 1977). Sickle haemoglobin also binds with increased affinity to the red-cell membrane (Bank et al., 1974; Fischer et al., 1975) and the accumulation of calcium may further increase haemoglobin interaction with the membrane (Palek et al., 1971).

Thus both the cell interior and the cytoplasmic membrane of the sickled cell are abnormal and clinical studies of erythrocyte deformability in vaso-occlusive crisis should therefore employ a measuring technique sensitive to both internal hyperviscosity and membrane rigidity.

11.2.3 Erythrocyte Deformability

The deformability of individual erythrocytes is a major factor in determining capillary flow and, despite the high shear rate of the human capillary bed, a reduction in membrane deformability is likely to impede flow, particularly at sites of capillary bifurcation (Klug et al., 1974). Sickled erythrocytes pass with difficulty through 5-μm Millipore filters (Jandl et al., 1961) and a similar technique, using 5-μm polycarbonate filters in a positive-pressure filtration system, was found to be more sensitive than viscometry in detecting changes in deformability under conditions of slight deoxygenation (Chien, 1977).

Micropipette elastimetry, performed at multiple sites on the membrane of single erythrocytes from normals and sickle-cell patients, has shown good correlation with a positive-pressure, 5-μm membrane, filtration system (Lessin et al., 1977). The ISC (Lessin et al., 1978), but not the oxygenated non-ISC (LaCelle, 1975), shows increased membrane rigidity when measured by micropipette elastimetry. The non-ISC does not show a decrease in deformability, using a 2-μm micropipette, until the pO_2 falls to 12-25 mm Hg (LaCelle, 1975). Micropipette techniques may theoretically miss localised membrane lesions if the major part of the ISC membrane has normal deformability (Mohandas and Bessis, 1979) and 5-μm membrane filtration techniques may therefore be more appropriate for clinical studies.

It is not yet established whether positive-pressure filtration through a 5-μm membrane (Lessin et al., 1977) gives comparable results to the negative pressure technique (Schmid-Schönbein et al., 1973; Reid et al., 1976). The sensitivity of the laser diffraction picture of the ektacytometer (Bessis and Mohandas, 1977; Mohandas and Bessis, 1979) in detecting sub-populations of rigid cells has also to be determined in comparison with filtration techniques. Thus clinical studies of sickle-cell crisis should ideally employ a variety of techniques for measuring deformability until the most appropriate method is determined.

11.3 Clinical Studies

11.3.1 Rheological Studies During Crisis

Despite the large amount of experimental work by rheologists on oxygenated and deoxygenated sickled cells there have been relatively few clinical studies of patients in sickle-cell crisis.

Richardson et al. (1976) measured whole-blood viscosity (WBV) at 230 and 23 s^{-1} in 8 children during sickle-cell crisis and found, compared with 26 children in the steady state, a significant increase in WBV, when adjusted to a standard haematocrit of 0.45, at both shear rates. A significant increase in plasma fibrinogen was also found. Since WBV and acute-phase proteins such as fibrinogen are likely to vary at different stages of crisis, a serial study was subsequently performed during nine crisis episodes in eight adult patients (Richardson et al., 1979). The results, compared with steady-state values for the same patients, showed significantly elevated WBV at both shear rates on days 1-2 and days 6-8 of crisis. The early (day 2) and late (days 6-8) peak in WBV are shown in Fig. 11.1 in comparison with the plasma fibrinogen, which increased progressively to a plateau after day 4. The day of admission to hospital was taken as day 1.

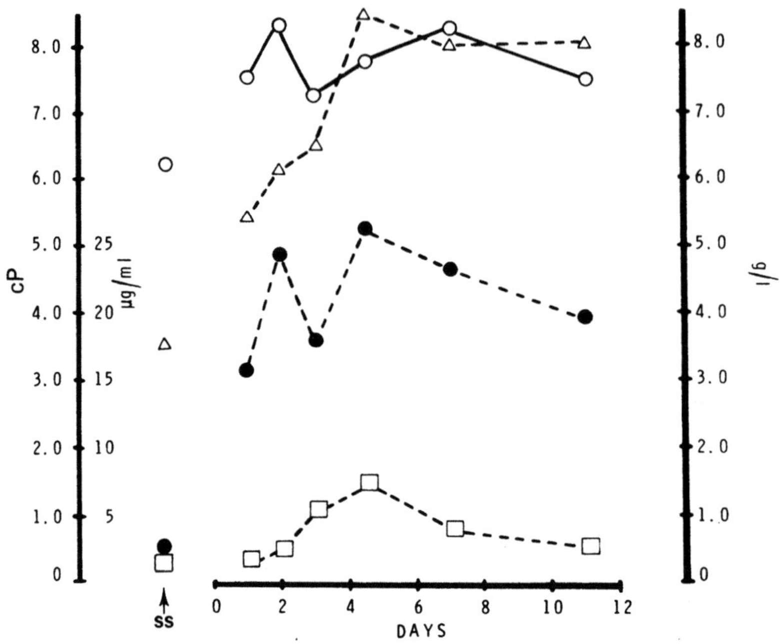

Fig.11.1. Serial mean values for whole-blood viscosity in centipoise (cP) measured at 23 s^{-1}(○), plasma fibrinogen in g/litre (△), fibrin degradation products in μg/ml (●), and cryofibrinogen in g/litre (□) in nine episodes of sickle-cell crisis compared with the mean of eight estimations in the steady state (*SS*). Based on the data of Richardson et al. (1979). (1 cP = 1 mPa.s)

The cause of the early peak in WBV is not clear. Serum fibrin degradation products (FDP) were elevated to almost 25 μg/ml by day 2 but this modest rise in small-molecular-weight fragments (D and E) would not be expected to increase red-cell aggregation (Marsh, 1978). Large-molecular-weight complexes of soluble fibrin monomer with fibrinogen do increase yield stress and WBV (Copley et al., 1968) but fibrin monomer, as detected by the cryofibrinogen method, did not show a significant increase until days 3-5. Other workers have reported a positive (ethanol gelation) test for monomer during crisis (Walsh et al., 1973; Ittyerah et al., 1976) but only Alkjaersig et al. (1976) have demonstrated, by fibrinogen chromatography, a rise in high-molecular-weight complexes at the onset of, and perhaps preceding, crisis. Further studies are required to determine the role of these complexes in causing hyperviscosity in the early stages of crisis.

Sickle-cell crisis is often accompanied, and may be precipitated, by bacterial or viral infection. The infection will also give rise to an acute-phase protein response by the liver. The corresponding increases in plasma fibrinogen and WBV for 20 young adults without sickle-cell disease, but with an acute infection requiring hospital admission, are shown in Fig. 11.2 (Richardson et al., 1979). The peak fibrinogen level coincided with the peak WBV on days 1-2 and, although the time scale for the two patient groups in Figs. 11.1 and 11.2 are not strictly comparable (hyperfibrinogenaemia occurred earlier in the infected group), the twin peaks of the sickle-cell viscosity curve were not seen in the infected patients. This further suggests that factors other than the acute-phase rise in fibrinogen were responsible for the early peak in WBV during crisis.

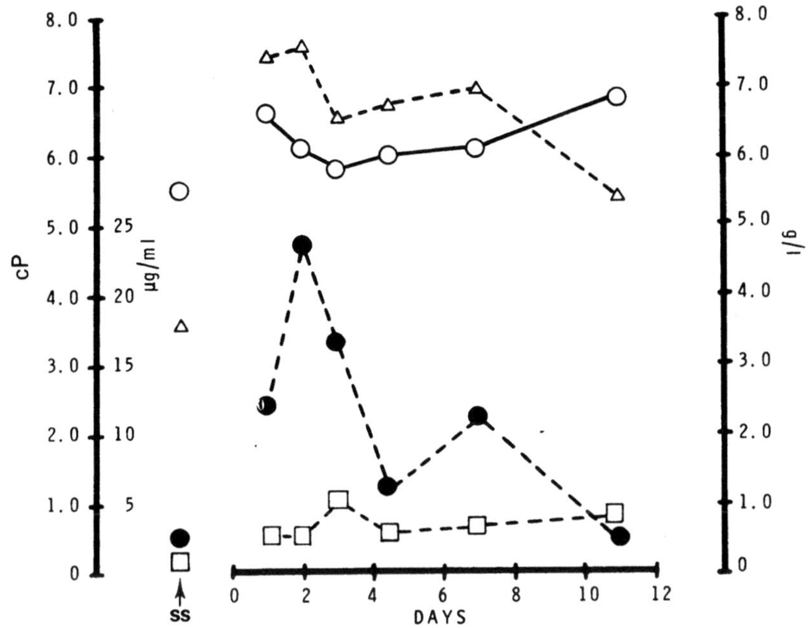

Fig. 11.2 Serial mean values for whole-blood viscosity at 23 s^{-1}, fibrinogen, fibrin degradation products, and cryofibrinogen in 20 patients with acute infection compared with 22 health controls (*SS*) representing steady-state values. Based on data of Richardson et al. (1979). Symbols as in Fig. 11.1. (1 cP = 1 mPa.s)

In a current study (unpublished) using a Contraves Low-Shear 30 viscometer operating at 37°, an early rise in viscosity, measured at low shear (0.2 s^{-1}) and at 23 s^{-1}, was again found on the day of admission (day 1). This early peak (Figs. 11.3 and 11.4) was still present when the WBV was not corrected for haematocrit. Although earlier studies (Diggs, 1965) found little consistent change in haematocrit with time during crisis, we have observed a small but progressive fall in Coulter S haematocrit (Fig. 11.5) which compensates for the effect of hyperfibrinogenaemia and causes a progressive fall in uncorrected WBV after the early peak. The corresponding rise in plasma fibrinogen (measured immunologically) and in plasma viscosity (Coulter Harkness viscometer) for the same patients is shown in Fig. 11.5.

The consistent early peak in WBV could have resulted from a decrease in erythrocyte deformability. It is known that the blood of sickle-cell patients in the steady state contains ISC which vary between 5% and 50% for different individuals but remain relatively constant within the same individual (Serjeant et al., 1969). The ISC appear to be removed rapidly from the circulation, with a T½ of 2 days compared with 5 days for whole blood from sickle-cell patients (Bertles and Milner, 1968), and the patient's ISC count correlates with his red-cell survival (Serjeant et al., 1969). Although Diggs (1965) and Richardson et al. (1979) reported no consistent change in the percentage of unfixed sickled cells during crisis, when the cells are fixed with formaldehyde then a higher percentage of sickled cells has been found at the onset of crisis with a subsequent fall during recovery (Barreras and Diggs, 1964; Rieber et al., 1977).

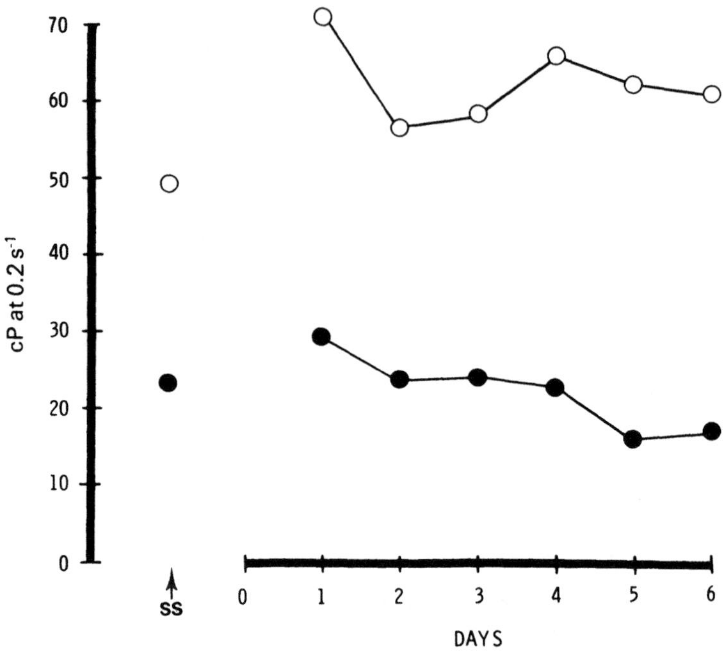

Fig.11.3. Serial mean values for whole-blood viscosity in centipoise (cP), measured at 0.2 s^{-1} and corrected (○) and uncorrected (●) for haematocrit, in eight episodes of sickle-cell crisis compared with the mean of 11 estimations in the steady state (*SS*). (cP = 1 mPa.s)

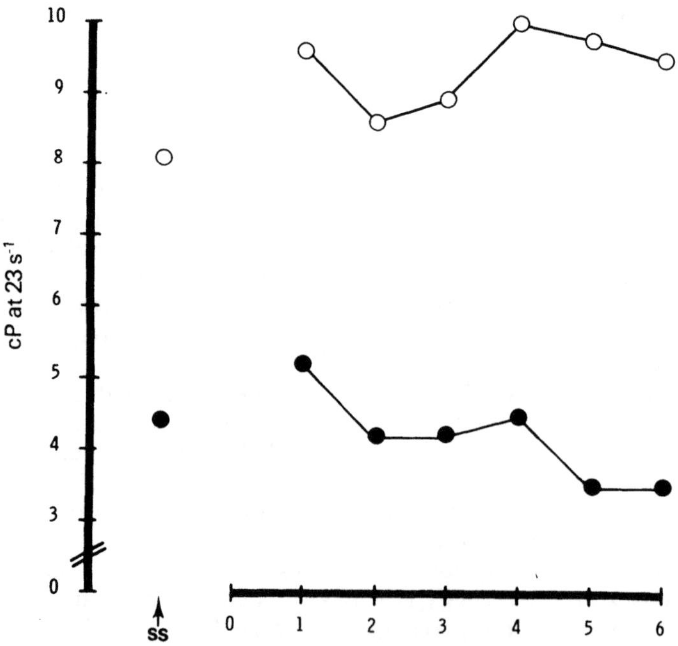

Fig.11.4. Serial mean values for whole-blood viscosity at 23^{-1} corrected (○) and uncorrected (●) for haematocrit in the same patients as Fig.11.3. (1 cP = 1 mPa.s)

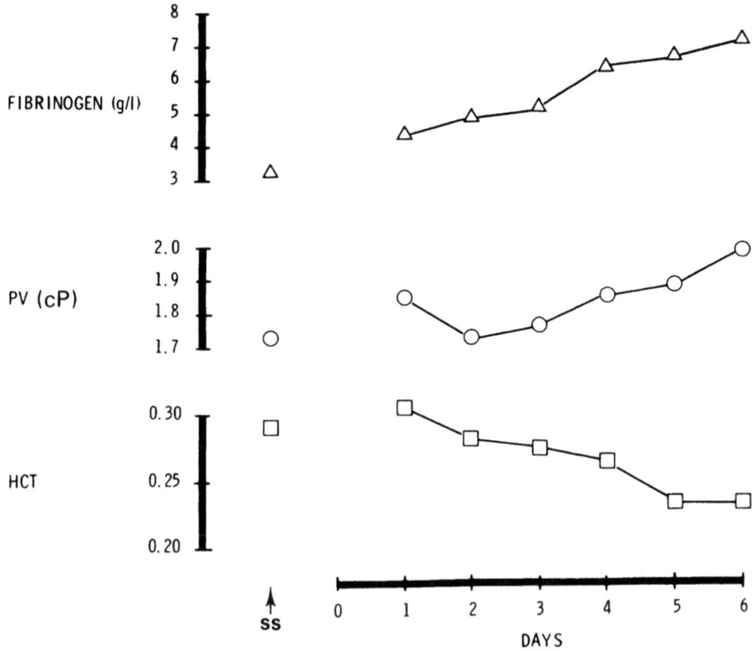

Fig.11.5. Serial mean values for fibrinogen, plasma viscosity and haematocrit for the same patients as Figs.11.3 and 11.4. (1 cP = 1 mPa.s)

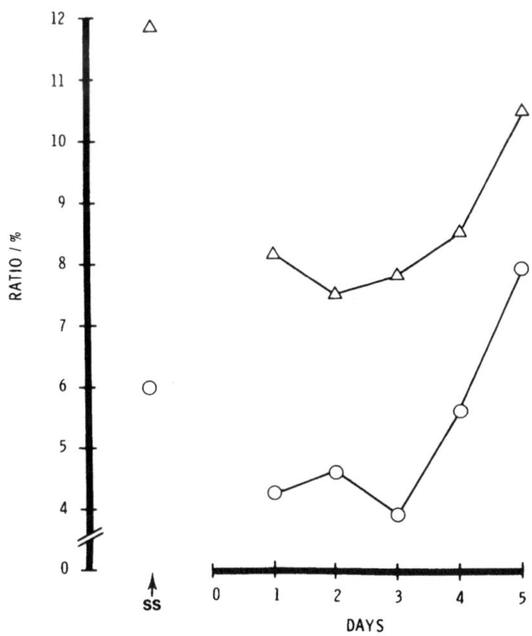

Fig.11.6. Serial mean values for percentage fixed sickled cells (△) in the same patients as Figs.11.3-11.5. Red-cell filterability ratio (○) determined in four crises

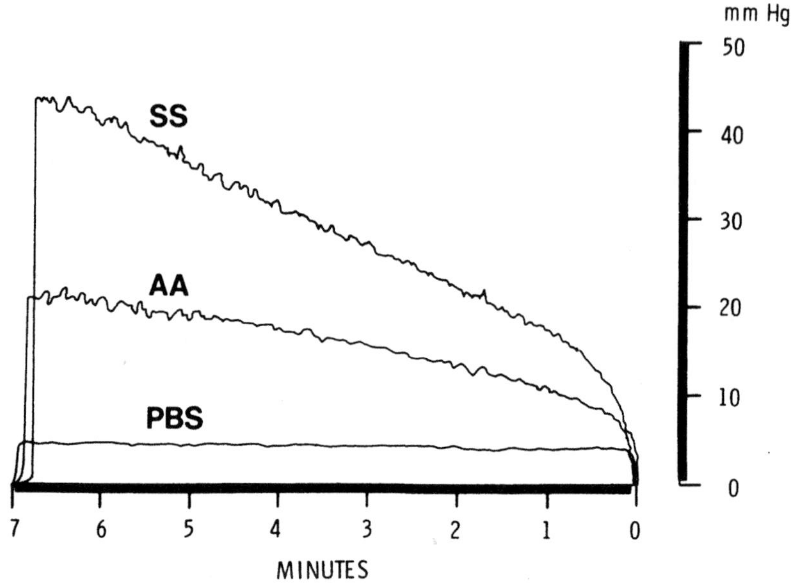

Fig.11.7. Positive-pressure filtration curves for oxygenated red cells from a patient with homozygous sickle-cell anaemia (*SS*), normal control blood (*AA*), and phosphate-buffered saline (*PBS*)

The change in percentage fixed sickled cells (FSC) with time for the same crisis episodes as in Figs. 11.3-11.5 is shown in Fig. 11.6. Venous blood was taken into a syringe containing 10% v/v formaldehyde in physiological saline (pH 7.3) and a wet preparation studied by interference microscopy. A fall in percentage FSC occurred during the first 3 days of crisis, compared with the steady-state value, with a subsequent rise towards the pre-crisis mean. This is consistent with the results of Rieber et al. (1977) who, in addition, showed an increased percentage of ISC 2 days prior to crisis in one patient.

Erythrocyte filterability was also measured using a positive-pressure system and 5-μm Nuclepore filters (based on Lessin et al., 1977). A filterability ratio was determined by dividing the area of the pressure curve obtained for a 0.1×10^{12}/litre red-cell suspension in phosphate-buffered saline (PBS) by the area of the pressure curve for PBS alone (Fig. 11.7). A higher ratio (decreased filterability) was obtained for oxygenated sickle-cell blood, obtained during the steady state, compared with normal blood. In a pilot study of four sickle-cell crises, the fall and subsequent rise in ratio closely followed the fall and rise in percentage FSC (Fig. 11.6). The more rigid erythrocytes appear to have been removed from the circulation during the first few days of crisis, possibly by trapping in areas of vaso-occlusion or by reticulo-endothelial cell activity in the marrow and liver.

Further studies are required at the onset of, and ideally preceding, crisis to determine whether the percentage FSC and the red-cell resistance to filtration increase from steady-state values, and cause a rise in WBV, before a subsequent fall occurs as crisis becomes established.

11.3.2 Role of Extracellular Factors

A number of extracellular factors which have an indirect effect on the erythrocyte have been studied in the belief that red-cell sickling at the onset of crisis is a secondary event resulting from an external change in homeostasis.

11.3.2.1 Plasma Viscosity. An increase in the larger molecular size plasma proteins causes an increase in plasma viscosity and will also, by forming rouleaux, increase WBV as measured at low shear. During the steady state, there is an increase in plasma viscosity, compared with normal controls, and this is related to an increase in globulin (Chien et al., 1976) but not fibrinogen (Richardson et al., 1979). During crisis, the progressive rise in plasma viscosity is related to the acute-phase rise in fibrinogen (Fig. 11.5).

The aggregating effect of plasma proteins on normal discocytes may be less apparent in sickle-cell disease since there is evidence that deoxygenated sickled cells aggregate poorly (Chien, 1977).

11.3.2.2 Alteration in Plasma Volume. Sicklers in the steady state have an expanded plasma volume compared with patients with other types of anaemia of a comparable degree of severity (Erlandson et al., 1960; Steinberg et al., 1977). The combined effect of pyrexia during crisis and the renal concentrating defect of sickle-cell disease (Buckalew and Someren, 1974) may cause an acute reduction in plasma volume. This haemoconcentration, together with the acute-phase hyperfibrinogenaemia of infection, might contribute significantly to hyperviscosity.

Erlandson et al. (1960) reported a 20%-25% reduction in plasma volume in two crisis episodes in one patient. Further studies are required to confirm this observation and to relate the reduction in plasma volume to the severity of the renal concentrating defect.

11.3.2.3 Platelet Activation. It has been postulated that platelet activation may contribute to vaso-occlusion at the onset of crisis. Platelet abnormalities reported during crisis have included a transient fall in platelet count (Gordon et al., 1974; Alkjaersig et al., 1976), altered platelet aggregation (Barnhart et al., 1973; Stuart et al., 1974; Gruppo et al., 1977), and a reduction in platelet survival (Haut et al., 1973). These changes are likely to be secondary phenomena resulting from vascular stasis or a direct effect of the infection on circulating platelets. The transient fall in platelet count, preceded by platelet release of heparin neutralizing activity (O'Brien et al., 1974), during crisis is shown in Fig. 11.8. A similar phenomenon was observed (Richardson et al., 1979) in patients with infection but without sickle-cell disease (Fig. 11.9) and it thus appears to be a non-specific secondary finding.

During the steady state, Barnhart et al. (1973) found microscopical evidence of platelet aggregation formation and platelet membrane shape change, and Gruppo et al. (1977) reported partial activation of platelets with resulting impaired aggregation with adenosine diphosphate. We have recently studied 16 adult sickle-cell patients on 29 occasions in the steady state in comparison with 23 age-matched healthy controls (Kenny et al., unpublished). The sicklers showed significant increases in platelet number, platelet microaggregate formation (Wu and Hoak, 1974) and a lowered threshold to aggregation with adenosine diphosphate. Although these findings suggested that chronic intravascular platelet activation occurs in the steady state, similar platelet changes were

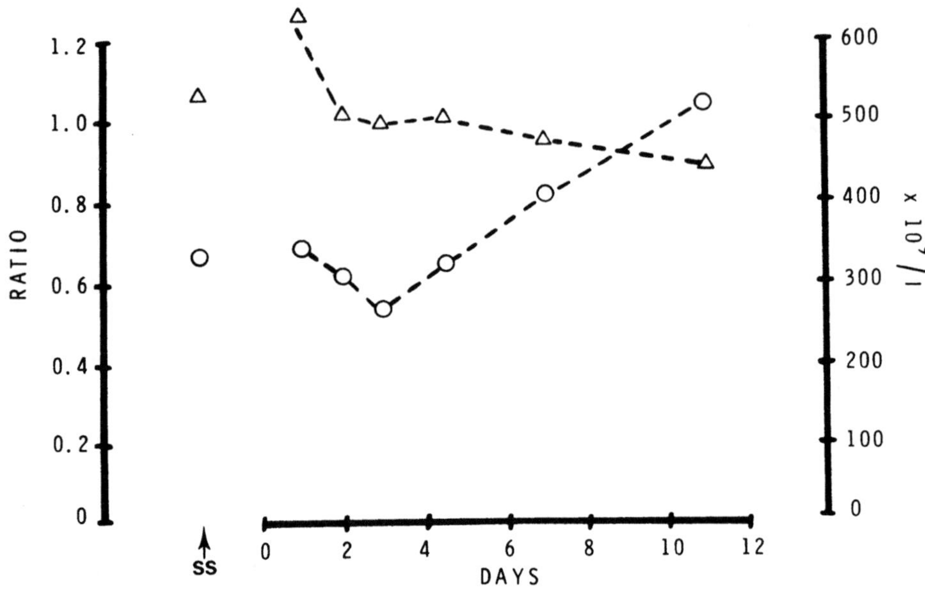

Fig.11.8. Serial mean values for heparin thrombin clotting time ratio (△) and platelet (○) count (x 10^9/l) in nine episodes of sickle-cell crisis compared with the mean of eight estimations in the steady state (*SS*). Based on the data of Richardson et al. (1979). HTCT ratio derived by dividing clotting time for normal plasma by that of test plasma

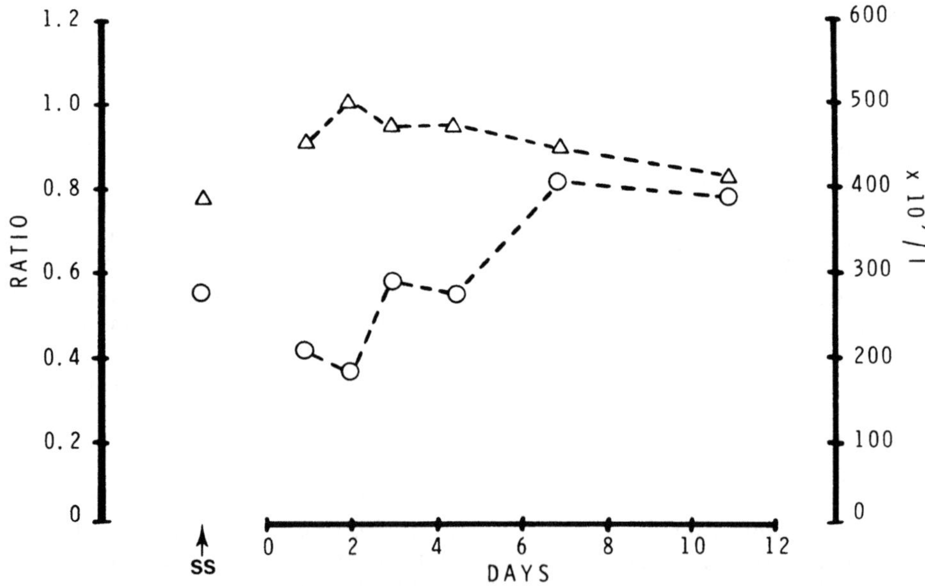

Fig.11.9. Serial mean values for heparin thrombin clotting time ratio (△) and platelet count (○) in 20 patients with acute infection compared with 22 healthy controls (*SS*) representing steady-state values. Based on data of Richardson et al. (1979)

obtained in 12 splenectomised young adults who did not have sickle-cell anaemia. Thus platelet changes in asymptomatic sicklers probably reflect an enlarged circulating population of young, metabolically active platelets as a consequence of childhood autosplenectomy. Despite a report that aspirin-dipyridamole prophylaxis may cause a slight reduction in symptomatic episodes (Alkjaersig et al., 1979), we doubt whether platelet activation is a primary factor in inducing the onset of vaso-occlusive crisis.

11.3.2.4 Coagulation Factors. Reports of coagulation activation in the steady state, with an exacerbation during crisis, may also be misleading. Asymptomatic patients show a significantly raised factor VIII level (Leslie et al., 1975; Richardson et al., 1979) but this has also been reported in other chronic haemolytic anaemias with increased reticulo-endothelial cell activity (Egeberg, 1962; Mannucci et al., 1969). Reduced levels of factor XIII (Ittyerah et al., 1976), reduced factor V (Leslie et al., 1975; Richardson et al., 1979) and raised antithrombin III (Richardson et al., 1979) in the steady state are similarly difficult to evaluate. When crisis develops, there are significant alterations in the levels of fibrinogen, factor VIII, antithrombin III, and fibrin degradation products consistent with coagulation activation, but similar changes were found as part of the acute-phase response to infection in 22 young adults without sickle-cell disease (Richardson et al., 1979). These changes therefore appear to represent secondary, non-specific events which are unlikely to be of aetiological significance in sickle-cell crisis.

11.3.2.5 Fibrinolytic Activity. A reduction in plasma fibrinolytic activity during crisis has been reported by several authors (Mahmood et al., 1967; Mahmood, 1969; Green et al., 1970; Walsh et al., 1973), particularly when there is accompanying infection (Green et al., 1970). Although reduced fibrinolytic activity is a well-recognised accompaniment to chronic atherosclerotic vascular disease (Nestel, 1959), patients with sickle-cell disease are able to produce significant shortening of the dilute whole-blood clot lysis time and euglobulin lysis time in response to exercise or the application of heat to the arm (Gordon et al., 1974). During crisis, fibrinolytic activity has been found to increase (Hilgartner et al., 1961), or not change significantly (Gordon et al., 1974), and patients given ancrod to achieve therapeutic defibrination during crisis also show the expected increase in zone of lysis on a fibrin plate (Mann et al., 1972). There is, therefore, little evidence that reduced fibrinolytic activity is a contributory factor to vascular occlusion during crisis.

11.4 Conclusions

The selection of an appropriate control group is extremely important in the study of sickle-cell crisis and the preceding steady state. Controls may require to be matched for age, asplenia, anaemia, reticulocytosis, reticulo-endothelial hyperplasia, race, or an acute-phase protein reaction. Very careful matching is particularly important in any study of rheology or thrombosis. It is also apparent that changes during vaso-occlusive crisis are dependent on the stage of crisis and a serial, daily study should ideally be made.

With these reservations, there would be considerable value in a serial study of erythrocyte deformability, whole-blood viscosity, and related factors such as percentage ISC during the early stages of crisis in comparison with the preceding steady state. While it is tempting to postulate that hyperviscosity, induced by the acute-phase rise in fibrinogen during infection, is the common factor which initiates vascular sludging in the early stages of crisis, our data suggest that a decrease in erythrocyte deformability may precede the fibrinogen effect. A serial study, by differentiating between primary causes and secondary events, should clarify which abnormality occurs first and thereby initiates the vicious circle of irreversible sickling. This information is vital to the development of effective methods for preventing crisis.

11.5 References

Alkjaersig N, Fletcher A, Joist H, Chaplin H Jr (1976) Hemostatic alterations accompanying sickle cell pain crises. J Lab Clin Med *88:* 440-449

Alkjaersig N, Chaplin H, Fletcher A P (1979) Experience with aspirin-dipyridamole prophylaxis in sickle cell disease. Thromb Haemostas *42:* 60

Bank A, Mears G, Weiss R, O'Donnell J V, Natta C (1974) Preferential binding of β^s globin chains associated with stroma in sickle cell disorders. J Clin Invest *54:* 805-809

Barnhart M I, Vettraino A, Lusher J M (1973) Microscopy as an aid in the evaluation and management of sickle cell anemia. Thromb Diathes Haemorrh (Stuttg) Suppl *53:* 193-214

Barreras L, Diggs L W (1964) Bicarbonates, pH and percentage of sickled cells in venous blood of patients in sickle cell crisis. Am J Med Sci *247:* 710-718

Bertles J F, Milner P F A (1968) Irreversibly sickled erythrocytes: a consequence of the heterogeneous distribution of the hemoglobin types in sickle-cell anaemia. J Clin Invest *47:* 1731-1741

Bessis M, Mohandas N (1977) Laser diffraction patterns of sickle cells in fluid shear fields. Blood Cells *3,* 229-239

Brownlee N R, Long L C, Balcerzak S P, Bromberg P A, Mills R L, Cornwell D G (1974) Relationship between erythrocyte lipid content and cell volume: regression equations for comparing normal and sickle cell populations. Physiol Chem Phys *6:* 479-487

Buckalew V M, Someren A (1974) Renal manifestations of sickle cell disease. Arch Intern Med *133:* 660-669

Chien S (1977) Rheology of sickle cells and erythrocyte content. Blood Cells *3:* 283-303

Chien S, Usami S, Bertles J F (1970) Abnormal rheology of oxygenated blood in sickle cell anemia. J Clin Invest *49:* 623-634

Chien S, Usami S, Jan K M, Smith J A, Bertles J F (1976) Blood rheology in sickle cell disease. In: Hercules J I, Cottam G L, Waterman M R, Schecter A N (eds) Proceedings of the Symposium on Molecular and Cellular Aspects of Sickle Cell Disease. US Department of Health, Education, and Welfare, Bethesda, Publication No. (NIH) 76-1007, pp 277-300

Copley A L, Luchini B W, Whelan E W (1968) On the rôle of fibrinogen-fibrin complexes in flow properties and suspension-stability of blood systems. In: Copley A L (ed). Hemorheology. Pergamon Press, Oxford, pp 375-387

Diederich D A, Trueworthy R C, Gill P, Cader A M, Larsen W E (1976) Hematologic and clinical responses in patients with sickle cell anemia after chronic extracorporeal red cell carbamylation. J Clin Invest *58:* 642-653

Diggs L W (1965) Sickle cell crises. Amer J Clin Path *44:* 1-19

Eaton J W, Skelton T D, Swofford H S, Koplin C E, Jacob H S (1973) Elevated erythrocyte calcium in sickle cell disease. Nature (Lond) *246:* 105-106

Eaton J W, Jacob H S, White J G (1979) Membrane abnormalities of irreversibly sickled cells. Semin Hematol *16:* 52-64

Egeberg O (1962) Blood coagulation and intravascular hemolysis. Scand J Clin Lab Invest *14:* 217-222

Erlandson M E, Schulman I, Smith C H (1960) Studies on congenital hemolytic syndromes. III Rates of destruction and production of erythrocytes in sickle cell anaemia. Pediatrics *25:* 629-644

Fischer S, Nagel R L, Bookchin R M, Roth E F, Tellez-Nagel I (1975) The binding of hemoglobin to membranes of normal and sickle erythrocytes. Biochim Biophys Acta (Amst) *375:* 422-433

Glader B E, Nathan D G (1978) Cation permeability alterations during sickling: relationship to cation composition and cellular hydration of irreversibly sickled cells. Blood *51:* 983-989

Glader B E, Lux S E, Muller-Soyano A, Platt O S, Propper R D, Nathan D G (1978) Energy reserve and cation composition of irreversibly sickled cells *in vivo*. Br J Haematol *40:* 527-532

Gordon P A, Breeze G R, Mann J R, Stuart J (1974) Coagulation fibrinolysis in sickle-cell disease. J Clin Path *27:* 485-489

Green D, Kwaan H C, Ruiz G (1970) Impaired fibrinolysis in sickle-cell disease; relation to crisis and infection. Thromb Diathes Haemorrh (Stuttg) *24:* 10-16

Gruppo R A, Glueck H I, Granger S M, Miller M A (1977) Platelet function in sickle cell anemia. Thromb Res *10,* 325-335

Harris J W, Brewster H H, Ham T H, Castle W B (1956) Studies on the destruction of red blood cells. The biophysics and biology of sickle-cell disease. Arch Intern Med *97:* 145-168

Haut M J, Cowan D H, Harris J W (1973) Platelet function and survival in sickle cell disease. J Lab Clin Med *82:* 44-53

Hilgartner M W, Horowitz H, Erlandson M, Ferguson A, Smith C H (1961) Studies of the coagulation mechanism in patients with sickle-cell anemia. Am J Dis Child *102:* 591-592

Ittyerah R, Alkjaersig N, Fletcher A, Chaplin H (1976) Coagulation factor XIII concentration in sickle-cell disease. J Lab Clin Med *88:* 546-554

Jandl J H, Simmons R L, Castle W B (1961) Red cell filtration and the pathogenesis of certain hemolytic anemias. Blood *18:* 133-148

Klug P P, Lessin L S, Radice P (1974) Rheological aspects of sickle cell disease. Arch Intern Med *133:* 577-590

Konotey-Ahulu F I D (1965) Sicklaemic human hygrometers. Lancet *i:* 1003-1004

LaCelle P L (1975) Pathologic erythrocytes in the capillary microcirculation. Blood Cells *1:* 269-284

Leslie J, Langler D, Serjeant G R, Serjeant B E, Desai P, Gordon Y B (1975) Coagulation changes during the steady state in homozygous sickle-cell disease in Jamaica. Br J Haematol *30:* 159-166

Lessin L S, Kurantsin-Mills J, Weems H B (1977) Deformability of normal and sickle erythrocytes in a pressure-flow filtration system. Blood Cells *3:* 241-262

Lessin L S, Kurantsin-Mills J, Weems H B (1978) Membrane alterations in irreversibly sickled cells (ISCs): ultrastructural, membrane protein and rheologic correlations. Abstracts, XVII Congress of the International Society of Hematology, Paris, p 64

Mahmood A (1969) Fibrinolytic activity and sickle cell crises. Br Med J *1:* 52-53

Mahmood A, Macintosh D M, Shaper A G (1967) Fibrinolytic activity in the clinical crisis of sickle cell anaemia. Br Med J *3:* 653-654

Mann J R, Deeble T J, Breeze G R, Stuart J (1972) Ancrod in sickle-cell crisis. Lancet *i:* 934-937

Mannucci P M, Lobina G F, Caocci L, Dioguardi N (1969) Effect on blood coagulation of massive intravascular haemolysis. Blood *33:* 207-213

Marsh N A (1978) The effect of fibrinogen and fibrinogen degradation products on erythrocyte sedimentation rate. Abstracts, XVII Congress of the International Society of Hematology, Paris, p 771

Mohandas N, Bessis M (1979) Red cell deformability changes in sickle cell anemia. In: Rosa J, Beuzard Y, Hercules J (eds) Development of therapeutic agents for sickle cell disease. North-Holland, Amsterdam, pp 15-29

Nestel P J (1959) Fibrinolytic activity of the blood in intermittent claudication. Lancet *ii:* 373-374

O'Brien J R, Etherington M, Jamieson S, Lawford P (1974) Heparin-thrombin clotting-time and platelet factor 4. Lancet *ii:* 656-657

Palek J (1977) Red cell calcium content and trans-membrane calcium movements in sickle cell anaemia. J Lab Clin Med *89:* 1365-1374

Palek J, Curby W, Lionetti F (1971) Effects of calcium and adenosine triphosphate on volume of human red cell ghosts. Am J Physiol *220:* 19-26

Reid H L, Barnes A J, Lock P J, Dormandy J A, Dormandy T L (1976) A simple method for measuring erythrocyte deformability. J Clin Pathol *29:* 855-858

Richardson S G N, Breeze G R, Stuart J (1976) Hyperfibrinogenaemia and hyperviscosity in sickle-cell crisis. J Clin Pathol *29:* 890-893

Richardson S G N, Matthews K B, Stuart J, Geddes A M, Wilcox R M (1979) Serial changes in coagulation and viscosity during sickle-cell crisis. Br J Haematol *41:* 95-103

Rieber E E, Veliz G, Pollack S (1977) Red cells in sickle cell crisis: observations on the pathophysiology of crisis. Blood *49:* 967-979

Rosenblate H J, Eisenstein R, Holmes A W (1970) The liver in sickle cell anemia. A clinical-pathologic study. Arch Pathol *90:* 235-245

Schmid-Schönbein H, Weiss J, Ludwig H (1973) A simple method for measuring red cell deformability in models of the microcirculation. Blut *26:* 369-379

Serjeant G R, Serjeant B E, Milner P F (1969) The irreversibly sickled cell; a determinant of haemolysis in sickle cell anaemia. Br J Haematol *17:* 527-533

Serjeant G R, Serjeant B E, Desai P, Mason K P, Sewell A, England J M (1978) The determinants of irreversibly sickled cells in homozygous sickle cell disease. Br J Haematol *40:* 431-438

Steinberg M H, Dreiling B J, Lovell W J (1977) Sickle cell anaemia: erythrokinetics, blood volumes, and a study of possible determinants of severity. Am J Hematol *2:* 17-23

Stuart M J, Stockman J A, Oski F A (1974) Abnormalities of platelet aggregation in the vaso-occlusive crisis of sickle cell anemia. J Pediatr *85:* 629-632

Usami S, Chien S, Scholz P M, Bertles J F (1975) Effect of deoxygenation on blood rheology in sickle cell disease. Microvasc Res *9:* 324-334

Walsh R T, Lusher J M, Barnhart M I (1973) Coagulation and fibrinolysis studies in sickle cell anemia. In: Mammen E F, Anderson G F, Barnhart M I (eds) Sickle cell disease. Schattauer Verlag, Stuttgart, pp 271-279

Wu K K, Hoak J C (1974) A new method for the quantitative detection of platelet aggregates in patients with arterial insufficiency. Lancet *ii:* 924-926

12 Circulatory Complications of Leukaemia and Paraproteinaemia

F.E. Preston

12.1 Macroglobulinaemia, Myeloma and the Hyperviscosity Syndrome (HVS)

The relationship between elevated serum viscosity levels and hyperproteinaemic states has been recognised for almost half a century. Interestingly, most of the earlier reports of this association related to patients with myelomatosis (Albers, 1937; Reimann, 1932; Wintrobe and Buell, 1933) and it was not until 1944 that increased serum viscosity levels were observed by Waldenström in patients with macroglobulinaemia (Waldenström, 1944).

Although Waldenström (1949) clearly recognised a relationship between macroglobulinaemia and various clinical signs and symptoms it was not until 1965 that Fahey and his colleagues defined the condition now known as the hyperviscosity syndrome (Fahey et al., 1965). These workers convincingly demonstrated that in patients with macroglobulinaemia many of the clinical features of microvascular insufficiency occurred as a direct consequence of an elevated serum viscosity. This, in turn, reflected the plasma concentration of the circulating paraprotein. These workers also showed that clinical improvement could be produced by removal of the patient's plasma by plasmapheresis.

Until very recently, it was considered that the hyperviscosity syndrome was confined almost exclusively to Waldenström's macroglobulinaemia, and it was believed that it occurred only very rarely in myelomatosis. Recent studies, however, have clearly indicated that this view is incorrect (Tuddenham et al., 1974; Preston et al., 1978a) and that HVS is a recognised complication of certain types of myelomatosis.

12.1.1 Clinical Features of HVS

It should be stressed that elevated blood viscosity levels are not necessarily associated with clinical signs and symptoms and for patients with a paraprotein there is considerable individual variation in the clinical effects of the raised blood viscosity. For any group of patients there is no correlation between measured blood or serum viscosity and the clinical symptomatology. Thus, some patients with relatively high blood viscosity levels may have minimal clinical sequelae whereas others may exhibit florid and debilitating clinical effects with only marginal elevation of their viscosity. This clearly indicates that factors other than blood viscosity influence the appearance of the signs and symptoms which constitute the hyperviscosity syndrome.

Although there is no correlation between the clinical effect and the blood viscosity between patients in a group, the measurement of the blood or serum viscosity of an individual is of considerable predictive value. This is because, for any single patient, there is remarkable reproducibility in the type and degree of symptoms for a given viscosity and for many patients the viscosity level may be accurately predicted from the accompanying clinical picture.

12.1.1.1 General Symptoms. Patients often complain of weakness, excessive fatigue, lethargy and drowsiness. Although affected patients are invariably anaemic, the symptoms are often greatly in excess of those normally associated with the degree of anaemia.

12.1.1.2 Visual Disturbances. Patients with HVS frequently complain of visual disturbances. These include haziness or blurring of vision, diplopia or even complete loss of vision. The onset is usually gradual and transient. Patients often complain that their vision becomes blurred whilst reading newspapers or books. Occasionally visual acuity may suddenly deteriorate and total blindness may also occur. The retina of patients with HVS is invariably abnormal and in this author's view the diagnosis of HVS is not tenable in the absence of some form of retinopathy. The commonest abnormality is congestion and pronounced tortuosity of the retinal veins. In advanced cases, sausage-like constrictions of the distended veins occur at sites of arterio-venous crossings. In some patients retinal haemorrhages and exudates may also occur and on occasion the fundal changes are indistinguishable from those observed in hypertensive encephalopathy.

12.1.1.3 Neurological Features. Neurological signs and symptoms are very commonly observed in patients with HVS. Patients often complain of headaches, but dizziness, vertigo, visual and auditory disturbances, tinnitus, nystagmus, ataxia and transient ischaemic attacks may also occur. Patients may complain of somnolence, and coma may be a presenting feature. On occasion there may be pyramidal signs and symptoms and, more rarely, peripheral neuropathies.

12.1.1.4 Other Complications. Cardiovascular insufficiency may also be observed in patients with HVS. It is not always possible to be certain that this is a direct consequence of HVS, particularly since many of these patients are elderly and are also profoundly anaemic. The frequency of response of the cardiac signs and symptoms to plasmapheresis does, however, suggest that at least some of the cardiac effects relate to the increased blood viscosity or the accompanying hypervolaemia. Similarly, renal impairment is also often seen in patients with severe HVS and again this may respond satisfactorily to methods aimed at reducing the blood viscosity. A few patients exhibit cryo-sensitivity which may be manifested by Raynaud's phenomenon and a tendency to peripheral gangrene, particularly in cold weather.

12.1.2 The Bleeding Diathesis of HVS

A bleeding diathesis is often present in patients with HVS and on occasion may even dominate the clinical picture. Epistaxis is common and may be particularly troublesome to control. Other haemorrhagic manifestations include bleeding from venepuncture sites, spontaneous bruising, and vaginal and genito-urinary haemorrhage. Gastro-intestinal haemorrhage may present as either haematemesis or frank melaena.

The routine investigations of coagulation are frequently abnormal in patients with HVS. However, one should be cautious in ascribing abnormalities of haemostatic function to this syndrome since similar abnormalities occur in the absence of hyperviscosity. In HVS the bleeding time is not infrequently prolonged even in those patients who are not thrombocytopenic. This may be accompanied by defective platelet aggregation. These abnormalities of platelet function presumably occur as a result of platelet coating by the paraprotein. The coagulation tests most frequently abnormal are the thrombin and reptilase clotting times. In the presence of a normal plasma fibrinogen this indicates inhibition of fibrin polymerisation by the circulating paraprotein (Lackner et al., 1970). Occasionally the prothrombin time or partial thromboplastin time may be unduly prolonged and again evidence of inhibitory activity against one or more clotting factors should be sought.

12.1.3 Hypervolaemia and HVS

It is now recognised that an increased plasma volume is an almost invariable concomitant of HVS and this is an important factor in the anaemia which accompanies the syndrome (Russell and Powles, 1978). In patients with myelomatosis and Waldenström's macroglobulinaemia the plasma volume is proportional to the serum viscosity which in turn is a function of the circulating paraprotein. Thus, at a given protein concentration the plasma volume is greater in patients with IgA myeloma and macroglobulinaemia than that observed in patients with IgG myeloma (see section 12.1.4). The increase in plasma volume is unlikely to relate to colloid osmotic pressure since this is generally regarded as being inversely related to molecular weight. An important question is whether the increased plasma volume contributes to the clinical or pathological features of HVS. Russell and Powles (1978) examined the relationships between plasma volume, serum viscosity, retinopathy and a bleeding tendency in patients with myeloma and macroglobulinaemia and concluded that the retinopathy of HVS was more closely related to the raised viscosity levels than to the hypervolaemia. The relationship between bleeding and viscosity levels, however, is more difficult to define. These same authors showed that although the correlation between bleeding and viscosity is strong there are, nevertheless, patients who bleed whose blood viscosity is insufficient to be associated with a retinopathy.

12.1.4 Incidence of HVS in Myelomatosis

The incidence of HVS in myelomatosis remains uncertain since it has only recently become recognised as a complication of IgA myeloma. Hobbs (1969) reported an incidence of 4% for patients with IgG myeloma but made no reference to IgA myeloma. More recently Tuddenham et al. (1974) reported HVS in 14% of their IgA myeloma patients whilst Preston et al. (1978a) reported an overall incidence of 9.6% with 25% of patients with IgA myeloma being affected.

12.1.5 Pathophysiology of HVS

For the majority of paraproteinaemic patients the increased serum viscosity is a reflection of (a) the concentration and (b) the intrinsic viscosity of the circulating paraprotein. In macroglobulinaemia the serum viscosity relates to the physical properties of the IgM molecule, which has a molecular weight of 1 million. In addition, it also has a high axial length/width ratio and this is associated with a high intrinsic viscosity of the IgM molecule. Consequently, for a given concentration of IgM the serum viscosity is higher than that produced by the same concentration of the smaller IgG molecule with a molecular weight of 160,000.

In all paraproteinaemic patients extremely high concentrations of the paraprotein, regardless of class or sub-type, will raise the blood viscosity to clinically significant levels. IgG_1 has a low intrinsic viscosity; consequently, HVS is not usually seen in patients with this type of myeloma until the paraprotein concentration exceeds 80 g/litre. At this concentration the serum viscosity may be sufficiently raised to produce symptoms. There are certain other immunoglobulins, notably IgA and IgG_3, which may significantly elevate the blood viscosity at relatively low concentrations. This relates to the tendency of both paraproteins to form complexes of high intrinsic viscosity.

Studies of the physico-chemical characteristics of IgA have clarified viscosity effects of this paraprotein (Preston et al, 1978a). Figure 12.1a and b illustrate the relationship between the relative serum viscosity and total protein concentration for a number of myeloma and non-myeloma sera. Both hyperconcentrated normal serum and an IgG_2 myeloma gave identical linear plots against the log of the relative serum viscosity (RSV), indicating the IgG globulins show no viscosity anomalies. When serum from a patient with Waldenström's macroglobulinaemia containing a 10 S IgG-IgG complex was similarly plotted, a similar linear relationship was seen up to 60 g/litre and no 10 S complex was demonstrable in the ultracentrifuge at these dilutions.

Above 70 g/litre, 10 S complexes were present and the slope of the log RSV/protein concentration plot changed (Fig. 12.1a). Both 7 S IgA myeloma and IgA myeloma with 10 S dimer but without higher polymers gave a linear but somewhat steeper plot with some indication of transition to a smooth curve above 80 g/litre for the myeloma with 10 S dimer. This possibly reflects the small

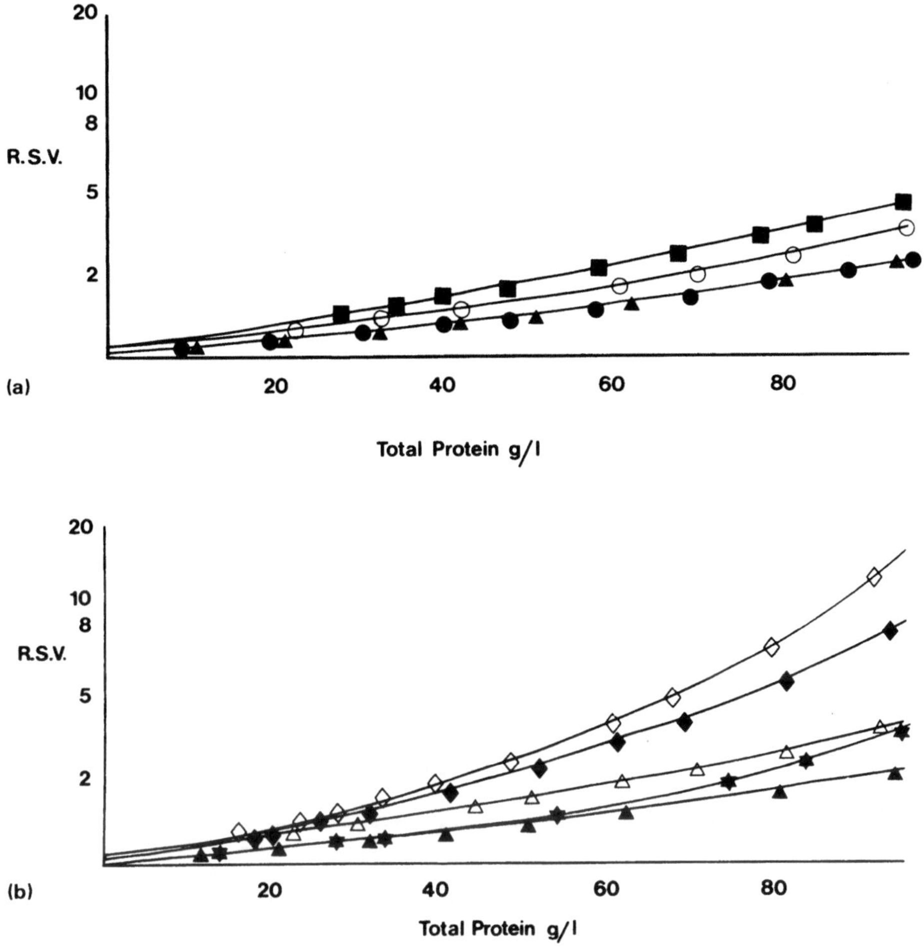

Fig. 12.1. Effect of serum dilution on relative serum viscosity in a group of hyperproteinaemia states likely to produce a hyperviscosity syndrome (a) ▲ Hyperconcentrated normal serum; ● IgG_2 myeloma; ○ Waldenström's macroglobulinaemia; ■ IgA myeloma (dimer) (b) ▲ Hyperconcentrated normal serum; ★ IgG_3 myeloma; △ IgA myeloma (monomer); ◆ macroglobulinaemia; ◇ IgA myeloma (dimer + polymer)

amounts of higher polymers. IgA myeloma with higher polymers gave a much steeper curvilinear response with no linearity even at low protein concentrations demonstrating the viscosity contribution of the higher IgA polymers (Fig. 12.1b) (Preston et al., 1978a). Macroglobulinaemia produced a curvilinear plot similar to that of polymeric IgA but above 30 g/litre the slope is less steep than for polymeric IgA (Fig. 12.1b).

In order to investigate the behaviour of IgA HVS serum further, Preston et al. (1978a) studied shear stress/shear rate plots for an IgA HVS serum, macroglobulinaemia serum and IgG_2 myeloma serum. These are shown in Fig. 12.2. IgG_2 serum behaved as a Newtonian fluid. Both macroglobulinaemia and

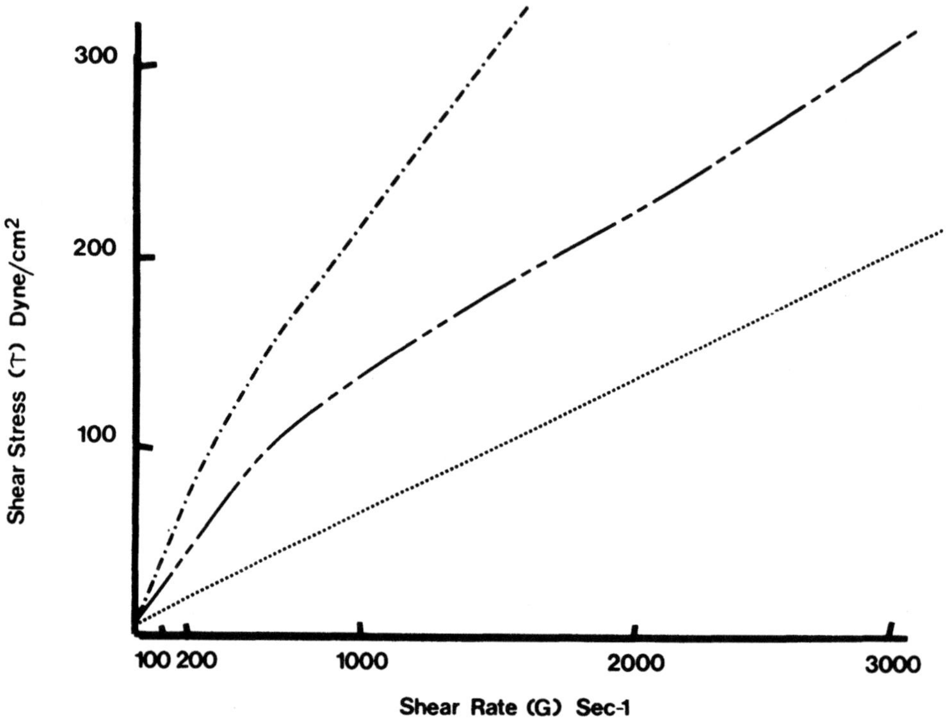

Fig.12.2. Shear stress-shear rate plots for three myeloma sera.
IgG_2 (Newtonian fluid) ....
IgM HVS (pseudoplastic behaviour) ——— – – ——— – – ——— –
IgA HVS (pseudoplastic behaviour) –·–·–·–·–·–·–·–·–·–·–
Viscosity (shear stress/shear rate) = 4.4 mPa.s (IgG_2), 19.4 mPa.s (IgM), 29.7 mPa.s (IgA)
Protein concentration in each sample 105 g/litre (paraprotein 70 g/litre)

IgA HVS sera showed marked pseudoplasticity with limited Newtonian behaviour at shear rates above 800 s^{-1} and viscosities of 19.4 and 29.7 mPa.s respectively. Pseudoplastic behaviour of this type is usually due to realignment of asymmetrical molecules. This interpretation was shown to be correct in further studies on IgA HVS serum.

It is this intrinsic property of the IgA molecule to form high-molecular-weight complexes such as dimers, trimers and tetramers with sedimentation coefficients of 10 S, 12 S and 15 S respectively which is so important in the production of the hyperviscosity syndrome. These complexes, particularly IgA trimers, are prominent in patients with IgA myeloma complicated by HVS (Preston et al., 1978a). In patients without HVS the IgA is mainly in the form of monomer and dimer. Further data obtained for purified IgA monomer, IgA dimer and IgA polymer show that the viscosity of the polymer is significantly greater than that of the larger IgM molecule. The intrinsic viscosity of IgA monomer and IgA dimer occupies an intermediate position between IgG and IgM. For this reason the whole-blood viscosity, for a given concentration of paraprotein, is higher in IgA myeloma than in IgG myeloma.

12.1.6 Diagnosis of HVS

The serum viscosity is invariably increased in all patients with paraproteinaemia other than those with light chain myelomas. The diagnosis of HVS should be confined to those patients in whom elevated blood viscosity is associated with the signs and symptoms of the syndrome.

Clinicians should be aware that there are certain groups of patients who are particularly susceptible to this complication. They include patients with the following:

1) Macroglobulinaemia

2) IgA myeloma

3) IgG_3 myeloma

4) All patients with serum paraprotein concentrations over 80 g/litre. HVS occurs particularly at initial clinical presentation, and during relapse when paraprotein concentrations are high.

The development of CNS signs and symptoms (particularly headaches, drowsiness, or visual disturbances) or a bleeding diathesis should suggest the possibility of the syndrome. Retinal examination is essential and the characteristic features represent valuable, if not essential, diagnostic support.

Although the serum viscosity is often raised in patients with paraproteinaemia irrespective of the presence or absence of HVS, some useful guidelines are available. Fahey et al. (1965) state that although there is great individual variation, no patient in their study with serum viscosities less than 4.0 mPa.s had symptoms. Similarly Russell and Powles (1978) report that the retinopathy of HVS is always associated with serum viscosity levels greater than 3.8 mPa.s. There are inherent dangers in relying solely on serum viscosity levels for the diagnosis of HVS since the whole-blood viscosity is influenced by the haematocrit and also by red-cell aggregation. For these reasons, it is perhaps more appropriate to measure whole-blood viscosity levels at various rates of shear.

12.1.7 Treatment of HVS

Plasmapheresis is an effective method of treating the hyperviscosity syndrome since large amounts of paraprotein can be removed relatively quickly. The fall in paraprotein concentration is paralleled by a fall in serum viscosity and this is often accompanied by a dramatic reversal of the patient's symptoms. The volume of plasma which requires to be removed depends upon its initial concentration and the type of paraprotein. In patients with Waldenström's macroglobulinaemia a single plasmapheresis may be sufficient to remove most of the IgM, since the majority of this immunoglobulin is confined within the intravascular space. In patients with IgG myeloma it may be necessary for a second or even third plasmapheresis since 60% of the total IgG may reside in the extravascular compartment. Consequently, following the first plasmapheresis a secondary rise may occur from the extravascular space. At clinical presentation symptoms are usually satisfactorily reversed by plasmapheresis and adequate control can be maintained with a standard chemotherapeutic regime. During relapse, however, the problem is much more difficult and the rapid production of new paraprotein may rapidly negate the beneficial effects of plasmapheresis.

It should be remembered that patients with HVS are invariably anaemic, with hypervolaemia an important contributory factor. Thus, blood transfusion may be hazardous since the improved haematocrit may severely aggravate the clinical features of HVS by its effect on whole-blood viscosity.

12.2 Leukaemia and HVS

The signs and symptoms of HVS are simply a reflection of impaired tissue perfusion consequent upon reduced blood flow. This in turn is a consequence of the increased blood viscosity. As already discussed, the symptom complex of neurological dysfunction, visual disturbances and a haemorrhagic tendency is usually associated with Waldenström's macroglobulinaemia and, more recently, myelomatosis. However, similar features can also be observed in patients whose viscosity is elevated as a consequence of alteration in the cellular content of the blood. For example, in patients with polycythaemia vera the abnormal viscosity resulting from the increased red-cell mass can produce signs and symptoms similar to those seen in the hyperviscosity syndrome associated with paraproteinaemia.

Neurological symptoms are not uncommon in patients with leukaemia. They are more usually considered to relate to leukaemic infiltration or else haemorrhage within the central nervous system. In some leukaemic states the clinical features may be accentuated by electrolyte imbalance, particularly in acute myeloblastic leukaemia and the blast crisis of chronic granulocytic leukaemia. There is now evidence that patients with leukaemia may also develop neurological signs and symptoms occurring as a result of the increased whole-blood viscosity consequent upon a grossly elevated white blood cell count (Preston et al., 1978b).

12.2.1 Clinical Features of Cellular Hyperviscosity

The syndrome is not confined to any specific leukaemic cell type and has been reported in chronic lymphocytic leukaemia, acute myeloblastic leukaemia and chronic granulocytic leukaemia. In chronic granulocytic leukaemia it may occur in the so-called benign phase or during blast-cell transformation. The syndrome appears most likely to occur in patients whose white-cell count is in excess of 200×10^9/litre. Patients with myeloid leukaemias appear to develop the signs and symptoms of cellular hyperviscosity at lower white-cell counts than those observed in patients with the lymphocytic type of leukaemia. Thus, in patients with chronic lymphocytic leukaemia the syndrome may not develop until the white-cell count exceeds 400 or 500×10^9/litre. When considering the viscosity effects of various types of leukaemic cell, individual cell size would appear to be an important factor but other variables such as the presence or absence of a nucleus will also influence the intrinsic viscosity of the white blood cells.

The clinical features of cellular hyperviscosity are similar to those observed in patients with malignant paraproteinaemias. Haemorrhagic manifestations, however, are not so commonly observed in this type of hyperviscosity syndrome. In the group of patients described by Preston et al. (1978b) patients presented with profound lethargy, unsteady gait, visual disturbances, deafness, headaches and coma. Significant hearing loss appears to be much more prominent a feature in patients with cellular hyperviscosity compared with the hyperviscosity effects associated with abnormal paraproteins. Fundal investigation reveals congested and tortuous retinal vessels with or without associated retinal haemorrhages.

12.2.2 *Blood Viscosity Measurements*

Since increased viscosity is produced by the circulating white cells it is necessary to measure the viscosity on whole blood and preferably at various rates of shear. The plasma viscosity in these patients is usually normal. As with patients with paraproteinaemia there is great individual variation between the level of the whole-blood viscosity and the occurrence of neurological signs and symptoms. It should be appreciated that most leukaemic patients have some degree of anaemia and, therefore, the reduction in the haematocrit will tend to counterbalance the viscosity effects of the increased white-cell count. It cannot be overstressed that blood transfusion, therefore, is potentially hazardous in such patients, since the increase in haematocrit following transfusion may raise the whole-blood viscosity sufficiently to produce coma or even death.

12.2.3 *Treatment of Cellular Hyperviscosity*

Leucopheresis presents an efficient and rapid method of reducing the circulating white-cell mass. The reduction in the total white-cell count will be accompanied by a similar reduction in whole-blood viscosity. In the six patients described by Preston et al. (1978b) the improvement in the laboratory results was accompanied by excellent clinical responses in three patients and a satisfactory response in one other. Two patients were not improved by leucopheresis. One of these patients was comatose when treated and died shortly afterwards without recovering consciousness. The general condition of the other patient improved for about 18 h after leucopheresis but then following the institution of chemotherapy he rapidly lost consciousness and died. Immediately prior to death there was haematological evidence of intravascular coagulation and this was subsequently confirmed by histological examination of the brain. This complication of chemotherapy in acute leukaemia almost certainly relates to the release of thromboplastins from damaged blast cells and these induce intravascular coagulation by activation of the extrinsic coagulation pathway.

Although the incidence of the cellular hyperviscosity syndrome in leukaemic states is not yet established it is obviously important to recognize it as a possible cause of neurological signs and symptoms in leukaemic patients since marked clinical improvement can be achieved by relatively simple means.

12.3 References

Albers D (1937) Multiple Myelome and Viscosität. Z Klin Med *132:* 807-811

Fahey J L, Barth W F, Solomon A (1965) Serum hyperviscosity syndrome. JAMA *192:* 464-467

Hobbs J R (1969) Immunochemical classes of myelomatosis including data from a therapeutic trial conducted by a Medical Research Council Working Party. Br J Haematol *16:* 599-617

Lackner H, Hunt V, Zucker M B, Pearson J (1970) Abnormal fibrin ultrastructure, polymerization and clot retraction in multiple myeloma. Br J Haematol *18:* 625-636

Preston F E, Cooke K B, Foster M E, Winfield D A, Lee D (1978a) Myelomatosis and the hyperviscosity syndrome. Br J Haematol *38:* 517-530

Preston F E, Sokol R J, Lilleyman J S, Winfield D A, Blackburn E K (1978b) Cellular hyperviscosity as a cause of neurological symptoms in leukaemia. Br Med J *1:* 476-478

Reimann H A (1932) Hyperproteinaemia as a cause of autohaemagglutination. JAMA *99:* 1411-1414

Russell J A, Powles R L (1978) The relationship between serum viscosity, hypervolaemia, and clinical manifestations associated with circulating paraprotein. Br J Haematol 39, 163-175

Tuddenham E G D, Whittaker J A, Bradley J, Lilleyman J S, James D R (1974) Hyperviscosity syndrome in IgA multiple myeloma. Br J Haematol *27:* 67-76

Waldenström J (1944) Incipient myelomatosis or "essential" hyperglobulinaemia with fibrinogenopenia — a new syndrome? Acta Med Scand *117:* 216-247
Waldenström J (1949) Recherches cliniques et physicocliniques sur les hyperglobulinemies. Presse Med *57:* 213-215
Wintrobe M M, Buell M V (1933) Hyperproteinaemia associated with multiple myeloma. Bull Johns Hopk Hosp *52:* 156-165

13 Occlusive Arterial Disease and Blood Rheology

G.D.O. Lowe, M.M. Drummond, C.D. Forbes and J.C. Barbenel

13.1 Introduction

A variety of common and important clinical events are thought to arise from organ ischaemia secondary to obstructed blood flow in the major arteries which supply the heart, brain and lower limbs. It is our purpose to review the possible contribution of blood rheology to these events. Recent interest in this topic probably arises for three reasons. In the last decade a number of studies have been reported which associate blood rheology factors with clinical arterial events, not only retrospectively but also prospectively. Secondly, changes in blood viscosity and its major determinants (haematocrit, plasma viscosity, plasma fibrinogen level) have been related to changes in cerebral and limb blood flow in man. Thirdly, alteration of the flow properties of blood appears to be a possible therapeutic approach to problems against which we currently have a very limited armamentarium. Most patients with angina pectoris obtain some relief from glyceryl trinitrate and beta-adrenergic blocking drugs, and arterial reconstructive surgery also gives symptomatic improvement to selected subjects who have angina or transient symptoms of leg or brain ischaemia. There is, however, little evidence that any medical or surgical measures alter the natural history of occlusive arterial disease — with the exception of stopping cigarette-smoking (Doll and Hill, 1964).

The epidemiology, pathology, and pathogenesis of "occlusive arterial disease" are complex. Patients with different clinical syndromes share certain factors which permit us to group them under this umbrella term. The first of these is a basic pathology, atherosclerosis — non-inflammatory, stenosing or occluding lesions of major arteries which contain smooth muscle, fibrous tissue, fibrinogen, fibrin and low-density lipoprotein. Lesions tend to occur at points of dynamic stress. Atherosclerosis starts early in life, and is progressive and universal. However, post-mortem studies show that persons who experience clinical coronary artery events (sudden death, angina pectoris, or major myocardial infarction) usually have more extensive coronary artery narrowing than do persons who have not suffered these events. Similarly, transient or permanent ischaemia of the lower limbs or brain is usually associated with narrowing of the caudal or the cranial arteries which is of above-average severity. There is also a tendency for the severity of atherosclerosis in one part of the arterial tree (coronary, cranial or caudal) to correlate with the severity of arterial narrowing at the other two sites (Mitchell and Schwartz, 1965).

Epidemiological studies support this pathological concept of arterial occlusion as a systemic disease. Thus, risk associations have been defined which synergically increase the likelihood of a person experiencing clinical arterial events. The major risk factors (increasing age, male sex, cigarette smoking, diabetes, blood pressure and serum cholesterol level) are common to all three major manifestations of arterial occlusion — coronary, cerebral and lower-limb events (Gordon and Kannell, 1972). The link between epidemiology and pathology may be that, in general, autopsy studies suggest that these risk factors also correlate with the extent of atherosclerosis. Whether less important risk factors (oestrogen consumption, obesity, physical activity, blood group, and the negative risk factor of alcohol consumption) are also common to the three main types of arterial event, and whether they also correlate with the extent of atherosclerosis, are questions which require further study.

If the major clinical syndromes of occlusive arterial disease have epidemiological and pathological features in common, they also exhibit dissimilarities. For example, the associations of male sex and cigarette-smoking with ischaemic events in the lower limbs are more pronounced than their associations with coronary or cerebral ischaemia (Gordon and Kannell, 1972). Arterial occlusion is not entirely due to atheroma: platelet-fibrin thrombi, platelet-fibrin emboli, and arterial spasm may all contribute to organ ischaemia, and there is evidence that the relative contribution of these other processes varies in different clinical syndromes. For example, in patients who die suddenly and in patients with "pure" angina there is usually severe coronary atherosclerosis, but not usually an occlusive thrombus (Roberts and Ferrans, 1976). On the other hand, patients who die after sustaining a major cardiac or cerebral infarction commonly have an occlusive thrombus in the appropriate artery (Chandler et al., 1974; Battacharji et al., 1967). Apart from emboli arising from the heart, atheromatous or platelet-fibrin emboli arising from proximal arterial lesions may give rise to transient cerebral or ocular ischaemia (Russell, 1961) and digital ischaemia (Slaney and Hamer, 1973). It has also been suggested that intravascular platelet aggregates might contribute to digital ischaemia (Vreeken and van Aken, 1971) and sudden coronary death (Haerem, 1972). Finally, coronary artery spasm may be associated with some episodes of angina (Oliva et al., 1973).

The contributions of these various phenomena to organ ischaemia remain controversial. For example, it has been argued that coronary artery thrombosis is a consequence, and not a cause of myocardial necrosis (Roberts and Ferrans, 1976). Similarly, intravascular platelet thrombosis may form as a response to ischaemic tissue injury as opposed to causing it (Petito, 1979). While there is evidence that deposition of fibrin and platelets on arterial walls not only increases arterial lumen occlusion but may also contribute to atherosclerosis (Roberts and Ferrans, 1976; Ross and Glomset, 1976), there is so far little evidence that drugs which reduce fibrin formation or which alter platelet behaviour significantly reduce the incidence of arterial events in intervention studies (Mackie and Douglas, 1978; Turpie and Hirsh, 1978).

To summarise, occlusive arterial disease is complex and our understanding of its pathogenesis is limited. Nevertheless, we can draw some conclusions which may direct our studies of blood rheology. (1) We can define with varying precision certain clinical syndromes of organ ischaemia, and relate blood rheology to these clinical events (preferably the more precise ones). (2) Extensive atherosclerosis appears to be a common pathological basis for most of these clinical events, with or without additional lumen occlusion by platelets and/or fibrin. Subjects with severe atherosclerosis at one site tend to have similarly extensive disease at other sites. The localisation of lesions suggests that fluid-dynamic forces may play a role in their formation. The relationship of blood rheology factors to the extent of arterial disease therefore merits study. (3) Certain well-recognised risk factors are associated in epidemiological studies with the development of clinical events and in morphological studies with the extent of arterial occlusion. The relationship of blood rheology factors to these risk factors for arterial disease is therefore also of interest. (4) In morphological studies, arterial occlusion by wall disease or by intraluminal masses derived from blood is commonly incomplete. A further reduction in blood flow which precipitates ischaemia may therefore exist. While arterial spasm, or a fall in blood pressure from any cause, might provide this additional impairment of blood flow, another possibility is the functional obstruction of impaired blood fluidity. (5) As there is evidence that aggregates or deposits of platelets and fibrin are associated with some clinical events, the relationship of blood rheology factors to these processes may also be of interest.

We shall now review studies relating blood rheology factors (blood viscosity, plasma viscosity, haematocrit, plasma fibrinogen, red-cell aggregation, red-cell deformability and platelet aggregates) to these various aspects of occlusive arterial disease.

13.2 Relation to Clinical Events

13.2.1 Acute Organ Infarction

Cardiac infarction. Several workers have studied changes in blood rheology factors after acute cardiac infarction. Jan et al. (1975), who performed serial studies, found that blood viscosity was elevated on admission to hospital and for more than 21 days afterwards: the cause of the elevated viscosity varied as a function of time. In the first 3 days, the viscosity increase was due mainly to an elevated haematocrit. This increased haematocrit has also been reported by others (Burch and DePasquale, 1962; Stables et al., 1967; Sedziwy et al., 1968;

Hershberg et al., 1972), but not by all (Conley et al., 1964; Ditzel et al., 1968). The causes of the increased haematocrit may include dehydration, reduction in plasma volume due to circulating catecholamines, and reduction in plasma volume due to pulmonary oedema formation (Sedziwy et al., 1968; Jan et al., 1975). Most workers have reported that the haematocrit subsequently falls, possibly due partly to resolution of these abnormalities of plasma volume, and in part due to "diagnostic phlebotomy" — at least in Boston (Hershberg et al., 1972).

Blood viscosity remains high despite the fall in haematocrit, due to increased plasma viscosity and increased red-cell aggregation. These abnormalities are related to increase in large acute-phase globulins:fibrinogen and α_2 globulin (Ditzel et al., 1968; Jan et al., 1975). While these changes are probably a response of the body to injury, rather than changes specific for cardiac infarction (Jan et al., 1975), Volger et al. (1979) have recently reported that the type of red-cell aggregates — irregular, "clump" forms which are more resistant to dispersion by shear forces — differs from the type of aggregates formed as a response to infection. This increased shear resistance of aggregates had previously been noted by Schmid-Schönbein and Wells (1968). Further work is required to establish the specificity of these red-cell aggregates for cardiac infarction. "Circulating" platelet aggregates, detected by a formalin-fixation technique, are increased following myocardial infarction (Wu and Hoak, 1974). While these are also non-specific markers of tissue injury, and possibly related (like red-cell aggregates) to plasma fibrinogen levels, we found that the level of platelet aggregates following myocardial infarction was higher than in patients with "non-vascular" illnesses (Lowe et al., 1979a).

Finally, decreased red-cell deformability has recently been reported, on the first day of cardiac infarction (Volger et al., 1979; Boyd et al., 1979): deformability increases to normal values within the subsequent 2 days. The cause of the reduced deformability remains to be established, but may also involve responses to stress or injury, as similar reductions in deformability are encountered following surgery, severe burns, septic shock and catecholamine infusion.

There is some evidence that these abnormalities of blood rheology are associated with complications of myocardial infarction. High plasma fibrinogen levels have been associated with incomplete recovery and thrombo-embolism (Eastham and Morgan, 1963; Fulton and Duckett, 1976). Volger et al. (1979) and Boyd et al. (1979) found that red-cell deformability was lowest in patients with cardiogenic shock or pulmonary oedema. Jan et al. (1975) found that patients in their series with shock, heart failure or pulmonary embolism had higher levels of blood viscosity and haematrocrit on admission than patients without complications. On the other hand, Hershberg et al. (1972), in a series of 190 patients admitted to a coronary care unit, found no correlation between admission haematocrit values and prognosis.

Cerebral infarction. Studies of blood rheology after acute cerebral infarction are few compared to studies of patients with cardiac infarction. The data are also less reliable, as "cerebral infarction" cannot be clearly separated on clinical grounds from small intracranial haemorrhages. While computerised coaxial tomography can now be used to define cerebral infarcts, it cannot determine whether they arise from embolism or from thrombosis on atheromatous plaques. Several workers have described increased blood viscosity, haematocrit and fibrinogen following clinically diagnosed "cerebral infarction" (Eisenberg, 1966;

Ott et al., 1974). Increased platelet aggregates (Wu and Hoak, 1974; Lowe et al., 1979a) and decreased red-cell deformability (Boisseau et al., 1979) have also been reported. While there is little information on the prognostic value of rheology factors in cerebral infarction, Fletcher et al. (1976) related increased plasma fibrinogen levels to the neurological score on admission and to the clinical outcome.

Problems with acute infarction. Interpretation of the above data is difficult, as the changes in blood rheology may be consequences of the infarct (due to altered haemodynamics, immediate secretion of hormones such as catecholamines or corticosteroids, or increases in "acute phase" proteins) rather than pathogenetic factors. Association with prognosis may reflect a true adverse haemodynamic effect of the abnormal rheology on the circulation, but the abnormality might equally well be merely a marker of some other variable which is of prognostic importance, such as the size of the infarct. For example, Dintenfass and Forbes (1974) found a correlation between blood and plasma viscosity and the level of serum aspartate transaminase (an indicator of myocardial damage) following cardiac infarction. While abnormal blood rheology may contribute to further ischaemia following cardiac or cerebral infarction, or predispose to thrombo-embolic complications by reducing blood flow, at present we can only speculate on the part that blood rheology might play in the precipitation of infarction.

13.2.2 Chronic Arterial Disease

Chronic coronary artery disease. McDonald and Edgill (1957) first described increased plasma fibrinogen levels in patients with angina pectoris and patients who had survived a previous cardiac infarction. Many workers have since confirmed a chronic increase in fibrinogen in subjects with coronary artery disease, and Hampton et al. (1966) described also increased variability in serial fibrinogen measurements in such persons compared to controls. Mayer (1964, 1965) and Nicolaides et al. (1977) described increased blood viscosity in men with chronic coronary artery disease, associated with increases in haematocrit, fibrinogen, and plasma viscosity. The increase in plasma viscosity is related to the increased plasma fibrinogen (Karppinen, 1970). On the other hand, Rozenberg (1968) could not confirm an increase in blood viscosity or haematocrit in survivors of a myocardial infarction.

Patients with angina were found to have increased platelet aggregates by Jonker and den Ottolander (1979) but not by Wu and Hoak (1974). Red-cell deformability has yet to be assessed by filtration methods in chronic coronary artery disease. Nicolaides et al. (1977) described increased deformability in patients with angina, using the centrifuge technique of Sirs (see Chapter 2) which correlates positively with the plasma fibrinogen level.

The prognostic significance of these abnormalities in chronic coronary artery disease is not known.

Chronic peripheral arterial disease. Dormandy et al. (1973a) first drew attention to increased blood viscosity in patients with chronic peripheral arterial disease, associated with high plasma fibrinogen levels; their findings have been confirmed by other workers (Stuart et al., 1979). Increased haemoglobin levels (Bouhoutsos et al., 1974), decreased red-cell deformability (Reid et al., 1976a; Drummond et al., 1979) and increased platelet aggregates (Lowe et al., 1979) may also contribute to the rheological disturbance in peripheral arterial disease.

Dormandy et al. (1973b) related increased blood viscosity and plasma fibrinogen to an adverse prognosis in patients with claudication. Increased plasma fibrinogen appears to be related also to increased risk of occlusion of peripheral arterial grafts (Hamer et al., 1973; Harris et al., 1978). Bouhoutsos et al. (1974) associated high pre-operative haemoglobin levels with complications following arterial surgery, while Bailey et al. (1979) found that failure of limited amputations in diabetics was associated with higher haemoglobin levels than successful operations. There is evidence, therefore, that blood rheology factors are of prognostic significance in chronic peripheral arterial disease.

13.2.3 Prospective Population Studies

Haemoglobin levels (which correlate closely with haematocrit levels) have been shown to be predictive for vascular events in prospective population studies. The Framingham study reported associations between high-normal haemoglobin levels and both cerebral infarction and myocardial infarction (Kannell et al., 1972; Dawber et al., 1977). Similar findings in relation to myocardial infarction were reported from Stockholm (Bottiger and Carlsön, 1973) and Wales (Elwood et al., 1974). In the Göteborg study, men who developed myocardial infarction or acute fatal coronary heart disease had higher haematocrits than men who did not, but the difference was not statistically significant (Wilhelmsen et al., 1973).

It is possible that plasma fibrinogen levels are also of predictive value for events related to coronary artery disease. Increased erythrocyte sedimentation was associated with coronary artery events in the Stockholm study (Bottiger and Carlson, 1973). It was recently reported from the Northwick Park study that men who had died from coronary artery disease had a significantly higher mean plasma fibrinogen level on admission to the study than men who were still alive (Meade, 1979).

13.3 Relation to Extent of Arterial Disease

We have previously referred to morphological studies which have shown that clinical arterial events are associated with extensive arterial narrowing. In view of the evidence for associations between clinical events and blood rheology factors, summarised in the previous section, it is of interest to examine the relationship of rheology factors to the extent of arterial disease, assessed either at autopsy or, *in vivo,* by arteriography.

Tohgi et al. (1978) reported a positive association between the haematocrit and the degree of cerebral atherosclerosis, graded at autopsy according to the World Federation of Neurology Code. Dormandy (1973a, 1975) found no relationship between blood viscosity and the extent of arterial narrowing seen at arteriography of the leg arteries in a series of patients with peripheral arterial disease. On the other hand, the highest viscosity values occurred in patients who also had clinical evidence of coronary artery disease — who presumably had more widespread arterial disease.

We have recently studied the relationship of blood viscosity, haematocrit and fibrinogen to the extent of coronary artery disease in 50 men, aged less than 56 years, who underwent coronary arteriography for the assessment of chest pain. We found that men with extensive coronary artery disease (stenosis of at least 50% of the lumen in at least two of the three major coronary vessels) had significantly higher blood viscosity, haematocrit and fibrinogen levels than men with less extensive coronary artery disease, or control subjects without chest pain (Lowe et al., 1980a). This preliminary evidence for an association between blood viscosity and the extent of coronary artery disease is open to a variety of interpretations. While the increased viscosity may play a role in atheroma formation, alternative explanations are that extensive arterial disease causes increased viscosity, or that the association is coincidental and arises from mutual association of viscosity and arterial disease with other factors.

13.4 Relation to Risk Factors

It was noted earlier that risk factors for arterial events have been defined in prospective epidemiological studies. The relationship of blood rheology to these risk factors has been assessed in selected groups as well as in unselected population studies.

Age. The incidence of clinical events and the extent of arterial narrowing both increase with age. Several studies show that the plasma fibrinogen level increases with age, and this may explain the increase in the ESR with age (Bottiger and Svedberg, 1967). Reported studies of blood and plasma viscosity changes with age show conflicting results: we have found significant rises with age in plasma viscosity and blood viscosity (corrected for haematocrit) in apparently healthy men, which parallel the increase in plasm fibrinogen (Lowe et al., 1980b). The haematocrit increases during childhood, but most studies find little change during adult life (Wintrobe et al., 1974), although McDonough et al. (1965) found a fall with age in men and a rise with age in women. Red-cell deformability, measured by filtration, has been reported to decrease with age (Reid et al., 1976b).

Sex. Women suffer fewer clinical arterial events than men prior to the menopause and for about 10 years thereafter. Mayer (1964) found that healthy females had lower blood viscosity than healthy males, especially before the menopause. This was related in part to the well-known sex difference in haematocrit due to menstruation. Mayer found also lower plasma viscosity in pre-menopausal women than men, but other workers have found no significant sex difference in plasma viscosity (Harkness, 1971). Neither do there appear to be sex differences in plasma fibrinogen, or in red-cell deformability (Drummond et al., 1979).

Cigarette-smoking. Blood viscosity, haematocrit, fibrinogen and plasma viscosity were found to be higher in middle-aged policemen who smoked, compared to their non-smoking colleagues, by Dintenfass (1975). We have confirmed these findings in a population of men studied over a wide range of age, and found increased viscosity to be present also in young smokers (Lowe et al., 1980b). These associations of haematocrit and fibrinogen with cigarette-smoking have been confirmed in population studies (McDonough et al., 1965; Kannell et al., 1972; Korsan-Bengtsen et al., 1972; Meade et al., 1979). The cause of the increase in haematocrit is probably the increased carboxyhaemoglobin level in smokers, which impairs blood oxygen-carrying capacity and increases blood-oxygen affinity. Presumably as a response to the resultant tissue hypoxia, smokers increase their red-cell mass and hence their haematocrit (Sagone et al., 1971). The mechanisms by which fibrinogen is elevated in smokers are not known. It does not appear to be due to their increased prevalence of bronchitis and reduced lung function (Korsan-Bengtsen et al., 1972). The effect of smoking on red-cell deformability is not yet established. Some workers have found a chronic reduction in deformability in smokers (Lagrue et al., 1979). We have found no chronic decrease in deformability in smokers, but we did observe an acute fall in deformability after smoking three cigarettes (Drummond et al., 1979).

Diabetes. Several studies, including our own study of young diabetics (Lowe et al., 1980c) have described abnormalities of blood rheology in diabetes mellitus. These are reviewed elsewhere in this book by Barnes (Chapter 14).

Blood pressure. Pickering (1936) found normal blood viscosity in hypertensive patients. Tibblin et al. (1966) described increased blood viscosity in young patients with essential hypertension, associated with increases in haematocrit, plasma viscosity and plasma proteins. While most workers have found that the majority of patients with uncomplicated hypertension have normal haematocrit or haemoglobin levels (Galluzi et al., 1965; Tarazi et al., 1966), epidemiological studies have reported correlations of haematocrit or haemoglobin level with both systolic and diastolic blood pressures within the normal ranges (McDonough et al., 1965; Kannell et al., 1972). This association may be explained by the contribution of haematocrit, through blood viscosity, to peripheral resistance and hence blood pressure. Tarazi et al. (1966) found an increased prevalence of high haematocrit values in patients who had renal artery stenosis (with or without high blood pressure) — an association which might be explained by the association of cigarette-smoking with renal artery stenosis (Mackay et al., 1979). Increased plasma fibrinogen levels were described in patients with malignant hypertension by Garvas et al. (1975), and a correlation between fibrinogen and blood pressure in a population study was reported by Korsan-Bengtsen et al. (1972). Seiffge (1979) found that spontaneously hypertensive rats had increased levels of blood and plasma viscosity, haematocrit and fibrinogen, but normal red-cell deformability.

Blood lipids. Most workers have found no effect of alimentary hyperlipaemia on blood or plasma viscosity (Charm et al., 1963). Dormandy et al. (1973a) found no correlation between blood viscosity and lipid fractions. We have found high fibrinogen levels as well as increased platelet aggregates in subjects with primary hypercholesterolaemia (Lowe et al., 1979b). Epidemiological studies have also reported correlations, within the normal ranges, of cholesterol with fibrinogen (Korsa-Bengtsen et al., 1972) and haematocrit (McDonough et al., 1965; Elwood et al., 1970).

Oestrogen consumption. Abnormalities in blood rheology in women taking oral contraceptives have been described by several workers, and are reviewed elsewhere in this book by Buchan and Macdonald (Chapter 16).

Obesity. Obesity appears to be associated with increased levels of haematocrit (McDonough et al., 1965) and fibrinogen (Bennett et al., 1966).

Physical activity. There is some evidence that increasing physical activity is associated with decrease in plasma viscosity (Charm et al., 1979).

Blood groups. Dintenfass (1976) has suggested that differences in blood rheology factors between ABO blood groups may be relevant to the epidemiological findings of excess arterial disease in blood group A compared to blood group O.

Alcohol consumption. Increasing alcohol consumption has been associated with decreasing risk of arterial occlusive events. Patients with alcoholic liver disease commonly have decreased levels of haematocrit and plasma fibrinogen (Wintrobe et al., 1974).

Geography and race. The incidence of clinical arterial events varies considerably worldwide. It has been suggested that the common occurrence of anaemia in "underdeveloped" countries may contribute to their apparent protection from arterial disease. Negroes in the southern United States have less coronary artery disease (Tejada et al., 1968) and lower haematocrits (McDonough et al., 1965) than whites living in the same area. Data on geographical and racial differences in other blood rheology factors are lacking.

13.5 Significance of Altered Blood Rheology in Occlusive Arterial Disease

It now appears that clinical arterial events, extensive occlusive arterial disease, risk factors and abnormal blood rheology are all related to each other. In which directions do the causal links run? The rheological enthusiast might suggest that risk factors cause abnormal blood rheology which then causes arterial disease. The rheological sceptic would argue otherwise. Might these associations of abnormal rheology with risk factors be purely coincidental? Do studies which associate abnormal rheology with arterial events or disease measure only the additive effects of the various risk factors? Do prospective studies which suggest that rheology factors (such as haematocrit or fibrinogen) predict clinical events, merely pick out these variables as "innocent" markers of multiple risk factors or extensive arterial disease?

Intervention studies — measuring the effect of improving the flow properties of blood on clinical events in controlled, prospective trials — are the only means of testing hypotheses concerning the pathogenetic significance of abnormal blood rheology. At present, the only practical methods of improving the ability of blood to flow appear to be venesection and the administraton of clofibrate, which lowers fibrinogen levels. No large-scale trial of haematocrit reduction has been reported. The results of primary and secondary prevention studies of clofibrate have been disappointing (Coronary Drug Project Research Group, 1975; Oliver et al., 1978). However clofibrate has a variety of actions apart from lowering fibrinogen, some of which may be harmful (Oliver et al., 1978). Oral agents which lower plasma fibrinogen levels more efficiently and selectively than clofibrate, or which effect a predictable reduction in haematocrit of about 5%, would be theoretically attractive for study.

Finally, it is of interest to consider possible mechanisms whereby increases in blood viscosity, haematocrit or fibrinogen might promote arterial events.

13.6 Mechanisms by Which Abnormal Rheology May Promote Ischaemia

13.6.1 Blood Flow Reduction

Sluggish blood flow has been observed in some parts of the circulation in patients with unequivocally high blood viscosity due to polycythaemia or paraproteinaemias. However, at first sight the relatively small increases in viscosity, haematocrit and fibrinogen levels associated with arterial disease might seem unlikely to cause a marked loss of perfusion, especially as blood in the arterial circulation normally has maximal fluidity (minimum viscosity) due to the high shear forces (and high shear rates) that it encounters. To this general statement we must add several qualifications which should make us keep an open mind as to the potential of commonly encountered viscosity alterations to reduce blood flow.

Firstly, resting blood rheology measurements do not reveal phasic alterations due to common events. Figure 13.1 shows the effect of 5 min strenuous exercise on high shear blood viscosity in 10 healthy young men studied by ourselves. A rise in the mean haematocrit of 7% was associated with an increase in high shear blood viscosity of 17%. Similar acute impairment in blood fluidity arises following acute illness or injury — due to increase in "acute phase" proteins, and after heavy cigarette-smoking — due to decreased red-cell deformability (Drummond et al., 1979). It is possible that such acute deteriorations in the flow properties of blood might at times be critical determinants of blood flow, particularly in arteries already narrowed by atherosclerosis. Increases in blood viscosity may well become more detrimental as the baseline viscosity level rises, and if increased basal blood viscosity is associated with increased severity of arterial narrowing (Lowe et al., 1980a) then the adverse rheological effects of strenuous exercise, intercurrent illnesses, or heavy smoking may be most pronounced in those persons with the most extensive arterial stenoses.

Secondly, extensive arterial disease may impair blood fluidity by reducing the shear rates in some parts of the circulation, and hence allowing blood viscosity to increase in the non-Newtonian manner observed many times *in vitro* in variable-shear rotational viscometers. Low shear forces may arise distal to arterial stenoses, and Roberts and Ferrans (1976) have suggested that some coronary thrombi observed after fatal cardiac infarction originate at such sites, and grow proximally due to slowed blood flow. Severe coronary artery disease impairs myocardial contractility and cardiac output, and a resulting fall in blood pressure may also reduce shear forces and hence increase blood viscosity in parts of the circulation. The importance of shear forces in maintaining blood fluidity has been emphasised by Schmid-Schönbein (1976).

What evidence do we have that blood rheology is a determinant of flow in the human heart, brain or limbs? Coronary artery flow is not easy to measure: nevertheless some workers, using indirect measures or estimates of myocardial blood flow, have found correlations with plasma or blood viscosity (Gordon et al., 1974; Lake and Dintenfass, 1975). Whether coronary blood flow is increased by viscosity reduction is not known. Low haematocrits are associated with increased coronary collateral vessels (Zoll et al., 1951) but this may be mediated by reduced oxygen-carrying capacity rather than reduced blood viscosity. Cerebral blood flow does appear to be related to blood viscosity and this relationship cannot be explained solely by changes in oxygen-carrying capacity (Pearson et al., Chapter 10, this book). Similarly, lower-limb blood flow relates to blood viscosity even after correction for oxygen transport (Yates et al., 1979). Red-cell deformability has yet to be correlated with blood flow in man, as has the level of "circulating" platelet aggregates.

13.6.2 Atherogenesis

Oppenheimer (1929) suggested that increased blood viscosity might favour atherogenesis by increasing mechanical stress on the vessel wall. Little evidence for this hypothesis has been produced in the subsequent 50 years. Fluid-mechanical factors are thought to be important in the localisation of atheromatous lesions to characteristic sites, and several theories have been produced which relate atherogenesis to increased endothelial shear stress, leading

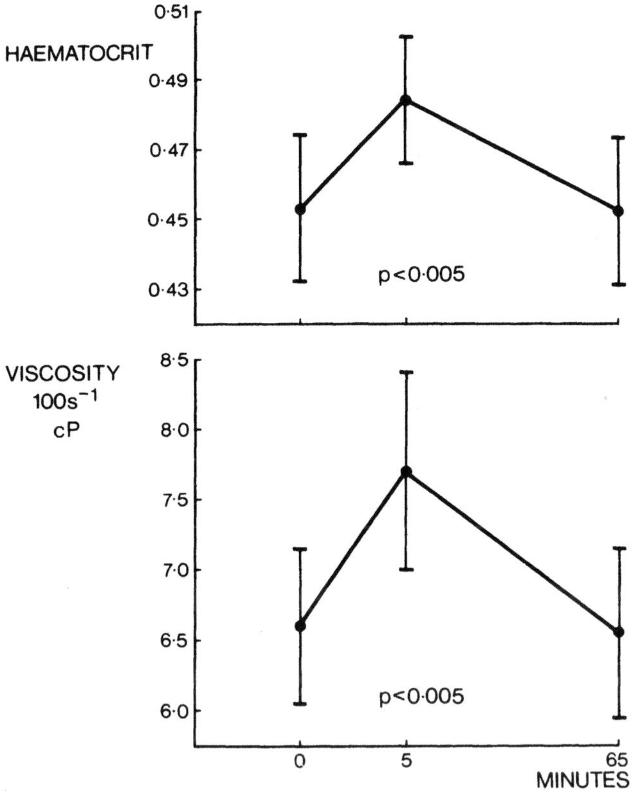

Fig.13.1. Increases in haematocrit and blood viscosity in 10 healthy young men following 5 min strenuous exercise (running up and down stairs) followed by 60 min rest. (mean ± one S.D.) 1 cP (centipoise) = 1 mPa.s

to increased flux of plasma proteins into the intima (Fry, 1968); decreased shear stress (Caro et al., 1971); or turbulence (e.g. Sako, 1962). The questions, which haemodynamic factors critically influence atherogenesis? and, what are the effects of blood rheology alterations on these factors? — require much further study to answer them. It should be noted that shear forces transmitted along the elastic arterial wall, which arise from cardiac contractions, are much greater than forces acting on the endothelium which arise from changes in the blood.

The recent demonstration by Smith et al. (1976) that much of the "fibrin" in developing atheromatous lesions is in fact fibrinogen, presumably derived from the plasma, may be relevant to the association of increased plasma fibrinogen levels with arterial disease and its risk factors. Vast amounts of effort have been expended over many years in attempts to validate the "lipid infiltration" theory that high plasma lipoprotein levels cause increased arterial wall lipids and hence increased occlusive arterial lesions. Some further investigation of the relationship between high plasma fibrinogen levels and fibrinogen accumulation in the arterial wall might be profitable. "Fibrinogen accumulation is relatively greater than lipoprotein accumulation in developing fibrous lesions. It is clearly a gross oversimplification to think of atherogenesis only in terms of lipid infiltration" (Smith et al., 1976).

13.6.3 Thrombo-embolism

It has long been realised that the deposition of platelets and fibrin on arterial walls is influenced by haemodynamics, and there is much current interest in the quantification of these influences. Shear forces affect platelet adhesion and aggregation, and fibrin formation and deposition; furthermore, some recent work suggests that changes in haematocrit and plasma fibrinogen levels within ranges associated with varying risk of arterial events, may influence platelet behaviour and fibrin deposition.

Red cells circulate in intimate contact with platelets, and are one source of adenosine diphosphate (ADP), which induces platelet adhesion and aggregation (Gaarder et al., 1961). Some "platelet adhesion tests" correlate with the haematocrit, probably due to increasing ADP release *in vitro* (Harrison and Mitchell, 1966). Schmid-Schönbein et al. (1979) have suggested that high shear stresses induce haemolysis, releasing ADP from erythrocytes and causing platelet aggregation and plug formation: in this *in vitro* system, release of ADP increased with increase of the haematocrit from 40% to 60%. Another chemical mechanism by which increasing haematocrit might increase platelet deposition is the removal of the platelet inhibitor prostacyclin (PGI_2) from the vessel wall by red cells (Willems et al., 1979).

Increases in haematocrit may increase platelet deposition by physical as well as chemical effects. Platelets are displaced towards the vessel wall with increasing haematocrit. In the Baumgartner preparation, platelet coverage of animal aortic subendothelium, exposed to whole blood flowing at defined shear rates, increases with increasing haematocrit within the normal range for humans. Above a haematocrit of 40%, platelet thrombus formation increases exponentially (Turitto et al., 1979). In another *in vitro* system for study of thrombus formation in whole blood under defined haemodynamic conditions, the haematocrit was related to the deposition of fibrin, as well as red cells, in the thrombus (Schultz et al., 1978).

There is also some evidence for a relationship of plasma fibrinogen to thrombus formation. Plasma fibrinogen has been correlated with deposition of platelets and fibrin in artificial thrombi (Schultz et al., 1978), platelet adhesion to glass (Hamer et al., 1973) and collagen (Lyman et al., 1971), and with the level of "circulating" platelet aggregates (Lowe et al., 1979a).

13.7 Conclusions

"Occlusive arterial disease" consists in a variety of pathological processes and has complex epidemiological associations. There is evidence that blood viscosity and its major determinants — haematocrit and plasma fibrinogen — are associated with clinical arterial events, many of the known risk factors for these events, and (possibly) the extent of arterial occlusion. Whether these associations are causal, consequential or coincidental is not known. However, there is some evidence that increased viscosity may reduce blood flow and on occasion this may contribute to clinical events. Increased haematocrit and fibrinogen levels may promote arterial disease in ways other than increasing blood viscosity. At present there is no evidence that alteration of blood rheology alters the natural history of arterial disease.

13.8 References

Bailey M J, Johnston C L W, Yates C P J, Somerville P G, Dormandy J A: (1979) Preoperative haemoglobin as predictor of outcome of diabetic amputations. Lancet *ii:* 168-170

Battacharji S K, Hutchinson E C, McCall A J (1967) Stenosis and occlusion of vessels in cerebral infarction. Br Med J *3:* 270-274

Bennett N B, Ogston C M, McAndrew G M, Ogston D (1966) Studies on the fibrinolytic enzyme system in obesity. Clin Pathol *19:* 241-243

Boisseau M R, Lorient M F, Bricaud H (1979) Red cell deformability and risk factors in one hundred patients with cerebrovascular thrombosis. Thromb Haemostas *42:* 107

Bottiger L E, Svedberg C A (1967) Normal erythrocyte sedimentation rate and age. Br Med J *2:* 85-87

Bottiger L E, Carlson L A (1973) The Stockholm Prospective Study 2. New events of coronary heart disease in men in relation to findings at initial examination. 9 year follow-up. In: Waldenstrom J, Larsson T, Ljungstedt N (eds) Early phases of coronary heart disease. The possibility of prediction. Nordiska Bokhandelns Förlag, Stockholm, pp 158-179

Bouhoutsos J, Morris T, Chavatzas D, Martin P (1974) The influence of haemoglobin and platelet levels on the results of arterial surgery. Br J Surg *51:* 984-986

Boyd M J, Dodds A J, Allen J, Bennett E D, Dormandy J A (1979) Decreased erythrocyte deformability after myocardial infarction. Clin Sci *58:* 12P

Burch G E, DePasquale N P (1962) The hematocrit in patients with myocardial infarction. JAMA *180:* 63-65

Caro C G, Fitzgerald J M, Schroter R C (1971) Atheroma and arterial wall shear — observation, correlation and proposal of a shear dependent mass transfer mechanism for atherogenesis. Proc Roy Soc, Series B, *177:* 109-159

Chandler A B, Chapman I, Erhardt L R, Roberts W C, Schwartz C J, Sinapius D, Spain D M, Sherry S, Ness P M, Simon T L (1974) Coronary thrombosis in myocardial infarction. Am J Cardiol *34:* 823-833

Charm S E, McComis W, Tejada C, Kurland G (1963) Effect of a fatty meal on whole blood and plasma viscosity. J Appl Physiol *18:* 1217-1220

Charm S E, Paz H, Kurland G S (1979) Reduced plasma viscosity among joggers compared with non-joggers. Biorheology *16:* 185-189

Conley C L, Russell R P, Thomas C B, Tumulty P A (1964) Hematocrit values in coronary artery disease. Arch Intern Med *113:* 170-176

Coronary Drug Project Research Group (1975) Clofibrate and niacin in coronary heart disease. JAMA *231:* 360-381

Dawber T R, Wolf P A, Colton T, Nickerson R J (1977) Risk factors. In: Zulch K J, Kaufmann W, Hossmann K A, Hossmann V (eds) Brain and heart infarct. Springer-Verlag, Berlin-Heidelberg-New York 1977, pp 226-252

Dintenfass L (1975) Elevation of blood viscosity, aggregation of red cells, haematocrit values and fibrinogen levels in cigarette smokers. Med J Aust *1:* 617-620

Dintenfass L (1976) Rheology of blood in diagnostic and preventive medicine. Butterworths, London, p 49

Dintenfass L, Forbes C D (1974) GOT, GPT and LDH levels and blood viscosity factors after recent myocardial infarction: effect of ABO blood groups. Haematologia *8:* 523-533

Ditzel J, Bang H O, Thorsen N (1968) Myocardial infarction and whole-blood viscosity. Acta Med Scand *183:* 577-579

Doll R, Hill A B (1964) Mortality in relation to smoking: ten years' observations of British doctors. Br Med J *1:* 1399-1410

Dormandy J A (1975) Blood: its viscosity and circulation. In: Harcus A W, Adamson L (eds) Arteries and veins. Churchill Livingstone, Edinburgh, pp 99-134

Dormandy J A, Hoare E, Colley J, Arrowsmith D E, Dormandy T L (1973a) Clinical, haemodynamic, rheological and biochemical findings in 126 patients with intermittent claudication. Br Med J *4,* 576-581

Dormandy J A, Hoare E, Khattab A M, Arrowsmith D E, Dormandy T L (19873b) Prognostic significance of rheological and biochemical findings in patients with intermittent claudication. Br Med J *4:* 581-583

Drummond M M, Lowe G D O, Belch J J F, Forbes C D, Barbenel J C (1979) An assessment of red cell deformability using a simple filtration method. Thromb Haemostas *42:* 106

Eastham R D, Morgan E H (1963) Plasma-fibrinogen levels in coronary-artery disease. Lancet *ii:* 1196-1197

Eisenberg S (1966) Blood-viscosity and fibrinogen concentration following cerebral infarction. Circulation *33-34* supp. II: 10-14

Elwood P C, Mahler R, Sweetnam P M, Moore F, Welsby E (1970) Association between circulating-haemoglobin level, serum-cholesterol, and blood-pressure. Lancet *i:* 589-591

Elwood P C, Waters W E, Benjamin I T, Sweetnam P M (1974) Mortality and anaemia in women. Lancet *i:* 891-894

Fletcher A P, Alkjaersig N, Davies A, Lewis M, Brooks J, Hardin W, Landau W, Raichle M E (1976) Blood coagulation and plasma fibrinolytic enzyme system pathophysiology in stroke. Stroke *7:* 337-348

Fry D L (1968) Acute vascular endothelial changes associated with increased blood velocity gradients. Circ Res *22:* 165-197

Fulton R M, Duckett K (1976) Plasma-fibrinogen and thromboemboli after myocardial infarction. Lancet *ii:* 1160-1164

Gaarder A, Jonsen J, Laland S, Hellem A, Owren P A (1961) Adenosine diphosphate in red cells as a factor in the adhesiveness of human blood platelets. Nature *192:* 531-532

Galluzi N J, Damato A, Bromer L (1965) Hypertension and hematocrit. Am J Med Sci *249:* 663-666

Gavras H, Oliver N, Aitchison J, Begg C, Briggs J D, Brown J J, Horton P W, Lee F, Lever A F, Prentice C, Robertson J I S (1975) Abnormalities of coagulation and the development of malignant phase hypertension. Kidney Int *8:* S-252-S-261

Gordon T, Kannell W B (1972) Predisposition to atherosclerosis in the head, heart and legs. The Framingham study. JAMA *221:* 661-666

Gordon R J, Snyder G K, Tritel H, Taylor W J (1974) Potential significance of plasma viscosity and haematocrit variations in myocardial ischaemia. Am Heart J *87:* 175-182

Haerem J W (1972) Platelet aggregates in intramyocardial vessels of patients dying suddenly and unexpectedly of coronary artery disease. Atherosclerosis *15:* 199-213

Hamer J D, Ashton F, Meynell M J (1973) Factors influencing prognosis in the surgery of peripheral vascular disease: platelet adhesiveness, plasma fibrinogen, and fibrinolysis. Br J Surg *60:* 386-389

Hampton J W, Mantooth J, Brandt E N, Wolf S G Jr (1966) Plasma fibrinogen patterns in patients with coronary atherosclerosis. Circulation *34:* 1098-1101

Harness J (1971) The viscosity of human blood plasma; its measurement in health and disease. Biorheology *8:* 171-193

Harris P L, Harvey D R, Bliss B P (1978) The importance of plasma lipid, glucose, insulin and fibrinogen in femoropopliteal surgery. Br J Surg *65:* 197-200

Harrison M J G, Mitchell J R A (1966) The influence of red blood-cells on platelet adhesiveness. Lancet *ii:* 1163-1164

Hershberg P I, Wells R D, McGandy R B (1972) Hematocrit and prognosis in patients with acute myocardial infarction. JAMA *219:* 855-860

Jan K-M, Chien S, Bigger J T Jr, (1975) Observations on blood viscosity changes after acute myocardial infarction. Circulation *51:* 1079-1084

Jonker J J C, den Ottolander G J H (1979) Platelet function studies in normal volunteers and patients with angina pectoris. Thromb Haemostas *42:* 144

Kannell W B, Gordon T, Wolf P A, McNamara P (1972) Hemoglobin and the risk of cerebral infarction: The Framingham study. Stroke *3:* 409-420

Karppinen K (1970) The electrophoretic mobility of red cells and platelets and the plasma viscosity in coronary heart disease. Acta Med Scand suppl 506

Korsn-Bengtsen K, Wilhelmsen L, Tibblin G (1972) Blood coagulation and fibrinolysis in a random sample of 788 men 54 years old. II. Relations of the variables to "risk factors" for myocardial infarction. Thromb Diathes Haemorrh *28:* 99-108

Lagrue G, Marcel G A, Faucher G, Grumel J M, Beiler D, Branellac A (1979) Red cell filterability and cardiovascular risk factors. Thromb Haemostas *42:* 106

Lake B, Dintenfass L (1975) Blood viscosity in the "ischaemic" exercise ECG. Lancet *ii:* 1092

Lowe G D O, Reavey M M, Johnston R V, Forbes C D, Prentice C R M (1979) Increased platelet aggregates in vascular and non-vascular illness: correlation with plasma fibrinogen and effect of ancrod. Thromb Res *14:* 377-386

Lowe G D O, Drummond M M, Third J L H C, Bremmer W F, Forbes C D, Prentice C R M, Lawrie T D V (1979b) Increased plasma fibrinogen and platelet-aggregates in type II hyperlipoproteinaemia. Thromb Haemostas *42:* 1503-1507

Lowe G D O, Drummond M M, Lorimer A R, Hutton I, Forbes C D, Prentice C R M, Barbenel J C (1980a) Relation between extent of coronary artery disease and blood viscosity. Br Med J *i:* 673-674

Lowe G D O, Drummond M M, Forbes C D, Barbenel J C (1980b) The effects of age and cigarette-smoking on blood and plasma viscosity in men. Scott Med J *25:* 13-17

Lowe G D O, Lowe J M, Drummond M M, Reith S, Belch J J F, Kesson C M, Wylie A, Foulds W S, Forbes C D, MacCuish A C, Manderson W G (1980c) Blood viscosity in young male diabetics with and without retinopathy. Diabetologia *18:* 359-363
Lyman, B, Rosenberg L, Karpatkin S (1971) Biochemical and biophysical aspects of human platelet adhesion to collagen fibers. J Clin Invest *50:* 1854-1863
McDonald L, Edgill M (1957) Coagulability of the blood in ischaemic heart-disease. Lancet *ii:* 457-460
McDonough J R, Hames C G, Garrison G E, Stubb S C, Lichtman M A, Hegelfinger D C (1965) The relationship of the hematocrit to cardiovascular states of health in the negro and white population of Evans County, Georgia. J Chronic Dis *18:* 243-257
Mackay A, Brown J J, Cumming A M M, Isles C, Lever A F, Robertson J I S (1979) Smoking and renal artery stenosis. Br M J *2:* 770
Mackie M J, Douglas A S (1978) Oral anticoagulants in arterial disease. Br Med Bull *34:* 177-182
Mayer G A (1964) Blood viscosity in healthy subjects and patients with coronary heart disease. Can Med Assoc J *91:* 951-954
Mayer G A (1965) Hematocrit and coronary heart disease. Can Med Assoc J *93:* 1151-1153
Meade T W (1979) The epidemiology of arterial thrombosis. Thromb Haemostas *42:* 260
Meade T W, Chakrabarti R, Haines A P, North W R S, Stirling Y (1979) Characteristics affecting fibrinolytic activity and plasma fibrinogen concentrations. Br Med J *1:* 153-156
Mitchell J R A, Schwartz C J (1965) Arterial disease. Blackwell Scientific, Oxford
Nicolaides A N, Bowers R, Horbourne T, Kidner P H, Besterman E M (1977) Blood viscosity, red-cell flexibility, haematocrit, and plasma-fibrinogen in patients with angina. Lancet *ii:* 943-945
Oliva P B, Potts D E, Pluss R G (1973) Coronary arterial spasm in Prinzmetal angina: documentation by coronary arteriography. N Eng J Med *288:* 745-751
Oliver M F, Heady J A, Morris J N, Cooper J (1978) A cooperative trial in the primary prevention of ischaemic heart disease using clofibrate. Br Heart J *40:* 1069-1118
Oppenheimer B S (1929) Vascular occlusion in polycythaemia vera. Trans Assoc Am Physicians *4:* 338-344
Ott E O, Lechner H, Aranibar A (1974) High blood viscosity syndrome in cerebral infarction. Stroke *5:* 330-333
Pettito C K (1979) Platelet thrombi in experimental cerebral infarction. Stroke *10:* 192-196
Pickering G W (1936) The peripheral resistance in persistent arterial hypertension. Clin Sci *2:* 209-219
Reid H L, Dormandy J A, Barnes A J, Lock P J, Dormandy T L (1976a) Impaired red cell deformability in peripheral vascular disease. Lancet *i:* 666-668
Reid H L, Barnes A J, Lock P J, Dormandy J A, Dormandy T L (1976b) A simple method for measuring erythrocyte deformability. J Clin Pathol *29:* 855-878
Roberts W C, Ferrans V J (1976) The role of thrombosis in the etiology of atherosclerosis (a positive one) and in precipitating fatal ischemic heart disease (a negative one). Semin Thromb Hemostas *2:* 123-135
Ross R, Glomset J A (1976) The pathogenesis of atherosclerosis. N Eng J Med *295:* 369-377
Rozenberg M C (1968) Blood viscosity and thrombosis *in vitro* in patients with previous myocardial infarction. Angiology *19:* 527-533
Russell R W R (1961) Observations on the retinal blood-vessels in monocular blindness. Lancet *ii:* 1422-1428
Sagone A L, Lawrence T, Balcerzak S P (1971) Smoking — a cause of "spurious" polycythaemia. Blood *38:* 826
Sako Y (1962) Effects of turbulent flow and hypertension on experimental atherosclerosis. JAMA *179:* 36-40
Schmid-Schönbein H, Wells R E (1968) The influence of quantified flow forces upon platelet and red cell aggregation in myocardial infarction. Circulation *37-38* Suppl 6: 174
Schmid-Schönbein H (1976) Microrheology of erythrocytes, blood viscosity, and the distribution of blood flow in the microcirculation. In: Guyton A C, Cowley A W (eds) International Review of Physiology, Cardiovascular Physiology II, volume 9, University Park Press, Baltimore, pp 1-62
Schmid-Schönbein H, Rohling-Winkel I, Blasberg P, Jüngling E, Wehmeyer A, Born G V R, Richardson P (1979) Release of ADP from erythrocytes under high shear stresses in tube flow. Thromb Haemostas *42:* 349
Schultz J S, Ciarkowski A A, Lindenauer S M, Penner J A (1978) Effects of hematological parameters on thrombus formation. Trans Am Soc Artif Intern Organs *24:* 565-567
Sedziwy L, Thomas M, Shillingford J (1968) Some observations on haematocrit changes in patients with acute myocardial infarction. Br Heart J *30:* 344-349
Seiffge D (1979) Hemorheological studies in the spontaneously hypertensive rat. Abstracts of Symposium Européen d'Hémorhéologie et Pathologie, Nancy, p 10-6

Slaney G, Hamer J D (1973) Arterial embolism. In: Birnstingl M (ed) Peripheral vascular surgery. Heinemann, London, pp 189-210
Smith E B, Alexander K M, Massie I B (1976) Insoluble "fibrin" in human aortic intima. Atherosclerosis *23:* 19-39
Stables D P, Rubenstein A H, Metz J, Levin N W (1967) The possible role of hemoconcentration in the etiology of myocardial infarction. Am Heart J *73:* 155-159
Stuart J, George A J, Blunt R J, Strachan C J L, Hurlow R A (1979) Hyperviscosity and coagulation activation in peripheral vascular disease. Thromb Haemostas *42:* 348
Tarazi R C, Frohlich E D, Dustan H P, Gifford R W Jr, Page I H (1966) Hypertension and high hematocrit. Am J Cardiol *18:* 855-858
Tejada C, Strong J P, Montenegro M R, Restrepo C, Solberg L A (1968) Distribution of coronary and aortic atherosclerosis by geographic location, race and sex. Lab Invest *18:* 509-529
Tibblin G, Bergent S E, Bjure J, Wilhelmsen L (1966) Hematocrit, plasma protein, plasma volume, and viscosity in early hypertensive disease. Am Heart J *72:* 165-176
Tohgi H, Yamanouchi H, Murakami M, Kameyama M (1978) Importance of the hematocrit as a risk factor in cerebral infarction. Stroke *9:* 369-374
Turpie A G G, Hirsh J (1978) Platelet suppressive therapy. Br Med Bull *34:* 183-190
Turitto V T, Weiss H J, Baumgartner H R (1979) The influence of shear rate and red cell concentration on platelet adhesion and thrombus formation in human blood. Thromb Haemostas *42:* 350
Volger E, Ostner K, Klein J, Wirtzfeld A (1979) Changes of red cell aggregation and deformability after acute myocardial infarction. Microvas Res *17:* S-153
Vreeken J, van Aken W G (1971) Spontaneous aggregation of blood platelets as a cause of idiopathic thrombosis and recurrent painful toes and fingers. Lancet *ii:* 1394-1397
Wilhelmsen L, Wedel H, Tibblin G (1973) Multivariate analysis of risk factors for coronary heart disease. Circulation *48:* 950-958
Willems C, van Aken W G, Peuscher-Prakke E M, van Mourik J a (1979) Synthesis of PGI_2 by cultured human vascular endothelial cells and stabilisation of PGI_2 by blood components. Thromb Haemostas *42:* 6
Wintrobe M M, Lee G R, Boggs D R, Bithell T C, Athens J W, Foerster J (1974) Clinical Hematology Seventh Edition, Lea and Febiger, Philadelphia
Wu K K, Hoak J C (1974) A new method for the quantitative detection of platelet aggregates in patients with arterial insufficiency. Lancet *ii:* 924-926
Yates C J P, Andrews V, Berent A, Dormandy J A (1979) Increase in leg blood-flow by normovolaemic haemodilution in intermittent claudication. Lancet *ii:* 166-168
Zoll P M, Wessler S, Schlesinger M J (1975) Interarterial coronary anastomoses in the human heart, with particular reference to anemia and relative cardiac anoxia. Circulation *4:* 797-815

Section D

Blood Rheology in Medicine and Surgery

14 Blood Viscosity in Diabetes Mellitus

A. J. Barnes

14.1 Introduction

It is well established that circulatory insufficiency plays a major pathological role in the long-term complications of diabetes. In the microcirculation of the retina, non-perfusion of capillaries is the earliest clinically detectable lesion in life and probably initiates neovascularisation (Kohner and Oakley, 1975). In larger vessels reduction or cessation of blood flow commonly leads to stroke, myocardial ischaemia and claudication of the legs (Jarrett and Keen, 1975). Of the three primary determinants of flow (vessel diameter, perfusion pressure and blood viscosity) abnormalities of vessel diameter have for long commandeered the attention of clinicians and research workers alike as the prime cause of circulatory insufficiency. However, in diseased vessels, which may be narrowed and unable to distend, the contribution of perfusion pressure and blood viscosity to flow changes may be enhanced. The present studies were therefore carried out to determine if blood viscosity (a major flow determinant) was increased in diabetic patients, whether blood viscosity differed in patients with and without different complications, and if viscosity was influenced by metabolic control.

14.2 Patients and Methodology

14.2.1 Patients

Rheological variables were measured in the following groups of individuals:
a) Sixty-one fasting non-diabetic subjects (30 males, 31 females) randomly

selected from ambulant hospital personnel and relations of patients (mean ± SEM, age 54 ± 2 years and weight 116% ± 2% of ideal).

b) Sixty-four fasting long-standing (> 20 years) ambulant diabetics (31 males, 33 females) of whom 53 were insulin treated and the remainder controlled with tablet therapy (age 56 ± 2 years and weight 111% ± 2% of ideal).

c) Ten patients (6 males, 4 females) presenting in diabetic ketoacidosis. Blood was taken within 1 h of admission to hospital and at 2, 4, 8, 25, 48 and 168 h after initiation of treatment with insulin and intravenous fluids. The mean age of the patients was 60 ± 8 years and body weight 108% ± 4% of ideal).

d) Thirty-one fasting newly diagnosed maturity-onset diabetics (15 males, 16 females). Blood was taken on the morning following their first clinic visit and at 1 and 4 months, at which time 14 patients were on diet alone, the remainder being treated with either tablets or insulin (not dealt with in this paper). The mean age of the patients was 52 ± 3 years and their weight 118% ± 10% of ideal at first visit. All subjects had a full clinical examination, electrocardiogram, chest X-ray and urinalysis.

All 61 non-diabetic subjects (group a) had fasting blood glucose levels below 5.8 mmol/litre. Following examination it was found that 14 subjects had some evidence of vascular disease, e.g. mild hypertension or an abnormality on their ECG or chest X-ray. The 64 long-standing diabetics (group b) were graded according to the presence and extent of different diabetic complications (Table 14.1) on the following criteria:

Retinopathy (based on observations of two independent observers using an ophthalmoscope, pupils dilated with 0.5% cyclopentolate hydrochloride) — 0 = no retinopathy, 1 = < a total of 15 microaneurysms in both eyes, 2 = > a total of 15 microaneurysms in both eyes, 3 = proliferative retinopathy.

Nephropathy — 0 = no albuminuria, 1 = intermittent albuminuria or persistent mild albuminuria at outpatients, 2 = persistent heavy albuminuria at outpatients.

Neuropathy — 0 = no neuropathy, 1 = absent ankle jerks and vibration sense in feet, 2 = further evidence of neuropathy including autonomic neuropathy.

Peripheral ischaemia — 0 = no evidence of peripheral ischaemia, 1 = absence of foot pulses in one leg or claudication in one leg, 2 = absence of foot pulses or claudication in both legs, 3 = gangrene in one or both legs.

Myocardial ischaemia — 0 = no evidence cardiac disease, 1 = either angina of effort, a previous history of myocardial infarction or ECG evidence of ischaemia or infarction, 2 = angina with ECG evidence of myocardial infarction or with extensive ischaemic changes on ECG.

Forty-two diabetics and 43 controls were smokers. Smoking habits did not differ between the two groups.

Two patients admitted in diabetic ketoacidosis (group c) had background retinopathy, 1 having evidence of an old myocardial infarct on the ECG. Of the 31 newly diagnosed diabetic outpatients (group d) none had good evidence of microvascular disease but 2 patients had peripheral vascular disease, 2 had coronary artery disease and 4 had mild hypertension (BP > 150/90).

Table 14.1. Grade and distribution of complications in 64 patients with long-standing diabetes (Group b)

Grade	Distribution	
	No. of patients	% of group affected
Retinopathy		
0	22	34%
1	11	17%
2	12	19%
3	19	30%
Nephropathy		
0	45	70%
1	12	19%
2	7	11%
Neuropathy		
0	26	41%
1	11	17%
2	27	42%
Peripheral ischaemia		
0	49	77%
1	4	6%
2	9	14%
3	2	3%
Myocardial ischaemia		
0	46	72%
1	9	14%
2	9	14%

14.2.2 Methods

Whole-blood viscosity was measured on 10 ml of fasting venous blood anticoagulated with 125 iu of lithium heparin. Blood viscosity was measured at shear rates of 2.62 and 0.77 s^{-1} using a Contraves low-shear 100 viscometer (Dormandy, 1974). Results were expressed either as measured viscosity or viscosity corrected to a standard PCV (packed cell volume) of 45% as previously described (Dormandy, 1970). The PCV was recorded after spinning blood at 12,000 rpm for 5 min in an Adams Autocrit centrifuge. Erythrocyte deformability was measured in long-standing diabetics and controls (groups a and b) by a filtration technique using Nuclepore polycarbonate sieves which had a pore diameter of 5 μm (Reid et al., 1976a). Filtration was carried out on whole

Table 14.2 Whole-blood viscosity, erythrocyte deformability and other variables (all mean ± SEM) in 61 non-diabetic subjects (group a) and 64 patients with long-standing diabetes (group b)

	Diabetics (64)	Controls (61)	Significance of difference
Viscosity at 0.77 s^{-1}			
(mPa.s) — measured	36.0 ± 1.1	32.2 ± 1.0	$p < 0.02$
— corrected to 45% PCV	36.4 ± 0.9	31.3 ± 0.5	$p < 0.001$
Viscosity at 2.62 s^{-1}			
(mPa.s) — measured	17.3 ± 0.5	15.7 ± 0.4	$p < 0.05$
— corrected to 45% PCV	17.4 ± 0.4	15.7 ± 0.2	$p < 0.001$
Erythrocyte deformability (ml/min)	0.38 ± 0.01	0.41 ± 0.02	N.S.
PCV (%)	44.5 ± 4.4	44.9 ± 3.6	N.S.
Fibrinogen g/litre)	5.15 ± 0.17	4.59 ± 0.09	$p < 0.01$
Total protein (g/litre)	69.6 ± 0.8	69.2 ± 0.6	N.S.
Albumin (g/litre)	45.0 ± 0.5	46.4 ± 0.4	$p < 0.05$
Total globulin (g/litre)	24.7 ± 0.6	22.7 ± 0.5	$p < 0.05$
Caeruloplasmin (g/litre)	0.41 ± 0.01	0.36 ± 0.01	$p < 0.01$
Triglyceride (mmol/litre)	1.64 ± 0.16	1.47 ± 0.11	N.S.
Cholesterol (mmol/litre)	6.76 ± 0.21	6.55 ± 0.13	N.S.
Phospholipid (g/litre)	2.82 ± 0.07	2.60 ± 0.06	$p < 0.05$
Creatinine (μmol/litre)	94.2 ± 3.52	88.0 ± 1.76	N.S.
Urea (mmol/litre)	7.23 ± 0.32	5.27 ± 0.17	$p < 0.001$
Glucose (mmol/litre)	13.6 ± 0.8	4.9 ± 0.1	$p < 0.001$

blood anticoagulated with edetic acid (2 mg/5 ml blood) within 2 h of venesection. Results were expressed as the volume of red blood cells (ml) filtered per minute. Plasma fibrinogen, caeruloplasmin and glucose and serum cholesterol, triglyceride and phospholipid were measured by standard methods (Barnes et al., 1977). Plasma urea and serum creatinine, albumin and total globulin were measured by Technicon Autoanalyser methods.

Statistical methods. Differences between the means of blood viscosity and other variables in diabetics and controls were assessed by Student's *t*-test. Blood viscosity and erythrocyte deformability were computed on log-transformed data to achieve normal distribution of these variables. The relation between blood viscosity and other variables was investigated by multiple regression analysis.

14.3 Results

14.3.1 Controls and Long-standing Diabetics (Groups a and b)

Mean ± SEM values for whole-blood viscosity, erythrocyte deformability and other possibly related variables in controls and long-standing diabetics are shown in Table 14.2. Whole-blood viscosity at the shear rates of 0.77 s^{-1} and 2.62 s^{-1}

Table 14.3. Whole-blood viscosity (shear rate $0.77\ s^{-1}$ and corrected to 45% PCV) in diabetes with different complications[a] (group b). Results expressed as mean ± SEM

	Blood viscosity (mPa.s) at shear rate $0.77\ s^{+1}$ in diabetics with complications	Comparison with 22 complication-free diabetics (blood viscosity 33.5 ± 1.4mPa.s at $0.77\ s^{-1}$)	Comparison with 61 non-diabetic controls (blood viscosity 31.3 ± 52 mPa.s at $0.77\ s^{-1}$)
Retinopathy n = 19	40.6 ± 2.1	$p < 0.01$	$p < 0.001$
Nephropathy n = 7	42.5 ± 2.3	$p < 0.01$	$p < 0.001$
Neuropathy n = 27	38.2 ± 1.2	$p < 0.02$	$p < 0.001$
Peripheral ischaemia n = 11	40.6 ± 2.8	$p < 0.05$	$p < 0.01$
Myocardial ischaemia n = 9	40.7 ± 3.0	$p < 0.05$	$p < 0.01$
Complication free n = 22	33.5 ± 1.4		N.S.

[a] Diabetics with the most severe grade or grades of a complication (see Table 14.1) compared with complication-free diabetics and with non-diabetic controls (group a)

was significantly higher in the diabetics than in the controls whether expressed as measured viscosity or as viscosity corrected to a standard PCV of 45%. Within the diabetic group the presence of retinopathy, nephropathy and other sequelae was associated with an increase in blood viscosity corrected for PCV (Table 14.3).

Erythrocyte deformability in the diabetics as a whole was not significantly different from the controls; however, deformability (0.31 ± 0.02 ml/min) was significantly less in 14 diabetics with the most extensive complications than in 22 patients with few or absent complications (0.39 ± 0.01 ml/min) ($p < 0.03$) or in the 61 non-diabetics (0.4 ± 0.02 ml/min) ($p < 0.01$).

Plasma fibrinogen, caeruloplasmin, urea and glucose and serum total globulin and phospholipid were significantly higher in the diabetics than in the controls (Table 14.2). By contrast serum albumin was significantly lower in the former. Although bivariate analysis showed several of these factors to correlate with blood viscosity, multivariate analysis showed the increase in viscosity in the diabetics to be significantly related only to plasma fibrinogen ($p < 0.01$) and to total globulin ($p < 0.05$).

14.3.2 Patients Presenting in Diabetic Ketoacidosis (Group c)

Mean ± SEM values of blood viscosity, PCV, plasma fibrinogen, osmolality and glucose and serum protein at zero, 4 h, 24 h and 1 week in 10 diabetics during recovery from ketoacidotic coma and pre-coma are shown in Table 14.4. All the variables measured fell significantly ($p < 0.05$ or less) between zero and 4 h and all but fibrinogen showed an additional significant fall by 24 h. There were no significant changes in any variable between 24 h and 1 week. Changes in viscosity

Table 14.4. Whole-blood viscosity and PCV, plasma concentrations of glucose, osmolality and fibrinogen, and serum concentrations of total protein at zero, 4 h, 24 h and 1 week in 10 diabetics during recovery from ketoacidotic coma and pre-coma. Results expressed as mean ± SEM

	zero	4 h	24 h	1 week
Measured viscosity (mPa.s)				
— 0.77 s^{-1}	59.5 ± 6.5	37.9 ± 4.0	26.9 ± 2.5	24.8 ± 3.0
— 2.62 s^{-1}	29.7 ± 3.3	18.9 ± 2.2	13.2 ± 1.1	12.7 ± 1.1
PCV %	42.7 ± 1.6	36.7 ± 1.2	35.0 ± 1.4	33.4 ± 1.9
Glucose mmol/litre	57.2 ± 11.1	25.9 ± 5.5	15.1 ± 1.9	12.9 ± 1.7
Osmality mOsm/kg	357.0 ± 18	339.0 ± 15	308.0 ± 12	289.0 ± 5
Fibrinogen g/litre	7.03 ± 0.6	6.48 ± 0.62	6.04 ± 0.50	6.21 ± 0.47
Total protein g/litre	70.3 ± 1.2	62.1 ± 1.2	58.9 ± 1.1	63.0 ± 1.5

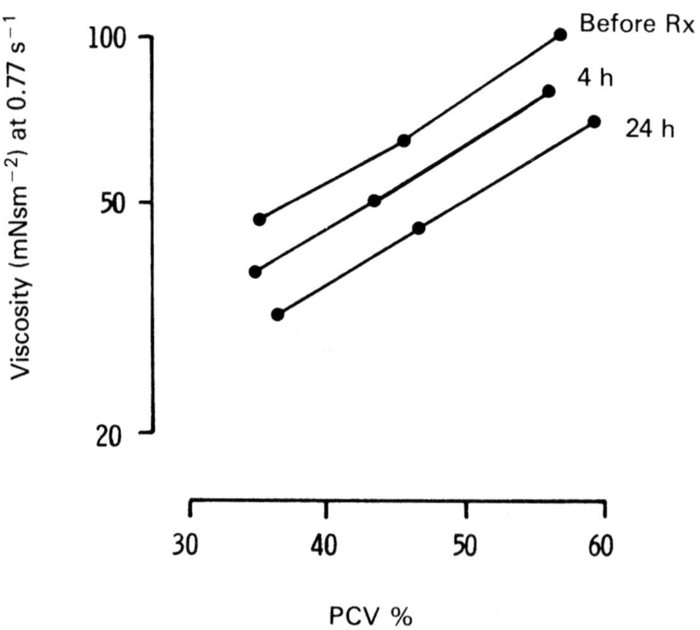

Fig. 14.1. Regression lines (mean of three diabetic patients) for viscosity against PCV before treatment and at 4 h and 24 h during recovery from ketoacidosis. Each regression line has been derived from a blood sample artificially made up to three separate haematocrits by addition or subtraction of plasma following centrifugation. 1 $mNsm^{-2}$ = 1 mPa.s

correlated significantly with changes in PCV during the first 24 h ($p < 0.01$ at both shear rates). In order to assess the contribution of factors other than PCV to blood viscosity in diabetic ketoacidosis, in three patients blood at zero, 4 h and 24 h was reconstituted to three separate haematocrits by addition or subtraction of plasma following centrifugation. Three point regression lines for viscosity (0.77 s^{-1}) against PCV were constructed for each of these times and are shown in Fig. 14.1. There was a progressive fall in blood viscosity with time that was independent of PCV. At a PCV of 45% blood viscosity fell from 68.1 mPa.s before treatment to 49.6 mPa.s at 24 h, a fall of 27%.

14.3.3 Newly Diagnosed Maturity-onset Diabetics (Group d)

Rheological variables in 31 newly diagnosed diabetics with more modest metabolic imbalance are shown in Table 14.5, where they are compared with those of the 61 non-diabetics (group a). The diabetics showed significant increases in measured and PCV-corrected blood viscosity, PCV, ESR and caeruloplasmin. Changes in these variables in 14 of the diabetics after 1 month of treatment with diet alone are given in Table 14.6. A significant fall in blood glucose was accompanied by reductions in measured viscosity and PCV but there were no changes in PCV-corrected blood viscosity, fibrinogen, proteins or lipids. Ten of these patients were studied again at 4 months when measured and PCV-corrected viscosity, PCV and fibrinogen fell significantly when compared with values at 1 month (Table 14.7).

Table 14.5. Mean ± SEM levels of blood viscosity, PCV and other variables in 31 newly diagnosed diabetics before initiation of treatment (group d) and in 61 non-diabetic controls (group a)

		Diabetics n = 31	Controls n = 61	Significance of difference
Measured viscosity	— 0.77 s^{-1}	41.9 ± 2.1	32.2 ± 1.02	$p < 0.001$
(mPa.s)	— 2.62 s^{-1}	20.6 ± 0.9	15.8 ± 0.4	$p < 0.001$
PCV corrected viscosity (45%)	— 0.77 s^{-1}	37.9 ± 1.2	31.5 ± 0.5	$p < 0.001$
(mPa.s)	— 2.62 s^{-1}	18.4 ± 0.5	15.8 ± 0.2	$p < 0.001$
PCV %		47.6 ± 0.7	45.0 ± 0.5	$p < 0.01$
ESR mm/h		16.0 ± 2.2	8.3 ± 0.3	$p < 0.01$
Fibrinogen g/litre		4.83 ± 0.19	4.58 ± 0.1	N.S.
Caeruloplasmin g/litre		0.42 ± 0.02	0.36 ± 0.09	$p < 0.01$
Total protein g/litre		70.3 ± 1.3	69.2 ± 0.6	N.S.
Albumin g/litre		46.4 ± 0.7	46.4 ± 0.5	N.S.
Globulin g/litre		24.0 ± 1.1	22.7 ± 0.5	N.S.
Glucose mmol/litre		12.2 ± 0.6	4.8 ± 0.1	$p < 0.001$
Cholesterol mmol/litre		6.37 ± 0.25	6.55 ± 0.13	N.S.
Triglycerides mmol/litre		1.4 ± 0.14	1.47 ± 0.11	N.S.

Table 14.6. Mean ± SEM levels of blood viscosity, PCV and other variables in 14 newly diagnosed maturity-onset diabetics before treatment and after 1 month's treatment with diet alone

	Before treatment	After 1 month	Significance of difference
Measured viscosity (mPa.s)			
— 0.77 s^{-1}	46.2 ± 3.6	40.3 ± 3.0	$p < 0.02$
— 2.62 s^{-1}	21.6 ± 1.5	19.8 ± 1.3	$p < 0.05$
PCV corrected viscosity (45%) (mPa.s)			
— 0.77 s^{-1}	38.4 ± 1.9	36.8 ± 1.5	N.S.
— 2.62 s^{-1}	18.2 ± 0.8	18.4 ± 0.7	N.S.
PCV %	48.9 ± 1.1	46.6 ± 1.2	$p < 0.01$
ESR mm/h	15.6 ± 4.2	14.6 ± 4.6	N.S.
Fibrinogen g/litre	4.81 ± 0.31	4.83 ± 24.0	N.S.
Caeruloplasmin g/litre	0.42 ± 0.03	0.42 ± 0.03	N.S.
Total protein g/litre	72.9 ± 2.4	71.2 ± 2.8	N.S.
Albumin g/litre	47.4 ± 0.9	47.3 ± 1.3	N.S.
Globulin g/litre	25.5 ± 2.0	23.9 ± 2.1	N.S.
Glucose mmol/litre	11.2 ± 0.8	8.7 ± 0.7	$p < 0.001$
Cholesterol mmol/litre	6.42 ± 0.26	6.2 ± 0.36	N.S.
Triglycerides mmol/litre	1.41 ± 0.18	1.49 ± 0.21	N.S.

Table 14.7. Mean ± SEM levels of blood viscosity, PCV, fibrinogen and glucose in 10 maturity-onset diabetics after 1 month and 4 months' treatment with diet alone

	Diet for 1 month	Diet for 4 months	Significance of difference
Measured viscosity (mPa.s)			
— 0.77 s^{-1}	41.6 ± 5.2	30.4 ± 3.3	$p < 0.02$
— 2.62 s^{-1}	21.0 ± 2.2	15.0 ± 1.0	$p < 0.02$
PCV corrected viscosity (45%) (mPa.s)			
— 0.77 s^{-1}	38.2 ± 3.0	31.0 ± 2.6	$p < 0.01$
— 2.62 s^{-1}	19.7 ± 1.2	15.6 ± 0.9	$p < 0.05$
PCV %	47.6 ± 2.4	45.8 ± 1.6	$p < 0.05$
Fibrinogen g/litre	4.83 ± 0.36	4.22 ± 0.1	$p < 0.05$
Glucose mmol/litre	7.8 ± 1.8	7.9 ± 2.3	N.S.

14.4 Discussion

The present studies indicate that the intrinsic flow properties of blood in diabetic patients are abnormal. Firstly, whole-blood viscosity at low shear rates was higher in long-standing diabetics than in matched non-diabetic subjects. This difference persisted when viscosity was corrected to a standard PCV of 45%, a manoeuvre often carried out to obviate changes in the viscosity caused by local alterations in haematocrit. Abnormalities of blood viscosity were more pronounced in patients with complications than in those without and in the most severely afflicted of these individuals erythrocyte deformability (possibly a more important determinant of blood flow in capillary-sized vessels) was also impaired. Previous studies using a variety of different viscometers have reported increases in blood, plasma or serum viscosity in diabetic patients although a correlation with complications has been more controversial (Cogan et al., 1961; Skovborg et al., 1966; Rees et al., 1967; Isogai et al., 1971; Labib et al., 1971; Zingg et al., 1971; McMillan, 1975; Hoare et al., 1976; Barnes et al., 1977; Dintenfass, 1977). A few workers have not found a difference in blood viscosity between diabetics and controls (Ditzel, 1968; Bollinger et al., 1969; Langer et al., 1971; Mosora et al., 1972) even in the face of higher levels of plasma fibrinogen in the former. Measurements of erythrocyte deformability have only more recently been made in diabetics, possibly because of the lack of general agreement regarding the measurement of this variable. However, the independent finding of more rigid red cells in diabetics by workers using filtration methods (Schmid-Schönbein and Volger, 1976; Barnes et al., 1977) a centrifugation method (Sirs et al., 1979) and a micropipette method (McMillan et al., 1978) strongly supports the case for an abnormality of red-cell rheology in this condition.

The aetiology of the observed rheological abnormalities in diabetic patients is only partially explained by the present and previous investigations. PCV was similar in long-standing diabetes (group b) and non-diabetics (group a) and plasma fibrinogen and to a lesser extent total globulin appeared to account for much of the difference in viscosity between the groups. The factors responsible for hyperfibrinogenaemia in diabetes are still to be established. Ferguson et al. (1975) demonstrated increased fibrinogen turnover in insulin-dependent diabetics with minimal clinical evidence of complications. Several factors including non-esterified fatty acids (Pilgeram and Pickart, 1968), growth hormone and cortisol (Jeejeebhoy et al., 1975; Miller, 1976) and prostaglandins (Carlson et al., 1978) are known to stimulate hepatic production of fibrinogen. Of these, fatty acids and growth hormone are commonly raised in diabetes and revert towards normal with improved metabolic control (Johanssen and Hansen, 1971). Clearly further studies are required in order to dissect out the relationships between the abnormal fibrinogen kinetics in diabetes on the one hand and different hormonal and metabolic factors on the other. Although abnormalities in erythrocyte deformability occurred in patients with widespread complications, we have previously not found a significant correlation between red-cell deformability and blood viscosity (Reid et al., 1976b) and more rigid red cells probably contributed little to the observed viscosity changes. Factors influencing erythrocyte deformability are complex, but a disturbance in the phospholipid and cholesterol content of the cell membrane might lead to alterations in the physical properties of red cells (Cooper, 1977) in diabetes. Furthermore, an increase in glycosylated haemoglobin might influence membrane properties by increasing haemoglobin binding to the membrane (Paulsen and Kory, 1976; McMillan et al., 1978). It has been pointed out by some workers that platelets and white cells influence the filtration of whole blood through polycarbonate sieves (Lingard, 1974). Although the use of EDTA as an anticoagulant probably minimised platelet aggregation in our filtration system, abnormalities in the physical properties of white cells in the diabetics might well have influenced filtration. It is quite possible that defective white cells in diabetes might interrupt capillary flow and predispose to areas of capillary non-perfusion in the retina.

The finding of high but partly reversible increases in blood viscosity in patients with diabetic ketoacidosis and also with lesser degrees of metabolic imbalance confirms and extends previous reports by Reubi (1953) and Skovborg and Nielsen (1969). This apparent relationship between raised blood viscosity and poor metabolic control might be of importance in two situations. Firstly in the acute situation (ketoacidosis or hyperosmolar non-ketotic coma or precoma) (group c) the very high levels of blood viscosity might participate in local circulatory stasis and predispose to venous and arterial thrombosis. Indeed cerebral and visceral thrombosis are recognised complications in patients with severe hyperglycaemia or ketoacidosis (Hobin and Maraist, 1973; Timperley et al., 1974; Tchertkoff et al., 1974; Ramon et al., 1975). Secondly, more modest degrees of metabolic imbalance (as shown in group d patients and typical of that seen in many diabetic outpatients) may also lead to increased blood viscosity albeit of a lesser degree. If raised blood viscosity were an important factor in the aetiology of diabetic microangiopathy or peripheral and other circulatory disease then hyperviscosity might at least in part explain the long-proposed association between poor metabolic control and diabetic complications (Stowers, 1975). The increase in whole-blood viscosity observed in ketoacidosis was about 70% accounted for by haemoconcentration, with hyperproteinaemia (especially hyperfibrinogenaemia) probably contributing to much of the remaining 30%. As both acidosis and hyperosmolality are known to affect the physical properties of red cells it might be predicted that erythrocyte deformability would be abnormal in diabetic ketoacidosis. Interestingly, Schmid-Schönbein and Volger (1976) have described such a defect which was reversed following treatment with insulin and intravenous fluids. The increase in blood viscosity in the maturity-onset diabetics (group d) was principally related to increased PCV, although PCV-corrected viscosity was also higher in these patients. In a similar fashion to the ketoacidotic patients the fall in blood glucose (this time from just above to just below the renal threshold for glucose) was accompanied by a fall in blood viscosity. This was initially related to a reduction in PCV but later on (4 months) also to a fall in plasma fibrinogen. Whether improvement in metabolic control in patients with long-standing diabetes (group b) would result in comparable reductions in viscosity has been repeatedly demonstrated in non-diabetics with circulatory disease.

The major question arising from the data presented is whether hyperviscosity in diabetes actually contributes to circulatory and microcirculatory complications or if it is merely a harmless manifestation of the metabolic disturbance, or perhaps secondary to vascular damage itself. Furthermore, viscosity is only one of several factors previously associated with diabetic complications. Abnormalities of growth hormone secretion, capillary fragility, platelet function, haemoglobin/oxygen dissociation and other factors are well described and many workers believe the aetiology of microvascular disease in diabetes to be multifactorial (McMillan, 1975). Careful longitudinal studies are therefore now required to define the evolution of possible aetiological factors and their chronological relationship with the onset and progression of different diabetic complications, so that the relative importance of each factor may be seen more objectively. The effect of specific therapeutic agents may then be assessed against the background of a detailed knowledge of the pathophysiological events leading up to and surrounding the development of diabetic sequelae. Such studies are likely to be costly, time consuming and tedious. Meanwhile, it is perhaps refreshing to know that improved metabolic control in diabetics may not only lower viscosity in the direction of normal but also favourably influence several other factors (Johanssen and Hansen, 1971; Johnson et al., 1979).

14.5 References

Barnes A J, Locke P, Scudder P R, Dormandy T L, Dormandy J A, Slack J (1977) Is hyperviscosity a treatable component of diabetic microcirculatory disease? Lancet *ii:* 789-791

Bollinger A, Berchtold P, Berger W (1969) Untersuchungen der Blutviskosität bei Diabetikern. Praxis *58:* 1104-1107

Carlson T H, Wentland S H, Leonard B D, Ruder M A, Reeve E B (1978) Effects of prostaglandin E_2, analogs, fatty acids, and indomethacin on fibrinogen level. Am J Physiol *235:* H223-230

Cogan D G, Merola L, Laibson P R (1961) Blood viscosity, serum hexosamine and diabetic retinopathy. Diabetes *10:* 393-395

Cooper R A (1977) Abnormalities of cell-membrane fluidity in the pathogenesis of disease. N Engl J Med *297:* 371-377

Dintenfass L (1977) Blood viscosity factors in severe non diabetic and diabetic retinopathy. Biorheology *14:* 151-157

Ditzel J (1968) Whole-blood viscosity and related components in diabetes mellitus. Dan Med Bull *15:* 49-53

Dormandy J A (1970) Clinical significance of blood viscosity. Ann Roy Coll Surg Engl *47:* 211-228

Dormandy J A (1974) Medical and engineering problems of blood viscosity. Bio-med Engng *9:* 284-303

Ferguson J C, Mackay N, Philip J A D, Sumner D J (1975) Determination of platelet and fibrinogen half-life with [^{75}Se] selenomethionine: studies in normal and in diabetic subjects. Clin Sci Mol Med *49:* 115-120

Hoare E M, Barnes A J, Dormandy J A (1976) Abnormal blood viscosity in diabetes mellitus and retinopathy. Biorheology *13:* 21-25

Hobin F P, Maraist F B (1973) Diabetic hyperosmolar non ketoacidotic coma complicated by mesenteric thrombosis. J. Fla Med Assoc *60:* 25-26

Isogai Y, Ichiba K, Iida A, Chikatsu I, Abe M (1971) Viscosity of blood and plasma in various diseases. In: Hartert H, Copley A L (eds) Theoretical and clinical hemorheology. Proc. 2nd Int. Conf. Hemorheology Heidelberg, 1969. Springer-Verlag, Berlin-Heidelberg-New York, pp 326-332

Jarrett R J, Keen H (1975) Diabetes and atherosclerosis. In: Keen J, Jarrett R J (eds) Complications of diabetes. Edward Arnold, London, pp 179-203

Jeejeebhoy K N, Ho J, Greenberg G R, Phillips M J, Bruce-Robertson A, Sodtke U (1975) Albumin, fibrinogen and transferrin synthesis in isolated rat hepatocyte suspensions. Biochem J *146:* 141-155

Johanssen K., Hansen A P (1971) Diurnal serum growth hormone levels in poorly and well controlled juvenile diabetics. Diabetes *20:* 239-245

Johnson M, Harrison H E, Raftery A T, Elder J B (1979) Vascular prostacyclin may be reduced in diabetes in man. Lancet *i:* 325-326

Kohner E M, Oakley N W (1975) Diabetic retinopathy. Metabolism *25:* 1085-1102

Labib M A M, Higazi A M, El-Ebrashy N, El-Ashmawy S, Madkoor M K, Barhooma G A (1971) Studies on diabetic retinal vascular changes with special reference to blood coagulation and viscosity. Bull Ophthalmol Soc Egypt *64:* 457-482

Langer L, Bergentz S E, Bjure J, Fagerberg S E (1971) The effect of exercise on haematocrit, plasma volume and viscosity in diabetes mellitus. Diabetologia *7:* 29-30

Lingard P S (1974) Capillary pore rheology of erythrocytes. 1. Hydroelastic behavior of human erythrocytes. Microvasc Res *8:* 53-63

McMillan D E (1974) Disturbance of serum viscosity in diabetes mellitus. J Clin Invest *53:* 1071-1079

McMillan D E (1975) Deterioration of the microcirculation in diabetes. Diabetes *24:* 944-957
McMillan D E, Utterback N G, La Puma J (1978) Reduced erythrocyte deformability in diabetes. Diabetes *27:* 895-901
Miller L L (1976) Direct effects of glucagon on protein and amino acid metabolism in the isolated perfused rat liver. Diabetes *25* (Suppl. 2): 865-871
Mosora N, Baciu T, Vincze J (1972) The viscosity of the serum, haematocrit and fibrinogen in diabetes mellitus and their relationship with diabetes mellitus. Diabetologia *8:* 59
Paülsen E P, Koury M (1976) Hemoglobin A_{Ic} levels in insulin-dependent and independent diabetes mellitus. Diabetes *25* (Supl. 2): 890-896
Pilgeram L O, Pickart L (1968) Control of fibrinogen biosynthesis: the role of free fatty acid. J Atheroscler Res *8:* 155-166
Ramon J S, Daroca J C, Perez J F C, Frances J M C, Gorina A B (1975) Complications cardiovasculares del coma diabético hiperosmolar. Estudio de cuatro observaciones. Rev Clin Española *137:* 521-526
Rees S B, Simon L, Peltier G A, Balodimos M, Gleason R, Marble A, Merrill E W (1967) Hemorheologic studies during the progression and remission of diabetic retinopathy. Biorheology *4:* 102-103
Reid H L, Barnes A J, Lock P J, Dormandy J A, Dormandy T L (1976a) A simple method for measuring erythrocyte deformability. J Clin Pathol *29:* 855-858
Reid H L, Dormandy J A, Barnes A J, Lock P J, Dormandy T L (1976b) Impaired red cell deformability in peripheral vascular disease. Lancet *i:* 666-667
Reubi F C (1953) Glomerular filtration rate, renal blood flow and blood viscosity during and after diabetic coma. Circulat Res *1:* 410-413
Schmid-Schönbein H, Volger E (1976) Red cell aggregation and red cell deformability in diabetes. Diabetes *25* (Suppl. 2): 897-902
Sirs J A, Boroda C, Rampling M W (1979) Rheological problems of diabetes and their treatment. Diabetologia (in press)
Skovborg F, Nielsen A V (1969) Variation in whole blood and plasma viscosity in acidotic diabetics. Bibl Anat *10:* 145-152
Skovborg F, Nielsen A V, Schlichtkrull J, Ditzel J (1966) Blood viscosity in diabetic patients. Lancet *i:* 129-131
Stowers J M (1975) Complications in relation to diabetic control — general review. In: Keen H, Jarrett R J (eds) Complications of diabetes. Edward Arnold, London, pp 1-5
Tchertkoff V, Nayak S V, Kamath C, Salomon M I (1974) Hyperosmolar nonketotic diabetic coma: vascular complications. J Am Ger Soc *22:* 462-466
Timperley W R, Preston F E, Ward J D (1974) Cerebral intravascular coagulation in diabetic ketoacidosis. Lancet *i:* 952-956
Zingg W, Sulev J C, Morgan C D, Ehrlich R M (1971) Blood viscosity in diabetic children. Diabetologia *7:* 461-462

15 Surgery and Post-operative Thrombosis

P. N. Matthews

15.1 Introduction

Venous thrombosis remains one of the commonest causes of mortality and morbidity following surgery. Since Virchow described his now classic triad over 100 years ago, many advances have been made in both diagnosis and prevention of venous thrombosis, and yet in some ways we are no nearer to a better understanding of the fundamental mechanism. While we recognise the fact that certain patients fall into "high risk" categories, we still do not really understand why one patient gets a thrombosis and another seemingly very similar patient does not.

Post-operative arterial thrombosis following reconstructive arterial surgery is one of the commonest causes of early failure and again we do not fully understand why this occurs in some patients and not others.

John Hunter (1794) was one of the first to describe some of the factors responsible for thrombosis in his paper entitled "A treatise on the Blood, Inflammation and Gunshot Wounds" first published in 1794. Anticipating some of Virchow's ideas, Hunter realised that stasis was an important but not the only cause of thrombosis. Virchow (1846) described the three classic causes of thrombosis and this is today still known as Virchow's triad.

At about the same time that Hunter was making his early observations, the Reverend Stephen Hales (1753), a minister at Teddington, made what was probably the first reference to blood viscosity as a cause of disease. In his book he stated that "the resistance which the blood meets in these capillary passages may be greatly varied, either by different degrees of viscosity or fluidity of the blood, or the several degrees of constriction or relaxation of those fine vessels". It was not until 1846 that Poiseuille (1846) performed his experiments on the flow of fluids through narrow tubes and proposed a formula relating blood flow with the driving force, the radius of the vessel and the viscosity of the blood.

While many advances have been made over the last 100 years, it is only recently that blood rheology has been shown to be related to circulatory insufficiency. There is now quite good evidence that blood viscosity may be a significant factor to consider in the pathogenesis of venous, and possibly also arterial, thrombosis.

This account will look at some of the rheological changes occurring during surgery and see if they can be linked with the development of arterial or venous thrombosis.

15.2 Changes in Blood Rheology Associated with Surgery

Major surgery and anaesthesia, with its associated intravenous infusions of large volumes of crystalloids, colloids and stored whole blood, causes an enormous number of physiological responses, and it would be surprising if a change in blood viscosity did not occur. The viscosity of blood is affected by the concentration and deformability of the red blood cells and also by the concentration of plasma proteins, chiefly fibrinogen.

15.2.1 Changes in Whole-blood Viscosity

One of the main determinants of blood viscosity is the haematocrit and this will vary considerably during and following surgery. To study the effect of other determinants of blood viscosity it is convenient to eliminate the effect of red-cell concentration by correcting to a standard haematocrit. Fig. 15.1 shows the typical changes occurring in whole-blood viscosity measured at two shear rates in a patient undergoing major vascular surgery. It can be seen that corrected whole-blood viscosity at lower shear rates rises to a maximum around the 4th day and then falls over the next 10 days. The blood viscosity at high shear rates changes very little. On the same graph it can be seen that the change in whole-blood viscosity at lower shear closely follows the post-operative change in plasma fibrinogen. It is now generally accepted that the most important plasma protein affecting blood viscosity is fibrinogen, mainly because of its shape and large size. The relatively higher viscosity of blood at low shear rates is due to the effect that fibrinogen has in causing aggregation of red cells. Harvey-Kemble and Hickman (1972) have measured fibrinogen levels in a group of 25 patients undergoing general surgical procedures and found that the fibrinogen levels rose by an average of 93%. The peak of this rise usually occurred around the 4th post-operative day.

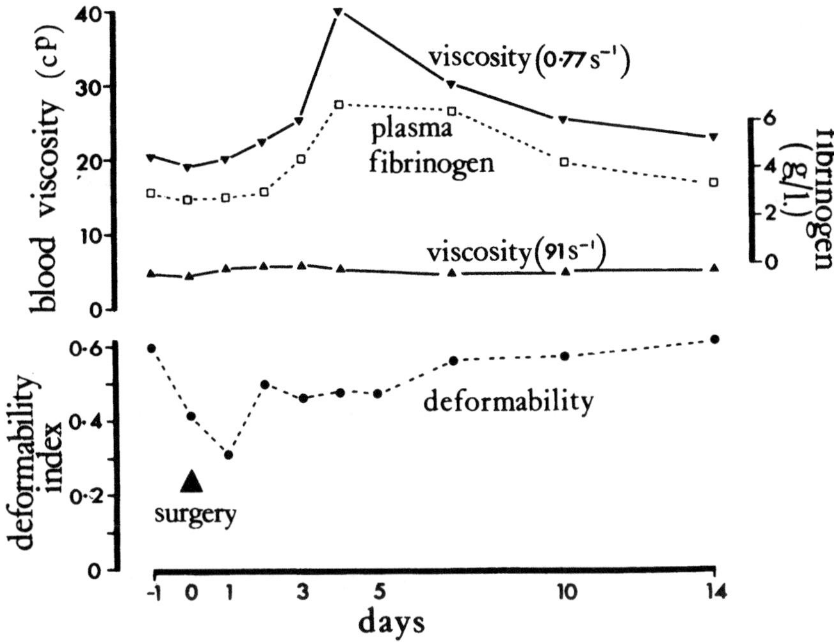

Fig.15.1. Typical changes in blood viscosity, plasma fibrinogen and red-cell deformability (RCD) following surgery. 1 centipoise (cP) = 1 mPa.s

15.2.2 Changes in Red-cell Deformability (RCD)

RCD has been defined as the geometric and physical characteristics which permit a cell, whose greater diameter may exceed 8 μm, to pass through 14 μm or longer segments of normal capillaries which range from 3-12 μm in diameter (Weed, 1970). As there is no single universally accepted method of measuring RCD and no conventional units of measurement, results have varied widely. The author has chosen to use a filtration technique with red cells resuspended in their own plasma because this seems to be more physiological and to give more reproducible results than some other methods.

15.2.2.1 Method Ten millilitres heparinised venous blood is centrifuged at 3000 rpm for 15 min. The plasma is removed and pre-filtered through a 5-μm Nuclepore filter (Nuclepore Corporation, Pleasanton, California, U.S.A.). The buffy coat is removed and discarded and a 0.2-ml aliquot of red cells is suspended in 4 ml of the filtered plasma to give a 5% suspension of red cells. Aliquots of 1 ml of this suspension are then filtered through high-grade, 5-μm Nuclepore polycarbonate sieves. The suspension is placed in a modified pop-top Nuclepore holder and allowed to filter under its own hydrostatic pressure for exactly 60 s. Filtrations are performed in triplicate on all blood samples using new filters. The volume of the cells filtered is compared with the volume contained in the original 1-ml suspension by comparing optical densities. From these measurements, a deformability index varying between 0 (when none of the suspension has filtered) and 1 (when all the suspension has filtered) can be calculated. One of the advantages with this method is that artefacts due to platelets or white cells are largely removed.

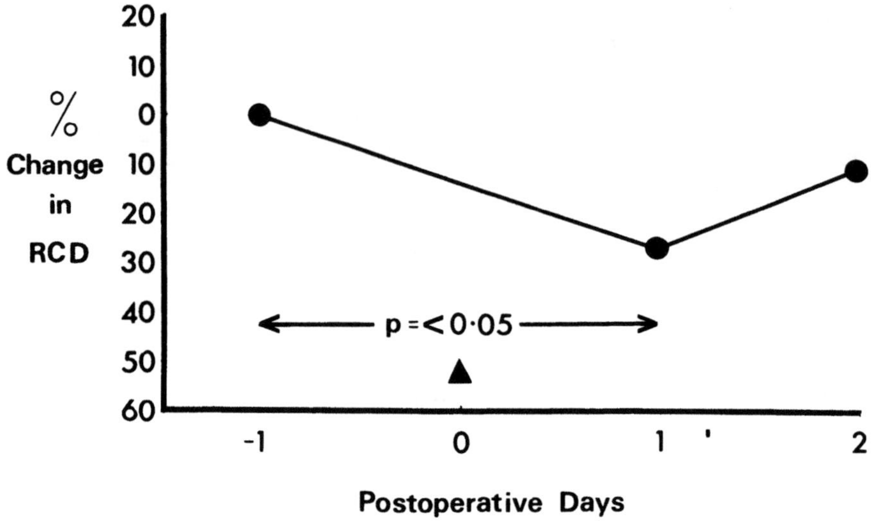

Fig.15.2. Change in RCD following varicose vein surgery under general anaesthetic (five patients)

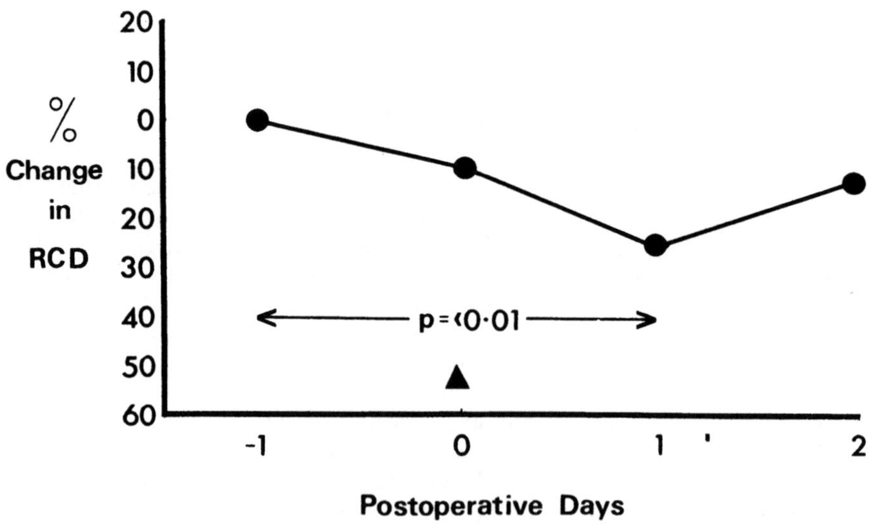

Fig.15.3. Change in RCD following inguinal hernia repair under general anaesthetic (12 patients)

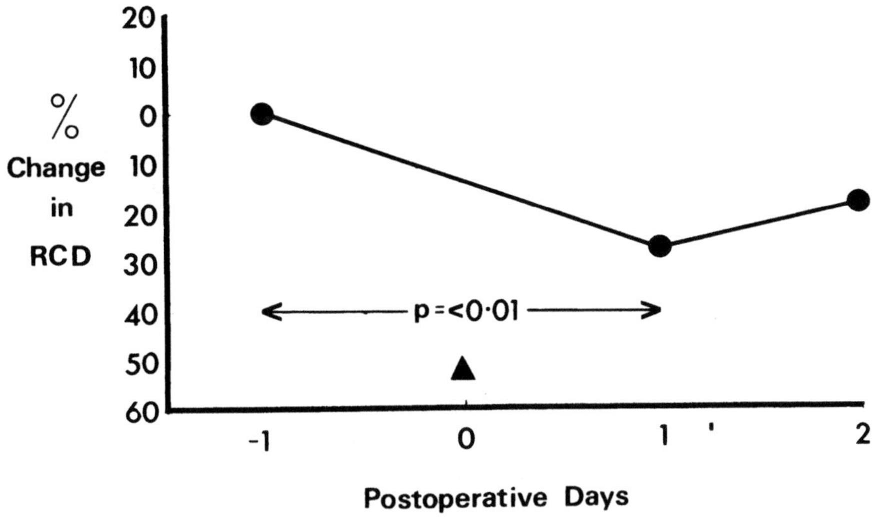

Fig.15.4. Change in RCD following inguinal hernia repair under epidural anaesthetic (five patients)

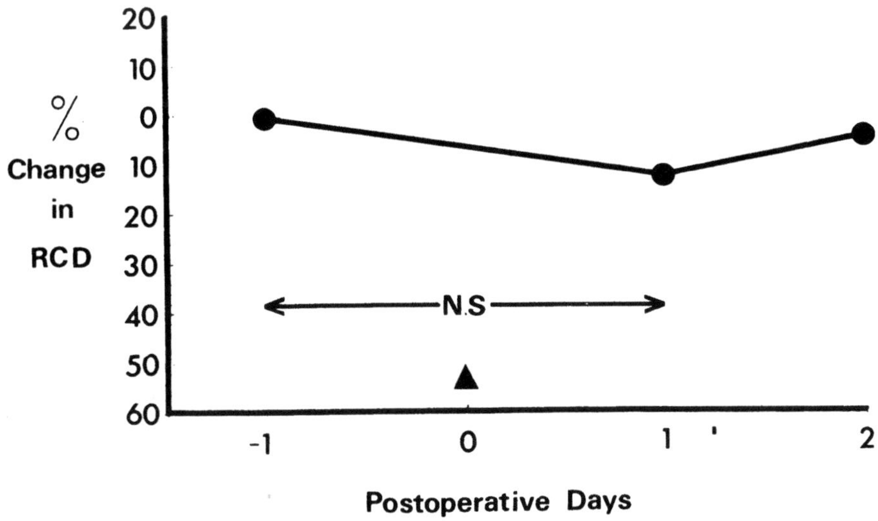

Fig.15.5. Change in RCD following cystoscopy under general anaesthetic (five patients)

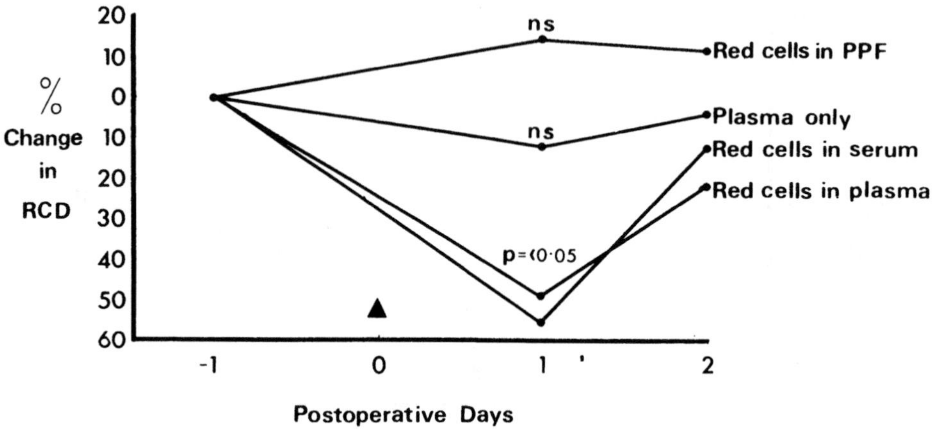

Fig.15.6. Change in RCD following hernia repair measured with red cells suspended in plasma, serum and PPF

15.2.2.2 Results Scholz et al. (1975) studied RCD following surgery, using a filtration technique and resuspending the red cells in Ringer-albumin solution. They could detect no change in RCD post-operatively except in one patient who was transfused 7 units of stored blood. We have studied RCD post-operatively in patients undergoing both reconstructive vascular surgery and varicose vein surgery and the typical changes that occur can be seen on Fig. 15.1. It can be seen that the maximum fall in RCD occurs 24 h post-operatively and has returned to normal by 5 days. It was interesting to note that there was not any significant change between the fall in RCD in arterial patients when compared with varicose vein patients. This was slightly surprising, considering the differences in operation time and transfusions.

Figures 15.2 to 15.5 show the post-operative changes in groups of patients having varicose vein operation, inguinal hernia repair under general anaesthetic and epidural anaesthetic, and check cystoscopies. It can be seen that patients having cystoscopy under general anaesthetic do not have any significant change of RCD, which tends to support the hypothesis that a decrease in RCD is associated with tissue trauma.

Figure 15.6 shows a similar change in RCD occurring in five patients having hernia repair, when the red cells are suspended in serum or plasma but not when the red cells are resuspended in purified protein fraction. This suggests that there is a factor in plasma, other than fibrinogen, which is causing the red cells to become less deformable. There was no significant change in plasma drip rate, showing that it is not a change in plasma filtration, with red cells playing a passive role, that is being measured.

In none of our clinical trials has there been any correlation between RCD and blood viscosity at high or low shear rates. It could be that the filtration method of measuring RCD may be more representative of events occurring in the microcirculation than the measurement of whole-blood viscosity. Baumgartner (1973) has demonstrated that red cells are necessary for platelets to be deposited on a subendothelial surface and Born (1977) has suggested that platelets may become activated by a chemical substance, probably ATP, that has been released from the red cells. It is conceivable that the decrease in RCD that occurs on the first post-operative day may predispose to local platelet aggregations in the venous system and hence increase the risk of thrombus formation.

15.3 The Relationship Between Blood Rheology and the Pathogenesis of Thrombosis

Post-operative venous thrombosis usually occurs in the absence of any recognisable lesion of the vein intima (McLachlin and Paterson, 1951), suggesting that blood flow and change in the blood coagulability must be the two factors of Virchow's triad that are the most important. Many studies have now shown that flow rates in the calf veins are reduced during anaesthetic and post-operative convalescence (McLachlin et al., 1960; Clark and Cotton, 1968). Sevitt (1974) has shown that a large proportion of thrombi begin in the valve pockets where the lowest flow rates are undoubtedly found. On the arterial side, Renwick et al. (1968) have studied flow rate following vascular reconstruction and found that flow is much slower in the first 4 h post-operatively and then gradually increases over the next 6 h. They also found that more than half of their vascular failures occurred in the first 24 h and were associated with an abnormally low flow state.

These studies and many others all seem to indicate that post-operative thrombosis, arterial or venous, is nearly always associated with reduced blood flow. Poiseuille's suggestion that blood flow is inversely related to blood viscosity has been verified in man in many studies (Dormandy, 1971; Dormandy et al., 1973).

The non-Newtonian behaviour of blood means that it has a higher apparent viscosity at low shear rates such as would tend to occur following surgical procedures. This higher apparent blood viscosity will lead to further reductions in blood flow and thus a progressive spiral of events may be initiated: lower flow rates causing increased apparent viscosity which in turn causes still lower flow rates.

The rather dramatic decrease in RCD that occurs in the first 24 h following surgery may be more important in the pathogenesis of venous thrombosis because it is known that most thrombi are formed, or have begun to form, soon after surgery. The later rise in blood viscosity may well cause propagation of the already formed thrombus.

15.4 Circumstantial Evidence Relating Increased Risk of Thrombosis with High Blood Viscosity

A large number of risk factors predisposing to post-operative venous thrombosis have been described and it is interesting that many of them are also associated with high blood viscosity.

15.4.1 Malignancy

It is now well established that malignant disease is associated with an increased risk of venous thrombosis (Kakkar et al., 1970) and that episodes of thrombosis may precede clinical evidence of some tumours. Tietjen et al. (1977) measured blood viscosity and plasma fibrinogen levels in 31 patients with known visceral cancers. They found that whole-blood viscosity, corrected to a standard haematocrit, was significantly higher in the cancer patients when compared with a control group. The plasma fibrinogen and globulin levels were also significantly elevated. This group concluded that the high apparent blood viscosity seen at low shear rates such as would be found in the venous circulation might well be contributing to the high incidence of venous thrombo-embolism seen in patients with neoplastic disease.

15.4.2 Haemoconcentration

The concentration of red cells in the blood is the single most important factor in determining its viscosity in large and medium vessels. The situation that occurs in the microcirculation is less clear and the relevance of bulk blood viscosity to flow is more doubtful.

The highest haematocrits occur in patients with polycythaemia vera. Wasserman and Gilbert (1963) reviewed 54 patients with polycythaemia vera who had undergone major surgery and found that both haemorrhagic and thrombotic complications were very common, especially in patients whose high pre-operative haematocrits had not been corrected. Reducing the haematocrit to normal levels significantly reduced the incidence of thrombotic complications. Bouhoutsos et al. (1974) reviewed 445 patients who had had arterial operations and found that the higher the pre-operative haemoglobin level, the more likely the risk of early post-operative complications. The most frequent complications were arterial thrombosis or bleeding and in most of the patients who developed these problems the haemoglobin levels were 17 g/dl or over. More recently, Bailey et al. (1979) studied a group of 59 diabetic patients having local amputations in the feet and found that the pre-operative haemoglobin level was very significantly lower in patients whose amputations healed when compared with those who failed to heal. All 30 patients who had haemoglobin levels above 13 g/dl had failed amputations. These and other reports strongly suggest that the increased blood viscosity resulting from an increased haematocrit leads to decreased arterial flow and thrombosis.

Due to the largely exponential relationship between blood viscosity and haematocrit, a small change in haematocrit, in the physiological range, will lead to a relatively large change in blood viscosity. Messmer et al. (1972a) suggested that an ideal PCV for oxygen delivery to the tissues is around 30%, which leads us to ask why we walk around with a PCV in the 45% to 50% range. All this leads to the rather interesting, seemingly paradoxical situation that reducing a patient's PCV would seem to improve tissue oxygenation.

15.4.3 Hyperfibrinogenaemia

It has already been shown how plasma fibrinogen, with its significant effect on blood viscosity at low shear rates, rises on about the 3rd or 4th post-operative day. Postlethwaite (1976) studied 28 patients undergoing major vascular surgery and found that prolonged elevation of plasma fibrinogen levels was associated with a significantly higher failure rate. He also studied the fibrinogen level pre-operatively in the successful and unsuccessful groups and found that patients undergoing successful surgery all started with a pre-operative plasma fibrinogen below 4.5 g/litre, while the patients whose surgery was unsuccessful had a wide scatter of pre-operative levels, mostly lying above that level.

Hamer et al. (1973) studied 76 patients having reconstruction surgery for peripheral vascular disease and found that a raised plasma fibrinogen level persisting 1 year after operation is often associated with failure.

15.5 Specific Studies Relating Venous Thrombosis and Blood Viscosity

Dormandy and Edelman (1973) studied 52 patients undergoing general surgical and gynaecological procedures in order to see if there was any relationship between high blood viscosity and the development of deep-vein thrombosis (DVT). They found that the 11 patients who developed DVT had a significantly elevated pre-operative corrected blood viscosity when compared with the pre-operative blood viscosity of those patients who did not develop a DVT. They concluded that the measurement of blood viscosity might be a useful physicochemical pre-operative predictive test to detect "high-risk" patients.

Humphreys et al. (1976) studied 28 patients undergoing laparotomy and found that the patients who subsequently developed a DVT had a significantly raised pre-operative index of yield stress. This correlation between yield stress and DVT can be explained when one considers the very low flow rates that occur in the calf veins during anaesthesia when the muscle pump is inactive. The relatively weak transmitted arterial pulsations may not be sufficient to overcome an increased attraction between red cells leading to prolonged periods of stasis and possible thrombus formation.

There is a known increased incidence of venous thromboembolism in women taking the contraceptive pill (Vessey and Doll, 1969). Aronson et al. (1971) studied a group of 47 healthy young women before and during administration of two different oral contraceptives. They found that in 8 patients there was an increase in blood viscosity that started within 2 months and increased to a maximum in 6 months. A striking finding was that 1 patient developed thrombophlebitis and the clinical signs were preceded by a marked rise in blood viscosity which dropped to normal within 2 months of stopping the pill.

15.6 Haemorheological Prevention of Thrombosis

The aim of the clinician must be to prevent the formation of a thrombus because once this has happened the sequelae, sometimes very serious, may occur before the thrombus has become clinically obvious. Nobody has yet found the ideal prophylactic against thrombosis. Haemorheological manipulation of the plasma, packed cell volume and RCD would seem to be a logical line of treatment and if it worked would provide further evidence for the involvement of blood viscosity in thrombosis.

15.6.1 Reduction of Plasma Viscosity

15.6.1.1 Plasma Exchange Removal of plasma that contains large quantities of abnormal proteins will effectively reduce plasma viscosity and could be used to treat pre-operatively patients with macroglobulinaemia and myeloma in order to reduce the known high incidence of thrombosis in these conditions.

15.6.1.2 Defibrination A rise in plasma fibrinogen is part of the physiological response to surgery and prolonged or excessive rises have been shown to correlate with failure of arterial surgery. This suggests that lowering plasma fibrinogen or preventing the normal post-operative rise might seem to be a reasonable method of lowering the incidence of arterial and venous thrombosis. Arvin is the purified extract of the venom of the Malayan pit viper and has been shown to be effective in reducing plasma fibrinogen levels. Postlethwaite et al. (1977) studied the effect of subcutaneous Arvin on the patency rate of Dacron grafts in rabbit inferior vena cavae and showed that Arvin significantly reduced the post-operative plasma fibrinogen, and this was associated with a significantly greater patency rate when compared with controls. Lowe et al. (1978) studied a group of patients undergoing surgery for fractured hips and found that Arvin reduced the incidence of DVT from a high 73% in the control group to 45% in the Arvin-treated group. The fact that post-operative thrombosis is significantly reduced by using a drug that lowers plasma fibrinogen is strong evidence that fibrinogen levels and therefore possibly blood viscosity are important in the genesis of thrombosis.

15.6.2 Haemodilution

The most important single determinant of blood viscosity is probably the concentration of red cells. Messmer et al. (1972b) have shown that acute normovolaemic haemodilution with a variety of plasma substitutes causes an increased cardiac output which was paralleled by decrease in blood viscosity. They have shown that the decreased oxygen-carrying ability of the blood, caused by reduction of the haematocrit, is more than compensated for by the increased cardiac output, provided blood volume remains the same. From these studies it would seem reasonable to attempt to lower haematocrit and therefore decrease blood viscosity and raise blood flow. Yates et al. (1979) have shown that blood flow to the legs of claudicants can be significantly improved by reducing the packed cell volume to 35% and they concluded that the probable mechanism was a reduction in blood viscosity.

Several studies have now shown that Dextran is effective in reducing post-operative thrombo-embolic complications. Kline et al. (1975) studied 831 patients undergoing elective surgery and showed that Dextran was more effective than saline in preventing pulmonary embolism, but did not cause any significant reduction in DVT. Gelin and Jansen (1975) studied 901 patients undergoing major surgical procedures and found that the incidence of both pulmonary emboli and DVT were significantly reduced in the Dextran group when compared with a control group given glucose infusions. They further studied blood flow through the leg veins by using ^{131}I Hippuran and found that flow in the deep veins was maintained much more effectively in the Dextran group. From these two studies it seems likely that Dextran works by having an anti-thrombotic effect and also by reducing blood viscosity and hence increasing flow.

15.6.3 Increased Red-cell Deformability

Red cells tend to become less deformable following surgery and a new line of anti-thrombotic treatment might be to prevent this from happening. At the time of writing there are no drugs that have definitely been shown to improve RCD. However early results of *in vitro* studies suggest that Flunarizine may increase the deformability of red cells. If these results are substantiated *in vitro* and *in vivo,* then a possible new line of prophylaxis against post-operative thrombosis may have been discovered.

15.6.4 Others

Low-dose heparin is now one of the most widely used prophylactic anticoagulants. The main effect of heparin is probably on coagulation; however, Erdi et al. (1976) have shown that low-dose subcutaneous heparin reduces the whole-blood viscosity at low shear rates and it is possible that this effect may contribute to the known effectiveness of heparin in preventing venous thrombosis.

15.7 Conclusion

This paper has tried to relate haemorheology and post-operative thrombosis and may therefore seem to have presented a rather biased view. It is not the author's intention to suggest that blood viscosity is the complete answer to this very important problem, but rather to bring to the reader's attention a previously rather neglected field. The relationship between blood viscosity and blood flow is now beyond dispute, as is the relationship between blood flow and thrombosis. It would therefore seem logical to suppose that blood viscosity would influence the development of both arterial and venous thrombosis, especially in the post-operative situation.

Acknowledgements. The author wishes to thank John Dormandy and Anthony Dodds for their help in preparing this paper; and Mrs. Celia Velarde for her typing. The author is grateful for the financial support of the St. George's Hospital Medical Research Committee.

15.8 References

Aronson H B, Magora F, Schenker J G (1971). The effect of oral contraceptives on blood viscosity. Am J Obstet Gynecol *110:* 997-1001

Bailey M J, Johnston C L W, Yates C J P, Somerville P G, Dormandy J A (1979). Preoperative haemoglobin as predictor of outcome of diabetic amputations. Lancet *ii:* 168-170

Baumgartner H R (1973). The role of blood flow in platelet adhesion, fibrin deposition and formation of mural thrombi. Microvasc Res *5:* 167-179

Born G V R (1977). Fluid, mechanical and biochemical interactions in haemostasis. Br Med Bull *33:* (3): 193-197

Bouhoutsos J, Morris T, Chevatzas D, Martin P (1974). The influence of haemoglobin and platelet levels on the results of arterial surgery. Br J Surg *61:* 984-986

Clark C, Cotton L T (1968). Blood flow in the deep veins of the leg. Br J Surg *55* (3): 211-214

Dormandy J A (1971). Influence of blood viscosity on blood flow and the effect of low molecular weight Dextran. Br Med J *4:* 716-719

Dormandy J A, Edelman J B (1973). High blood viscosity: an aetiological factor in venous thrombosis. Br J Surg *60* (3): 187-190

Dormandy J A, Hoare E, Colley J, Arrowsmith D E, Dormandy T L (1973). Clinical, haemodynamic, rheological and biochemical findings in 126 patients with intermittent claudication. Br Med J *4:* 576-581

Erdi A, Thomas D P, Kakkar V V, Lane D A, Dormandy J A (1976). Effect of low dose subcutaneous heparin on whole blood viscosity. Lancet *ii:* 342-344

Gelin L, Jansen H (1975). Moderate preoperative haemodilution. Mortality and thrombus formation in general surgery. In: Messmer K, Schmid-Schönbein H (eds) Intentional haemodilution, Biblthca. Haemat. No.41. Karger, Basel, pp 239-247

Hales S (1733) Statistical Essays Vol. 2, Innys and Manby, London

Hamer J D, Ashton F, Meynell M J (1973) Factors influencing prognosis in the surgery of peripheral vascular disease: platelet adhesiveness, plasma fibrinogen and fibrinolysis. Br J Surg *60* (5): 386-389

Harvey-Kemble J V, Hickman J A (1972) Post-operative changes in blood viscosity and the influence of haematocrit and plasma fibrinogen. Br J Surg *59* (8): 629-632

Humphreys W V, Walker A, Charlesworth D (1976) Altered viscosity and yield stress in patients with abdominal malignancy: relationship to DVT. Br J Surg *63:* 559-561

Hunter J (1794) A treatise on the blood, inflammation and gunshot wounds. John Richardson, Chapter I.

Kakkar V V, Howe C T, Nicolaides A N, Renney J T G, Clarke M B (1970) Deep vein thrombosis of the leg. Is there a "high risk" group? Am J Surg *120:* 527-530

Kline A, Hughes L E, Campbell H, Williams A, Zlosnick J, Leach K G (1975) Dextran 70 in the prophylaxis of thrombo-embolic disease after surgery: a clinically orientated, randomised, double blind trial. Br Med J *2:* 109-112

Lowe G D O, Campbell A F, Meek D R, Forbes C D, Prentice C R M, Cummings S (1978) Subcutaneous ancrod in prevention of deep-vein thrombosis after fractured neck of femur. Lancet *ii:* 698-700

McLachlin J, Paterson J C (1951) Some basic observations on venous thrombosis and pulmonary embolism. Surg Gynecol Obstet *93* (1): 1-8

McLachlin A D, McLachlin J A, Jory T A, Rawling E G (1960) Venous stasis in the lower extremities. Ann Surg *152:* 678-685

Messmer K, Sunder-Plassmann L, Klovekorn W P, Holper K (1972a) Circulatory significance of haemodilution. In: Harders H (ed) Advances in microcirculation. Karger, Basel, Vol.4, 1-77

Messmer K, Lewis D H, Sunder-Plassmann L, Klovekorn W P, Mendler N, Holper K (1972b) Acute normovolaemic haemodilution. Europ Surg Res *4:* 55-70

Poiseuille J L (1846) C R Acad Sci Paris *9:* 961-967, 1041-1048

Postlethwaite J C (1976) The importance of plasma fibrinogen in vascular surgery. Ann Roy Coll Surg Eng *58:* 457-464

Postlethwaite J C, Goyle K B, Dormandy J A, Hynd J W (1977) Improvement in experimental vascular graft patency by controlled defibrination. Br J Surg *64:* 28-30

Renwick S, Gabe I T, Shillingford J P, Martin P (1968) Blood flow after reconstructive arterial surgery measured by implanted electromagnetic flow probes. Surgery *64* (3): 544-553

Scholz P M, Kinney J M, Chien S (1975) The effects of major abdominal operations on human blood rheology. Surgery *77* (3): 351-359

Sevitt S (1974) The structure and growth of valve pocket emboli in femoral veins. J Clin Pathol *27:* 517-528

Tietjen G W, Chien S, Scholz P, Gump F E, Kinney J M (1977) Changes in blood viscosity and plasma proteins in carcinoma. J Surg Oncol *9:* 53-59

Vessey M P, Doll R (1969) Investigation of relation between the use of oral contraceptives and thrombo-embolic disease — a further report. Br Med J *2:* 651-657

Virchow R (1846) Beitr Exp Pathol *2:* 227-380

Wasserman L R, Gilbert H S (1963) Surgery in polycythaemia vera. New Eng J Med *269* (23): 1226-1230

Weed R I (1970) The importance of erythrocyte deformability. Am J Med (Editorial) *49* (2): 146-150

Yates C J P, Berent A, Andrews V, Dormandy J A (1979) Increase in leg blood flow by normovolaemic haemodilution in intermittent claudication. Lancet *ii:* 166-168

16 Rheological Studies in Obstetrics and Gynaecology

P. C. Buchan and H. N. Macdonald

16.1 Introduction

Because of the strong connection between arterio-venous thrombo-embolism and oral contraception (Vessey and Mann, 1978) much interest has been focussed on the alterations in rheological parameters caused by hormonal ovulation suppression. A full understanding of these effects of oestrogens and progestagens must be based on a sound appreciation of the physiological background upon which they are imposed but little attention has been paid, as yet, to the rheological changes in the menstrual cycle and pregnancy. In common with most biological measurements rheological parameters vary in response to circadian (Seaman et al., 1965), menstrual (Dintenfass and Yu, 1968) and reproductive (Hamilton, 1950) cycles. This chapter is a review of the existing literature on rheological changes during ovulatory cycles, oral contraception and pregnancy, supplemented by our own data drawn from serial and cross-sectional studies of women in these various states.

16.2 Methodology

A detailed description of the methodology will appear elsewhere and only a brief account is included at present. Viscosity measurements were made with a Deer Rheometer (Rheometer Marketing Ltd., Leeds) at a shear stress of 0.574 dyne/cm^2 at 37°C. Packed cell volume was obtained by use of a Coulter Counter-S; plasma fibrinogen was measured in the menstrual and oral contraceptive studies by the method of Ratnoff and Menzie (1951) and in the pregnancy study by the method of Ellis and Stransky (1961); erythrocyte deformability, expressed as an index, was estimated by a microfiltration method using 5-μm Nuclepore filters (Buchan, 1980).

Our own data on non-pregnant subjects was obtained from a serial study of 20 young women, aged between 22 and 34 years, 12 of whom were experiencing regular spontaneous menstruation and 8 of whom had been taking a combined oral contraceptive containing ethinyloestradiol 30 μg and norgestrel 250 μg for at least 6 months. Each subject had blood sampled at the same time each day on four separate occasions approximately a week apart during the cycle. A statistical summary of the rheological data appears in Table 16.1, which also includes data from 30 healthy women who had been taking the same oral contraceptive combination for at least 6 months and who contributed a single blood sample between the 7th and 21st day of their cycle. The data from the serial studies during spontaneous and oral contraceptive suppressed cycles are depicted graphically in Figures 16.1 to 16.6. The graphs are drawn against a diagrammatic representation of the appropriate hormone profiles. Serial data were analysed statistically using a paired t-test on the mean differences between the stages.

16.3 Normal Menstrual Cycle

16.3.1 Plasma Fibrinogen

The literature is conflicting as regards the behaviour of plasma fibrinogen levels during the normal menstrual cycle. Howie et al. (1970) noted a pre-menstrual rise which was short of statistical significance in a small study. Turksoy et al. (1961) also demonstrated a pre-menstrual rise preceded by a more impressive ovulatory peak. Our own data followed the trend of Howie et al. (1970) but with a statistically significant pre-menstrual peak and no suggestion of an ovulatory rise which remains to be confirmed (Fig. 16.1).

16.3.2 Plasma Viscosity

We have found a downward trend in plasma viscosity throughout the cycle with a significant rise preceding menstruation. This higher level is maintained until the next follicular phase is well established (Fig. 16.1). These findings require confirmation as we can find no other published study of plasma viscosity throughout the menstrual cycle. This pattern reflects the pre-menstrual rise in

Table 16.1. Mean values of rheological parameters in 'users' and 'non-users' of oral contraceptives

	12 'Non-users'				8 'Users'				30 'Users'
Days of cycle	1-7	12-16	18-21	26-28	1-7	12-16	18-21	26-28	7-21
Plasma fibrinogen g/l	2.4**	2.45**	2.35**	2.64^{ns}	2.72	2.8	2.82	2.73	2.79
Standard error	0.09	0.08	0.08	0.1	0.09	0.08	0.08	0.1	0.17
Packed cell volume	0.382**	0.397*	0.39**	0.399*	0.407	0.405	0.405	0.404	0.417
Standard error	0.005	0.006	0.007	0.005	0.008	0.007	0.008	0.009	0.008
Erythrocyte deformability index	0.776^{ns}	0.783^{ns}	0.907*	0.756^{ns}	0.861	0.647	0.788	0.956	0.772
Standard error	0.047	0.075	0.007	0.054	0.064	0.048	0.066	0.059	0.092
Plasma viscosity mPa.s	1.74^{ns}	1.69*	1.68*	1.74^{ns}	1.75	1.78	1.76	1.71	1.77
Standard error	0.024	0.023	0.025	0.023	0.033	0.031	0.03	0.049	0.022
Whole-blood viscosity mPa.s	9.13**	11.05^{ns}	10.01*	12.63^{ns}	11.24	10.62	12.48	10.49	11.9
Standard error	0.375	1.23	1.17	0.88	0.79	0.717	1.83	0.837	1.210

Comparison of 30 'users' with the 'non-users' at each stage of cycle. ns = $p>0.05$ * = $p<0.05$ ** = $p<0.01$

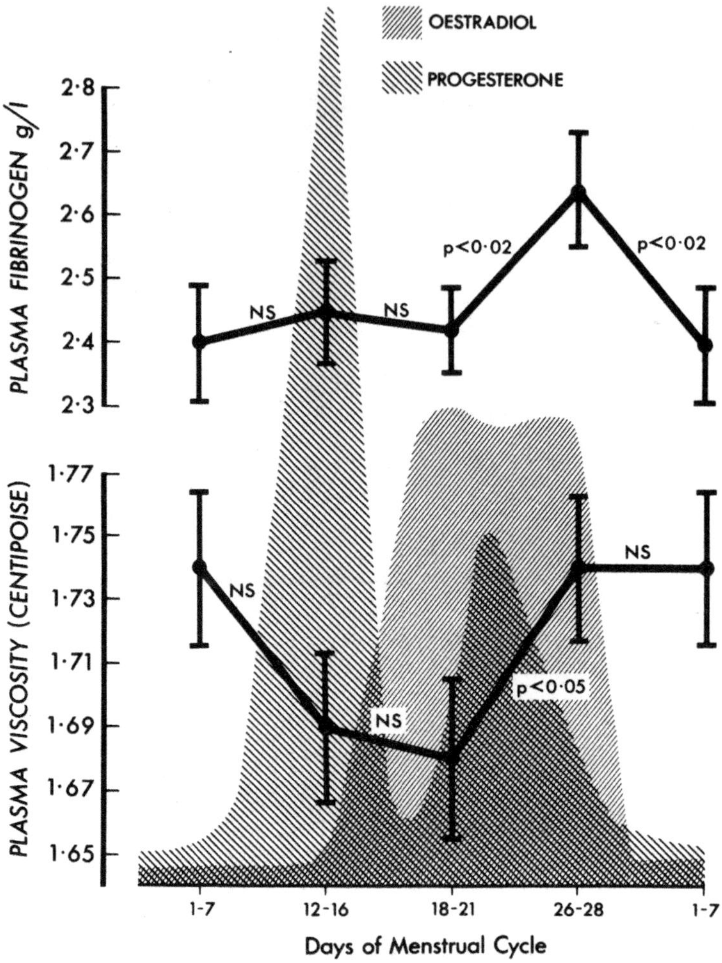

Fig.16.1. Variations in plasma fibrinogen and plasma viscosity during the menstrual cycle (mean ± SE, n = 12). 1 centipoise = 1 mPa.s

fibrinogen which is regarded as one of the main determinants of plasma viscosity. Surprisingly, however, plasma viscosity does not follow the fibrinogen level downwards during menstruation nor do these two parameters follow similar trends during the follicular phase of the cycle.

16.3.3 Erythrocyte Deformability

Merke and Lundh (1976) observed that erythrocyte deformability was lower in the luteal than in the follicular phase of the cycle. Although confirming a decrease in the luteal phase, our results suggest that this is reaction from post-ovulatory rise rather than continuation of a fall from higher pre-ovulatory levels (Fig. 16.2). The recorded methodology from the earlier study does not permit closer examination of the reasons for this dissimilarity.

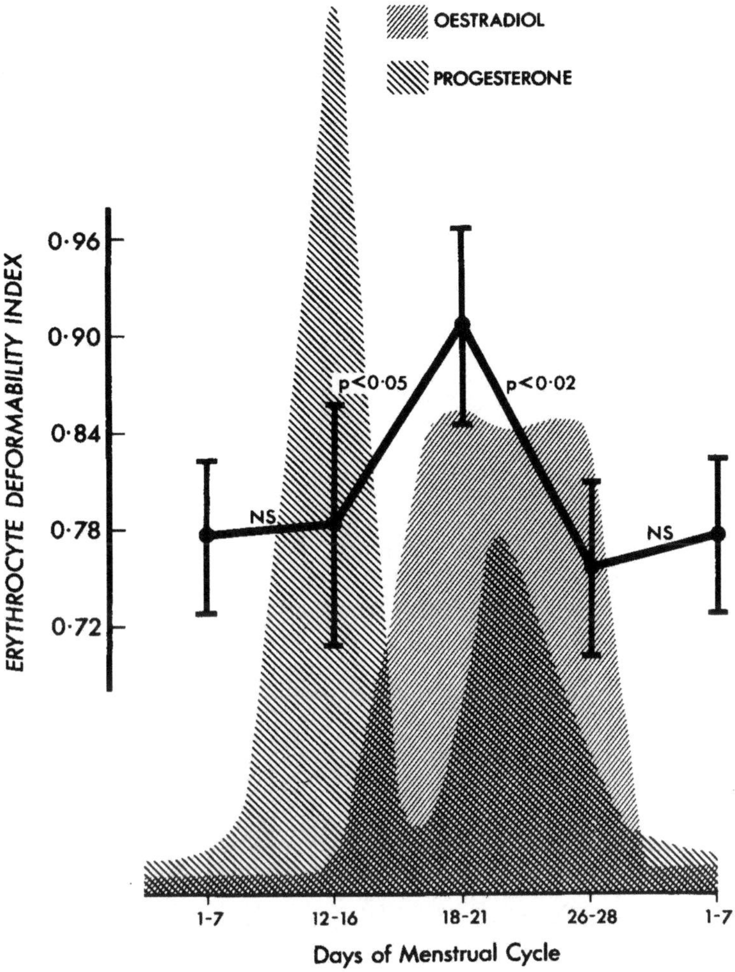

Fig.16.2. Variations in erythrocyte deformability during the menstrual cycle (mean ± SE, n = 12)

16.3.4 Packed Cell Volume

There is agreement that the packed cell volume decreases during menstruation and rises gradually to a peak at the 18th day (Vellar, 1974) or the 24th day (Danforth et al., 1946). Our data exhibit the same decrease from a pre-menstrual high to a post-menstrual low point but returns to a higher level at ovulation (Fig. 16.3) which is rather earlier than in the other two studies.

16.3.5 Whole-blood Viscosity

Dintenfass and his colleagues from serial studies of normally menstruating women (Dintenfass et al., 1966; Dintenfass and Yu, 1968) described a pre-menstrual rise which was reversed during menstruation with little subsequent change. Our findings were essentially the same (Fig. 16.3) as a pre-ovulatory peak

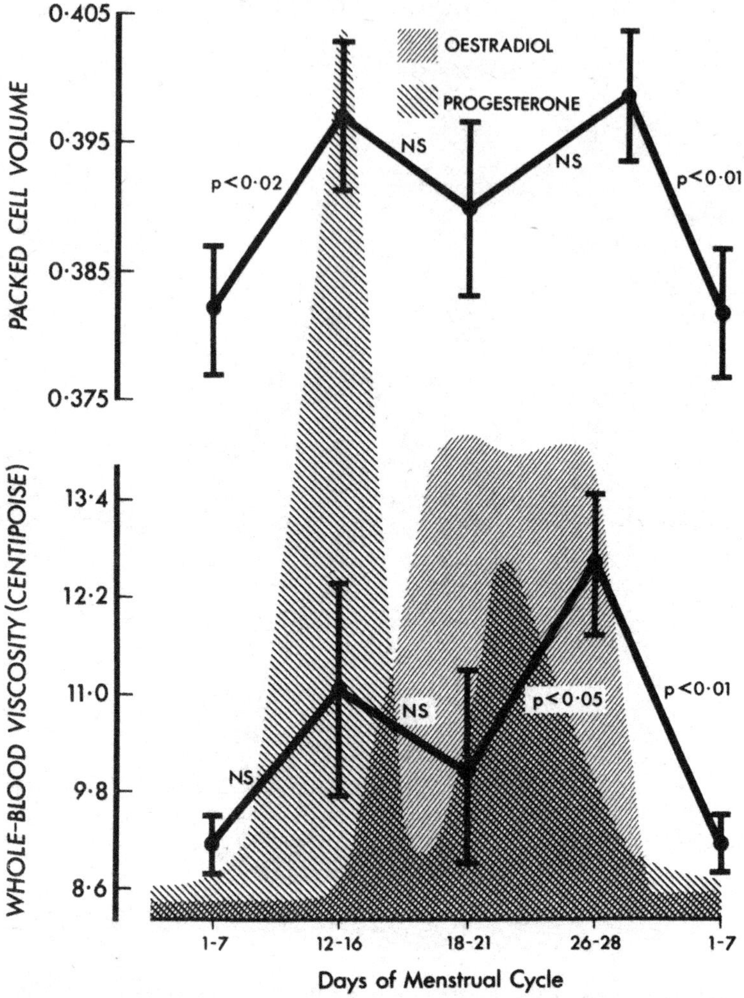

Fig.16.3. Variations in packed cell volume and whole-blood viscosity during the menstrual cycle (mean ± SE, n = 12). 1 centipoise = 1 mPa.s

proved statistically insignificant. This trend matches that of packed cell volume and plasma fibrinogen more closely than the other parameters studied.

One may conclude that menstruation coincides with a transient fall in packed cell volume which may be initiated before the onset of bleeding and is associated with decreasing whole-blood viscosity. Plasma viscosity on the other hand, although rising with fibrinogen levels during the luteal phase, does not decrease again until quite late in the ensuing follicular phase and so may moderate the extent to which whole-blood viscosity might otherwise follow packed cell volume back to pre-menstrual levels. Plasma viscosity is presumed to reflect changes in fibrinogen levels; however, in this situation it would appear to be modified significantly by additional unknown factors. Erythrocyte deformability correlates best with the combined secretion of progesterone and oestrogen and, as might be expected, clearly reciprocates the increase in whole-blood viscosity.

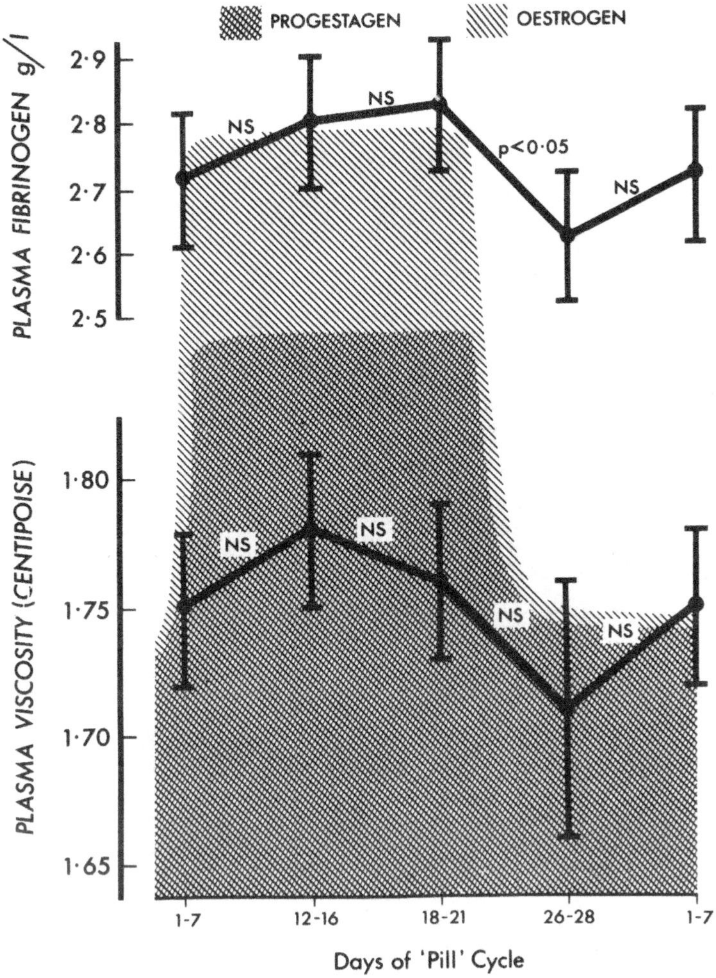

Fig.16.4. Variations in plasma fibrinogen and plasma viscosity on the contraceptive pill (mean ± SE, n = 8). 1 centipoise = 1 mPa.s

16.4 Oral Contraception

16.4.1 Plasma Fibrinogen

Poller (1978), Lowe et al. (1979), Howie et al. (1970) and several other groups all have reported an elevation of plasma fibrinogen levels associated with oral contraceptives and a similar effect is apparent in our data (Table 16.1). The only significant difference noted in the serial study was that between the high point noted after the hormonal preparation had been taken for 2 weeks and the premenstrual low point immediately after completion of the course (Fig. 16.4). The stimulatory effect of oral contraceptives was also demonstrated by a cross-sectional study of pill "users" in whom the fibrinogen level was significantly above those found at each of the first three stages of the normal menstrual study (Table 16.1).

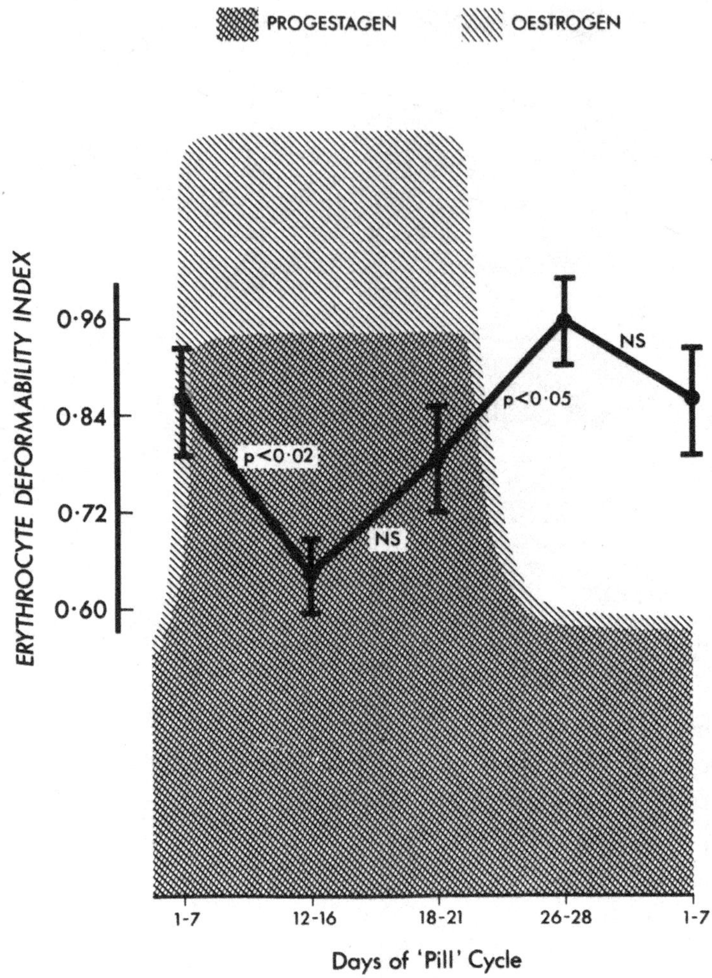

Fig.16.5. Variations in erythrocyte deformability on the contraceptive pill (mean ± SE, n = 8)

16.4.2 Plasma Viscosity

We failed to find any significant alteration in plasma viscosity from serial study of oral contraceptive users (Fig. 16.4). The level in the cross-sectional data was significantly above that noted around ovulation in the spontaneous menstrual cycles but otherwise not different. Two earlier studies of plasma viscosity during oral contraception (Leonhardt et al., 1978; Lowe et al., 1979) also found no difference from the control group. There is no study which isolates the effects of exogenous oestrogens or progesterones on plasma viscosity.

16.4.3 Erythrocyte Deformability

Our data suggest that oral contraceptives influence this parameter more than any of the others measured. The effect of the strongly progestagenic preparation

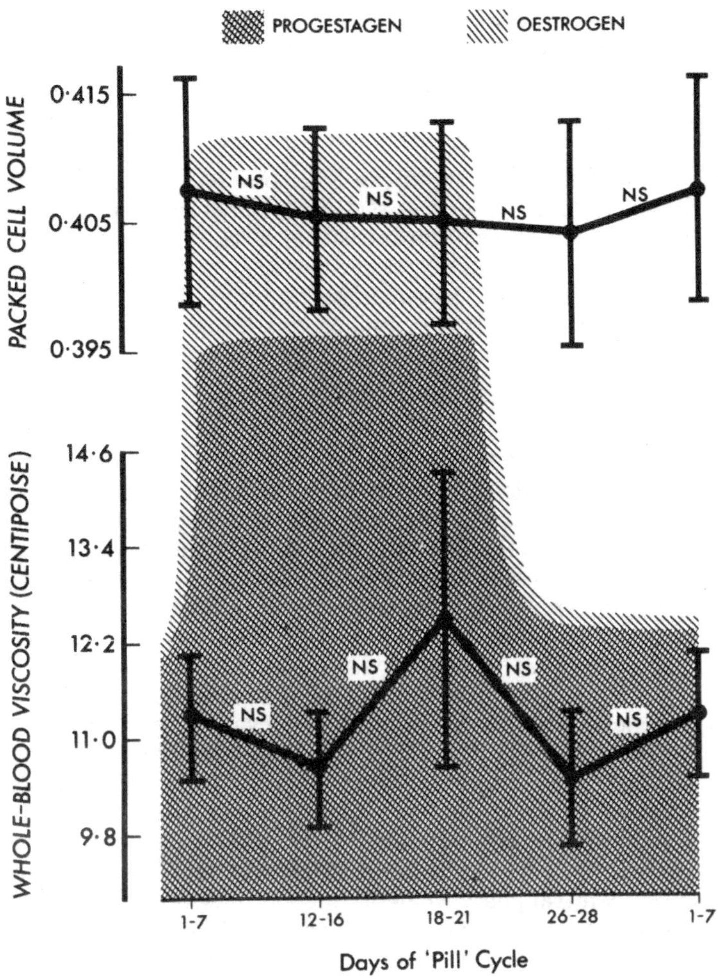

Fig.16.6. Variations in packed cell volume and whole-blood viscosity on the contraceptive pill (mean ± SE, n = 8). 1 centipoise = 1 mPa.s

studied was to increase erythrocyte deformability from a low point shortly after starting the 3-week course to a high point immediately after its completion, followed by a rapid return to the initial low level (Fig. 16.5). This conflicts with Oski et al. (1972), who reported decreased deformability whilst on a combination of 100 μg mestranol and 2 mg norethisterone, which is considerably more oestrogenic than the preparation taken by our subjects. In contrast Durocher et al. (1975) and Leonhardt et al. (1978) examined groups of women taking various oestrogen-progesterone preparations and found no difference from the controls. The level of erythrocyte deformability found in our cross-sectional survey of pill users was significantly less than the luteal phase peak noted in the normal menstrual cycle but not different from the levels at other stages. It may well be that in considering the effect of oral contraception on erythrocyte deformability, account must be taken of the relative oestrogen and progesterone balance and of the effect these hormones may have on fibrinogen and other plasma proteins which affect deformability.

16.4.4 Packed Cell Volume

Packed cell volume did not alter significantly in our serial study of pill users (Fig. 16.6) but the level found in the cross-sectional study (Table 16.1) was significantly above that seen at any stage of the menstrual cycle. This association of oral contraception and elevated packed cell volume has been reported (Aronson et al., 1971; Lowe et al., 1979) and could be due in part or whole to reduced menstrual loss or a stimulatory effect upon erythrocyte production.

16.4.5 Whole-blood Viscosity

The combined effect of the changes in these several rheological parameters was an unvarying level of whole-blood viscosity in the serial study of pill users (Fig. 16.6). The whole-blood viscosity in the cross-sectional study of pill users was significantly above the early follicular and early luteal phase troughs in whole-blood viscosity seen in normally menstruating women, but this is difficult to interpret as these stages represent rather different hormonal environments. Our data, therefore, support Aronson et al. (1971), Leonhardt et al. (1978) and Lowe et al. (1979) to the extent that whole-blood viscosity in pill users is elevated in comparison with certain stages of the normal menstrual cycle.

In summary it appears that plasma viscosity is elevated slightly by a strongly progestagenic combined pill in comparison to the oestrogen-dominated follicular phase of the cycle — largely, it seems, as a result of higher levels of fibrinogen. Despite the rise in plasma viscosity and packed cell volume, whole-blood viscosity changes little during oral contraception and differs significantly from the normal menstrual cycle only in the early follicular phase, when packed cell volume is at an ebb, and in the early luteal phase when plasma viscosity is at its lowest. Erythrocyte deformability appears to increase on ingestion of oestrogen and a strong progestagen, thus reflecting the physiological change seen in the luteal phase of the cycle. The inverse relationship between erythrocyte deformability and whole-blood viscosity is less striking in the serial study of pill users, but when the cross-sectional data are compared with those for the normal menstrual cycle (Table 16.1) there is suggestion of a reciprocity in the luteal phase. Despite these alterations in rheological parameters during oral contraception, none of the values fall outside the range seen in the normal menstrual cycle.

6.5 Pregnancy

The patients reported in this study were 27 primigravidae and 13 multiparae, aged 22 to 30 years, of whom 12 regularly smoked more than 10 cigarettes per day. None had any serious medical illness or was on any regular medication, apart from the routine ferrous sulphate and folic acid prescribed at the antenatal clinic. Blood samples were taken at regular intervals throughout pregnancy, each sample being taken between 9 and 11 a.m. Statistical analysis was by means of a paired t-test on the mean differences between compared sets of data.

Table 16.2. Plasma fibrinogen levels during pregnancy

Author		Mean plasma fibrinogen (g/litre)												
			Weeks of pregnancy											
		Six weeks post-partum	12	16	20	24	28	30	32	34	36	38	40	40+
Phillips and Skrodelis, 1958					3.2	3.5	3.3		3.6		3.8		4.0	
Talbert and Langdell, 1964			5.0		4.5		4.4			4.9	5.1		5.5	
Bonnar et al, 1969		2.41	3.6		3.8	3.9		4.25			4.45		4.5	
Condie and Ogston, 1976		3.85	3.38		3.41			4.24				4.63		
Buchan and MacDonald 1980	Mean	2.45	3.25	3.16	3.33	3.45	3.47	3.47	3.68	3.66	3.88	3.73	3.5	3.2
	SD	0.2	0.44	0.21	0.39	0.34	0.43	0.52	0.36	0.46	0.32	0.3	0.51	0.42
	n	30	27	10	32	27	26	13	14	20	20	12	15	4

16.5.1 Plasma Fibrinogen

There is a general agreement that plasma fibrinogen levels increase during pregnancy (Table 16.2). Our own study demonstrates this trend with a steady rise which is significant by 12 weeks ($p<0.001$) and maximal at 36 weeks; the subsequent decline is not statistically significant.

16.5.2 Plasma Viscosity

Only two cross-sectional studies have been reported (Table 16.3). Hamilton (1950), using an Ostwald viscometer, found plasma viscosity consistently reduced during pregnancy in contrast to Eastham (1965), whose measurements with a Harkness viscometer suggested an increase. Our experience indicates an unequivocal rise ($p<0.001$) in plasma viscosity as early as 12 weeks and a subsequent steady decline to the non-pregnant level at 38 weeks and even lower by term.

16.5.3 Erythrocyte Deformability

The literature is confused as regards the effect of pregnancy on erythrocyte deformability, for there are reports of no change (Murphy, 1968), a decrease (Durocher et al., 1975) and most recently an increase in erythrocyte filterability (Buchan, 1980). All these investigators used microfiltration methods, but Murphy and Durocher et al. used washed erythrocytes suspended in artificial media, whereas Buchan measured deformability of erythrocytes in their native plasma, which is clearly more physiological. The difference in technique may explain the discrepancy, as pregnancy causes an increase in the plasma levels of globulin and fibrinogen, both of which are known to affect erythrocyte deformability (Rampling and Sirs, 1977). The serial measurements of erythrocyte deformability in our study (Table 16.4) demonstrate an increase in deformability which is established by 12 weeks ($p<0.05$) and increases to a maximum around 34 weeks ($p<0.01$); despite subsequent decrease, the level at term remains significantly above the non-pregnant value ($p<0.05$).

16.5.4 Packed Cell Volume

The tendency to "physiological" anaemia in pregnancy is well recognised (Table 16.5) and appears in our study as a steady decline in packed cell volume which is significant by 12 weeks ($p<0.05$) and continues to reach a minimum around 24 weeks ($p<0.001$), to be followed by a slow increase until term.

16.5.5 Whole-blood Viscosity

All reported studies suggest that whole-blood viscosity is decreased in pregnant women (Pellisier, 1912; Esiaschwili, 1933; Cohen and Thomson, 1936; Hamilton, 1950). The two most comprehensive studies are those of Cohen and Thomson (1936), who derived whole-blood viscosity from the haematocrit with the formula described by Nygaard et al. (1935) and Hamilton (1950), who made direct measurements with an Ostwald viscometer. Their results are compared with those from the present study in Table 16.6. Although quantitatively different, due to measurement at a different shear rate, our data follow the same pattern of

Table 16.3. Plasma viscosity during pregnancy

Author		Non-pregnant	Mean plasma viscosity (mPa.s) (Weeks of pregnancy)										
			(6-9)	(10-13)	(14-17)	18-21)	(22-25)	(26-29)	(30-33)	(34-37)	(38-40)		
Hamilton, 1950		1.74	1.59	1.61	1.62	1.58	1.63	1.63	1.66	1.69	1.68		
				(9-19)			(20-32)			(33-40)			
Eastham, 1965		1.6		1.66			1.64			1.66			
		Six weeks post-partum	(12)	(16)	(20)	(24)	(28)	(32)	(34)	(36)	(38)	(40)	(40+)
Buchan and Macdonald, 1980	Mean	1.71	1.99	1.88	1.94	1.81	1.81	1.8	1.77	1.71	1.75	1.67	1.69
	SD	0.09	0.21	0.11	0.27	0.17	0.27	0.12	0.12	0.13	0.19	0.18	0.12
	n	30	34	8	33	26	27	14	13	15	16	15	4

Table 16.4. Erythrocyte deformability during pregnancy

	Erythrocyte deformability index (ml/min) Six weeks post partum	Weeks of pregnancy 12	16	20	24	28	32	34	36	40	40+
Mean	0.83	0.98	1.04	1.08	1.09	1.13	1.15	1.18	1.12	1.06	0.98
SD	0.11	0.12	0.12	0.13	0.13	0.14	0.14	0.13	0.14	0.11	0.1
n	30	30	8	33	25	28	16	12	14	15	4

decrease, which is significant by 16 weeks ($p<0.01$) and continues until 34 weeks, when the trend reverses so that the level of whole-blood viscosity at 38 to 40 weeks is not significantly different from the non-pregnant level.

16.6 Conclusions

The rises in plasma fibrinogen and viscosity seen in the luteal phase of the menstrual cycle appear to continue throughout the first trimester of pregnancy. Plasma viscosity then decreases to less than the non-pregnant level by term, in contrast to fibrinogen, which peaks at 36 weeks. Thus a rising fibrinogen level appears to have more immediate effect on plasma viscosity than does a decrease. Possibly there is a point beyond which rising plasma viscosity triggers additional mechanisms for its control. As increases appear restricted to the luteal phase and pregnancy, progesterone, superimposed on oestrogen, seems the important hormonal stimulus to plasma fibrinogen and consequently viscosity. The levels of each of these parameters in patients on the more strongly progestagenic contraceptive pill were in general above that found in the normal menstrual cycle (Table 16.1).

Changing levels of fibrinogen might also be expected to affect erythrocyte deformability and these two parameters do indeed follow similar patterns of change in the luteal phase of the normal cycle and in pregnancy. During the menstrual cycle, however, the peak for erythrocyte deformability anticipates that of fibrinogen by almost 1 week. Similarly the increase occurring whilst on the progestagenic pill is reversed swiftly on its cessation, whereas the rise in fibrinogen levels is slower, possibly even delayed, although extended observations are needed to establish this. One might therefore suspect that erythrocyte deformability increases not as a direct result of elevated fibrinogen levels but rather as the swifter response to a common causative factor as, for example, progesterone. Whatever the reason for increased erythrocyte deformability, there seems no doubt of its effect in reducing whole-blood viscosity, as changes in each tend to be mirror images especially in the luteal phase and during pregnancy. It is intriguing that increasing concentrations of progesterone may provoke more viscous plasma but simultaneously increase erythrocyte deformability, thus reducing whole-blood viscosity.

Packed cell volume is an even more important determinant of whole-blood viscosity than erythrocyte deformability. From the pre- to post-menstrual phase and throughout pregnancy, whole-blood viscosity and packed cell volume behave almost identically. Both tend to be higher in pill users than at any stage of the normal menstrual cycle expect pre-menstrually, when erythrocyte deformability is low and plasma fibrinogen and viscosity are at a peak. The precise reason for changing packed cell volume will vary even in the situations presently studied. Certainly reduced menstrual loss is a factor in the high packed cell volume associated with oral contraception. A disproportionate increase in plasma volume dictates most of the decrease in packed cell volume seen in pregnancy, and similar factors may influence the cyclical fluctuations of the normal menstrual cycle. The rheological changes consequent upon regular oral contraception are quantitatively within the range found in the normal menstrual cycle. Thrombogenic sequelae of the pill cannot therefore be attributed to unusual levels of viscosity, but loss of cyclical fluctuation associated with spontaneous ovarian activity could play a part. The rheological changes following conception also display this loss of

Table 16.5. Packed cell volume during pregnancy

Author		Mean packed cell volume											
		Six weeks post-partum	Weeks of pregnancy										
			12	16	20	24	28	32	34	36	38	40	40+
Rath et al, 1950			0.388	0.372	0.37	0.367	0.372	0.337		0.37		0.379	
Lund, 1951			0.4	0.39	0.37	0.355	0.352	0.342		0.344		0.357	
Benstead and Theobald, 1952				0.398	0.383	0.371	0.355	0.346		0.355		0.368	
Edgar and Rice, 1956				0.378	0.362	0.344	0.342	0.33		0.33		0.363	
	Mean	0.394	0.371	0.366	0.351	0.346	0.348	0.348	0.356	0.354	0.359	0.356	0.378
Buchan and Macdonald, 1980	SD	0.018	0.018	0.018	0.018	0.024	0.016	0.018	0.026	0.026	0.029	0.026	0.042
	n	30	26	11	33	27	26	14	20	21	12	17	4

Table 16.6. Whole-blood viscosity during pregnancy

Author		Non-pregnant	Whole-blood viscosity (mPa.s) (Weeks of pregnancy)										
				(9-12)	(13-16)	(17-20)	(21-24)	(25-28)	(29-32)	(33-36)	(37-40)		
Cohen and Thomson, 1936		4.8		4.68	4.46	4.33	4.31	4.26	4.21	4.2	4.42		
			(6-9)	(10-13)	(14-17)	(18-21)	(22-25)	(26-29)	(30-33)	(34-37)	(38-40)		
Hamilton, 1950		4.61	4.2	4.19	4.11	3.92	3.84	3.95	4.32	4.16	4.49		
		Six weeks post-partum	(12)	(16)	(20)	(24)	(28)	(32)	(34)	(36)	(38)	(40)	(40+)
Buchan and Macdonald, 1980	Mean	10.82	9.95	8.93	9.46	9.06	8.25	8.37	8.04	8.7	10.14	9.42	9.87
	SD	2.58	2.65	1.37	2.63	4.27	2.69	2.92	2.09	2.27	3.96	3.41	3.07
	n	30	30	8	33	25	28	16	12	14	16	15	4

fluctuation and this may contribute to the increased incidence of thrombo-embolic disease in pregnancy.

Even more interesting are the implications of abnormalities in rheological parameters in high-risk pregnancies. Heilman et al. (1977) reported that, compared with normal pregnancy, patients with foetal intra-uterine growth retardation and pre-eclampsia had elevated whole-blood viscosity. Matthews and Mason (1974) reported elevated plasma viscosity in patients with pre-eclampsia. Our own, as yet incomplete, data on heavy smokers and patients with intra-uterine growth retardation seem to confirm these observations. Clearly the place of hyperviscosity in abnormal pregnancy requires extensive study which, hopefully, could direct attention along previously unsuspected therapeutic avenues of value.

16.7 References

Aronson H B, Magora F, Schenker J G (1971) Effect of oral contraceptives on blood viscosity. Am J Obstet Gynecol *110:* 997-1001

Benstead N, Theobald G W (1952) Iron and the 'physiological' anaemia of pregnancy. Br Med J *1:* 407-410

Bonnar J, McNicol G P, Douglas A S (1969) Fibrinolytic enzyme system and pregnancy. Br Med J *3:* 387-389

Buchan P C, MacDonald H (1980 in press) Evaluation and modification of whole blood filtration in the measurement of erythrocyte deformability in pregnancy and the newborn. Br J Haematol

Cohen M E, Thomson K J (1936) Studies on the circulation in pregnancy. I. The velocity of blood flow and related aspects of the circulation in normal pregnant women. J Clin Invest *15:* 607-625

Condie R G, Ogston D (1976) Sequential studies on components of the haemostatic mechanism in pregnancy with particular reference to the development of pre-eclampsia. Br J Obstet Gynaecol *83:* 938-942

Danforth D N, Boyer P K, Graff S (1946) Fluctuations in weight, hematocrit and plasma protein with the menstrual cycle. Endocrinology *39:* 188-193

Dintenfass L, Yu J S (1968) Changes in the blood viscosity and the consistency of artificial thrombi in 17-year-old girls during the menstrual cycle. Med J Aust *1:* 181-183

Dintenfass L, Julian D G, Miller G (1966) Viscosity of blood in healthy young women. Lancet *i:* 234-235

Durocher J R, Weir M S, Lundblad E G, Patow W E, Conrad M E (1975) Effect of oral contraceptives and pregnancy on erythrocyte deformability and surface charge. Proc Soc Exp Biol Med *150:* 368-371

Eastham R D (1965) Plasma viscosity in the improved haematological screening of antenatal patients. J Obstet Gynaecol Br Commonw *71:* 763-764

Edgar W, Rice H M (1956) Administration of iron in antenatal clinics. Lancet *i:* 599-602

Ellis B C, Stransky A A (1961) A quick and accurate method for the determination of fibrinogen in plasma. J Lab Clin Med *58:* 477-488

Esiaschwili I (1933) Blutgerinnung und Viskosität in der Schwangerschafts und Nachgeburtsperiode. Zentralbl Gynaekol *57:* 2712-2728

Hamilton H F H (1950) Blood viscosity in pregnancy. J Obstet Gynaecol Br Emp *57:* 530-538

Heilman L, Mattheck C, Kurz E (1977) Rheologische Veranderungen des Blutes in der normalen und pathologischen Schwangerschaft und deren Einfluß auf die Sauerstoffsdiffusion. Arch Gynaekol *223:* 283-298

Howie P W, Prentice C R M, Mallinson A C, Horne C H W, McNicol G P (1970) Effect of combined oestrogen-progestogen oral contraceptives, oestrogen, and progestogen on antiplasmin and antithrombin activity. Lancet *ii:* 1329-1332

Leonhardt H, Grigoleit H G, Reinhardt I (1978) Auswirkungen von Zigarettenrauch und oralen Kontrazeptiva aud die Plasma und Vollblutviskosität. Med Welt *29:* 880-883

Lowe G D O, Reavey M M, Forbes C D, Prentice C R M (1979) Increased blood viscosity in young women taking oral contraceptives. Clin Sci Mol Med *56:* 21P

Lund C J (1951) Studies on the iron-deficiency anaemia of pregnancy, including plasma volume, total haemoglobin, erythrocyte protoporphyrin in treated and untreated normal and anaemic patients. Am J Obstet Gynecol *62:* 947-954

Matthews J D, Mason T W (1974) Plasma viscosity and pre-eclampsia. Lancet *ii:* 409

Merke C, Lundh B (1976) Erythrocyte filterability and heme catabolism during the menstrual cycle. Ann Intern Med *85:* 322-324

Murphy J R (1968) Hemoglobin CC disease:- Rheological properties of erythrocytes and abnormalities in cell water. J Clin Invest *47:* 1483-1495

Nygaard K K, Wilder M, Berkson J (1935) The relationship between viscosity of blood and the relative volume of erythrocytes (hematocrit value). Am J Physiol *114:* 128-136

Oski F A, Lubin B, Buchert E D (1972) Reduced red cell filterability with oral contraceptive agents. Ann Intern Med *77:* 417-419

Pellisier M P (1912) De la viscosité du sang total chez la femme enceinte et accouchée et de ses rapports avec la tension artérielle. Arch d'Obstet *11:* 306-320

Phillips L L, Skrodells V (1958) The fibrinolytic enzyme system in normal, hemorrhagic and disease states. J Clin Invest *37:* 965-973

Poller L (1978) Oral contraceptives, blood clotting and thrombosis. Br Med Bull *34:* 151-156

Rampling M W, Sirs J A (1977) Fibrinogen, fibrinogen degradation products and erythrocyte flexibility. Thromb Haemostas *38:* 668-676

Rath C E, Caton W, Reid D E, Finch C A, Conroy L (1950) Hematological changes and iron metabolism in normal pregnancy. Surg Gynecol Obstet *90:* 320-327

Ratnoff O, Menzie C (1951) A new method for the determination of fibrinogen in small samples of plasma. J Lab Clin Med *37:* 316-320

Seaman G V F, Engel R, Swank R L, Hissen W (1965) Circadian periodicity in some physiological parameters of circulating blood. Nature *207:* 833-835

Talbert L M, Langdell R D (1964) Normal values of certain factors in the blood clotting mechanism in pregnancy. Am J Obstet Gynecol *90:* 44-50

Turksoy R N, Phillips L L, Southam A L (1961) Influence of ovarian function on the fibrinolytic enzyme system. I. Ovulatory and anovulatory cycles. Am J Obstet Gynecol *82:* 1211-1215

Vellar O D (1974) Changes in hemoglobin concentration and hematocrit during the menstrual cycle. 1. A cross-sectional study. Acta Obstet Gynecol Scand *53:* 243-246

Vessey M P, Mann J I (1978) Female sex hormones and thrombosis. Br Med Bull *34:* 157-162

17 Blood Rheology in the Newborn

C.H.M. Walker

17.1 Introduction

The importance of disorders of microcirculation in the pathogenesis of disease in the newborn was highlighted when Chu et al. (1965) coined the synonym "The Pulmonary Hypoperfusion Syndrome" for hyaline membrane respiratory distress syndrome (RDS) of the newborn. Yet, despite the supportive evidence of others (Strang, 1966), blood rheology as such has not been the subject of much investigation in this age group, emphasis being directed more toward cardiopulmonary pathophysiology. More recently, with the improved management of RDS and the increasing survival of low-birth-weight babies in general, and with the persistence of intracranial haemorrhage as a major threat to life (Anderson et al., 1976) more attention has been paid to blood flow, especially cerebral blood flow. However, the emphasis again has been in cerebrovascular dynamics and haemostasis rather than on the red cell or the physical properties of plasma in the newborn. While not suggesting that this emphasis has been entirely misplaced, what follows is an attempt to put into perspective the numerous inter-relating factors of importance in the aetiology of various disorders in the newborn, especially those of the central nervous and respiratory systems.

17.2 Effects of Increased Blood Volume in the Newborn

Several factors work together to provide the normal baby with a high blood cell content (haematocrit ratio, Hct). Erythropoietic activity shortly before term is such that the Hct at birth is normally around 0.50. Owing to the loss of fluid from plasma to extracellular space there is haemoconcentration in the order of 2 - 6 g/dl haemoglobin (Hb) with a proportionate rise of about 20% in Hct. The early infusion of albumin has been shown to prevent this rise, presumably by maintaining oncotic pressure and plasma volume (Bland et al., 1973).

In the foetus subjected to intrapartum asphyxia there is an intrauterine transfer of blood from placenta to foetus resulting in an even higher Hct in the newborn (Chou and Ackerman, 1973; Flod and Ackerman, 1971; Philip et al., 1969; Turnbull and Walker, 1955) except when there is a tight nuchal cord (Linderkamp et al., 1978b). Two mechanisms have been proposed. First, it has been suggested that the foetus responds to hypoxia before and during labour by producing an excess of erythropoietin with a subsequent release of extra red cells into the circulation (Finne, 1966; Gross et al., 1973; Turnbull and Walker, 1955; Walker and Turnbull, 1953). Second, and equally probable, is the stress effect of asphyxia causing release of catecholamines, which are high even in the normal newborn (Messow-Zahn et al., 1978), and which cause vasoconstriction of the umbilical artery and "trapping" of extra blood within the foetus.

The metabolic needs of the normal newborn demand a resting oxygen uptake which is approximately twice (per kilogram body weight) that of the adult and it is therefore essential that oxygen-carrying power and blood circulation are adequately maintained. It was probably this concern which encouraged the vogue for allowing large volumes of blood to pass from placenta to baby before cord clamping and yet this very manoeuvre can place the infant's microcirculation in danger (Yao and Lind, 19754). The advice presented in the literature has been conflicting, not only about the passage of blood from placenta to newborn, but also concerning the value of post-natal donation of blood or colloid to the pre-term baby (Bland et al., 1973; Secher and Karlberg, 1962), following which a strikingly hyperdynamic state may be induced (Oh et al., 1966). It is well known that the baby's circulation can accommodate a sudden infusion of up to one-third of the blood volume without producing cardiac failure. However, the micro-vasculature immediately after birth is both more fragile (Walker and Balf, 1954) and more permeable (Secher and Karlberg, 1962; Strang, 1966) than later in life. This is particularly true of the pre-term infant, in whom the capillary network is highly susceptible to injury, both traumatic and hypoxic.

The greater the placental transfer of blood, the greater are the chances of severe haemoconcentration, since it has been shown that the number of capillary fenestrations in the microcirculation in babies who have had a large placental transfer of blood is considerably greater than those who have not (Pietra et al., 1968) so transcapillary fluid transfer is enhanced. Increase in pulmonary and systemic vascular resistance (Fouron and Hebert, 1973) and raised venous pressure may occur with subsequent increase in the escape of albumin (Friedman, 1976), a phenomenon also noted following birth asphyxia (Ingomar and Klebe, 1974) which could explain why the early infusion of albumin seems beneficial (Bland et al., 1973).

The clinical significance of this was comprehensively reviewed by Moss and Monset-Couchard (1967), and it is now accepted that there may be reduced pulmonary function as measured by lung compliance, functional residual capacity (Oh et al., 1967) and blood gases (Ingomar and Klebe, 1973), and even breast feeding may be impaired (Cort and Pribylova, 1964). It has been a mistake, therefore, to assume that foetal red cells behave like adult cells, to imagine that the autoregulation of blood flow is always efficient in the newborn and to expect all babies to benefit from such transfusions of blood.

17.3 Foetal Blood

17.3.1 The Foetal Red Cell

Since the loss of intravascular fluid can cause an increase of up to 20% in Hct within a few hours of birth (Usher et al., 1963),. the microcirculation becomes even more dependent on the deformability of the red cell. It might seem a little surprising, therefore, that just at this time the foetal red cells have reduced capacity for deformation and that their reduced flexibility can cause the vascular resistance to rise even in the absence of significant change in Hct (Greene et al., 1973). Gross and Hathaway (1972) produced evidence to suggest that no less than four different cellular factors could reduce the deformability (and filterability) of the foetal cell. These include:

a) Reduced stability of adenosine triphosphate (ATP), (ATP being an important factor in the maintenance of shape and flexibility of the red cell) together with diminution of ouabain-sensitive ATPase and phosphate incorporation.

b) Changes in cell membrane lipid diminished ability of the foetal red cell to tolerate oxidant stress.

c) Abnormalities in water and ion transport and in the action of the cation pump.

d) Structural differences such as microcytosis, pyknocytosis, the presence of Heinz bodies and pitting of the red cell surface.

Foetal red cells vary greatly in size and shape. On the whole they are larger and more spherical than adult cells (Turnbull and Walker, 1955). Fortunately this does not appear to reduce their overall filterability; in fact Weed et al. (1969) have shown that ATP-depleted cells get less filterable as they become smaller.

17.3.2 The Plasma and Plasma Factors Which Influence Red Cells

The evidence to date would suggest that plasma factors other than mere volume relative to red-cell number probably play a part in determining the adequacy of blood flow. Foetal cells have a life which is approximately two-thirds that of adult cells and the more pre-term the infant is, the shorter it becomes. The fact that placing foetal cells in adult plasma does not appear to prolong their life (Bratteby et al., 1968) suggests that cell life and possibly other factors which determine cell function are governed by inherent factors, but the observation that adult cells incubated in premature infant plasma may become pyknotic (Oski and Naiman, 1965) suggests that a plasma factor does exist.

It has been known for some time that α-tocopherol deficiency in the newborn alters red-cell membrane properties (McWhirter, 1975; Oski and Barness, 1967) and hyperbilirubinaemia may also affect deformability (Powell et al., 1968). The levels of plasma proteins, especially fibrinogen and IgM, are likely to be of

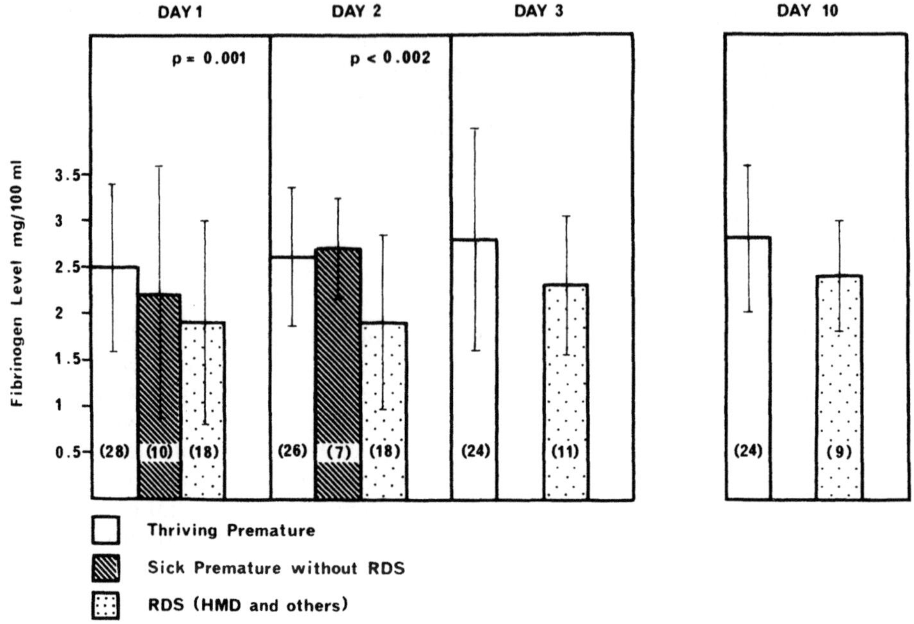

Fig.17.1. Fibrinogen levels in premature infants during first 10 days of life. (mean ± 1 SD; ( ) number of cases) *RDS:* Respiratory distress syndrome. *HMD:* Hyaline membrane disease. (Courtesy of Jensen et al., 1973)

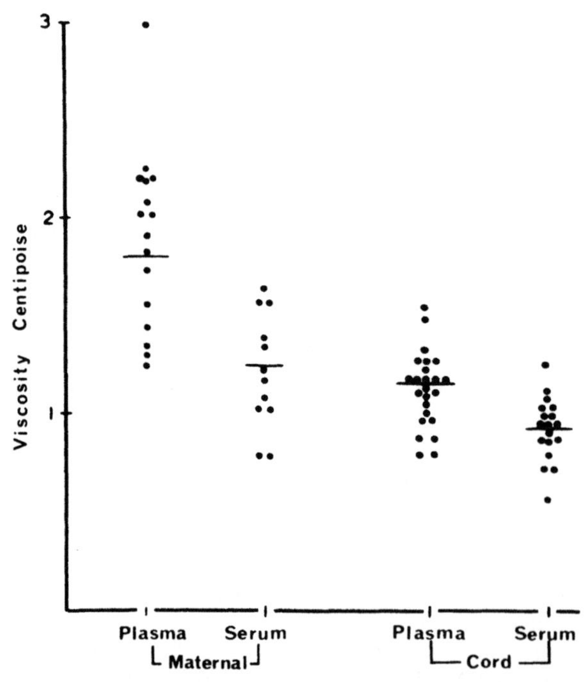

Fig.17.2. Viscosity (shear rate 1.993 s^{-1}) of maternal and cord plasma and serum. Mean values shown. (Courtesy of Foley et al., 1978) 1 centipoise = 1 mPa.s

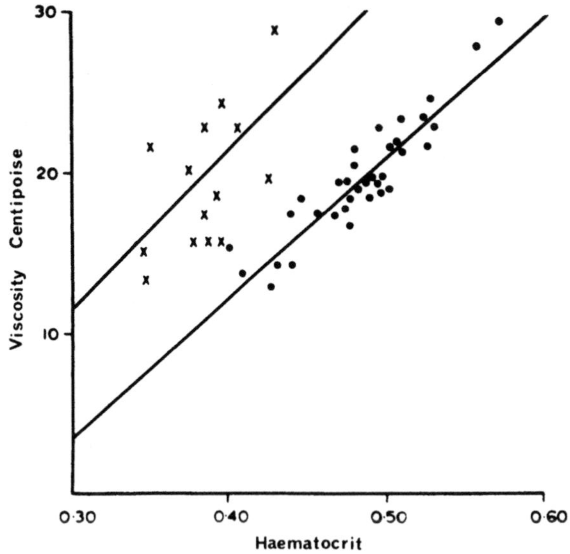

Fig. 17.3. Viscosity (shear rate 1.993 s^{-1}) and haematocrit in maternal (x) and cord (●) blood. (Courtesy of Foley et al., 1978) 1 centipoise = 1 mPa.s

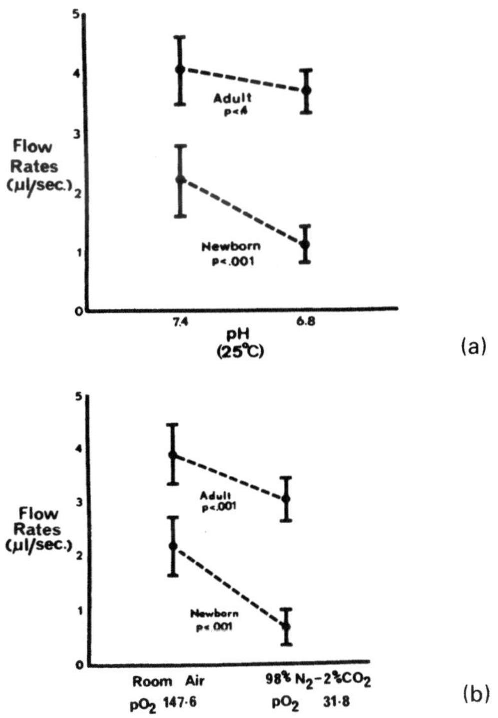

Fig. 17.4. Decreased filtration flow rate of 10 subjects: comparison of adult and foetal cells under conditions of (a) acidosis and (b) hypoxia (mean ± 1 SD). (Courtesy of Gross and Hathaway, 1972)

importance since they are low, especially in the ill pre-term baby (Jensen et al., 1973) (Fig. 17.1). This has the effect of lowering plasma viscosity, both plasma and serum viscosity in cord blood being lower than that in maternal blood (Fig. 17.2), but this may also reduce deformability of the red cell as it does in adults. However, Foley et al. (1978) found cord whole-blood viscosity to be the same as that of the mother despite the higher Hct (Fig. 17.3). Newborn whole-blood viscosity is higher than that of the mother, even in the face of a lower plasma viscosity (Gaehtgens and Schickendantz, 1975). Equally or even more important are the common changes in blood gases at birth. All babies are born with a mixed metabolic and respiratory acidosis resulting in an average blood pH of about 7.25. Most babies can sustain this well and correct it within a few hours, but those who have sustained significant asphyxia or who have failed to establish adequate ventilation after birth are greatly at risk of developing levels of $p0_2$ and pH, which Gross and Hathaway (1972) have shown can reduce red-cell filterability considerably (Fig. 17.4). They discuss further the evidence which suggests that ATP is again likely to be involved since it is known that ATP binding to Hb is increased and glutathione stability reduced in a low $p0_2$ environment.

17.4 The Neonatal Hyperviscosity Syndrome and Associated Conditions

There is now abundant evidence that the viscosity of blood of the newborn rises dramatically when the Hct exceeds 0.65 (Baum, 1966; Kontras et al., 1970; MacIntosh and Walker, 1973); while Linderkamp et al. (1978a) have confirmed the importance of blood volume and viscosity in the regulation of blood flow it is uncertain to what extent hypoperfusion due to hyperviscosity contributes to the signs and symptoms frequently presenting in the first few days of life. It is relevant, however, that Bergqvist and Zetterström (1975) found normal blood flow in the resting newborn even with high Hct, but under maximal demand circumstances both flow and capillary filtration coefficient were indeed related to Hct.

The list of babies prone to this syndrome (plethora, cyanosis, respiratory distress, cardiomegaly and central nervous system signs) is extensive (Kontras, 1972; Ongkie, 1974; Wessenberg, 1978). It includes those who are pre-term (Bergqvist and Zetterström, 1974); post-term or small for gestational age (Fig. 17.5; Bergqvist, 1974; Humbert et al., 1969; Walker and MacIntosh, 1972); those who have suffered from asphyxia (Dunn, 1970); those whose mothers have diabetes (Klebe and Ingomar, 1974), pre-eclamptic toxaemia, or placenta praevia; and any baby of any weight or gestational age whose cord was clamped late or who for some reason received a large transfusion from mother or co-twin. To these must be added babies with inherent red-cell abnormalities (sickle-cell disease) or plasma abnormalities which in themselves increase viscosity. Cardio-respiratory distress (Danks and Stevens 1964; Gatti et al., 1966; Oh et al., 1967), cardiomegaly and pulmonary congestion (Lanning et al., 1977), disturbed renal function (Aperia et al., 1974), hypoglycaemia and hyperbilirubinaemia have all been asociated with Hct. More recently, hyperviscosity has been reported as an associated feature of other neonatal conditions, e.g. necrotising enterocolitis (Hakanson and Oh, 1977; Leake et al., 1975), neonatal thyrotoxicosis (Bussman et al., 1977), gangrene (Papageorgiou and Stern, 1972), Down's syndrome (Weinberger and Oleinick, 1970), hyperplastic visceromegaly (Beckwith syndrome); in the older child with Duchenne muscular dystrophy (Tillman et al.,

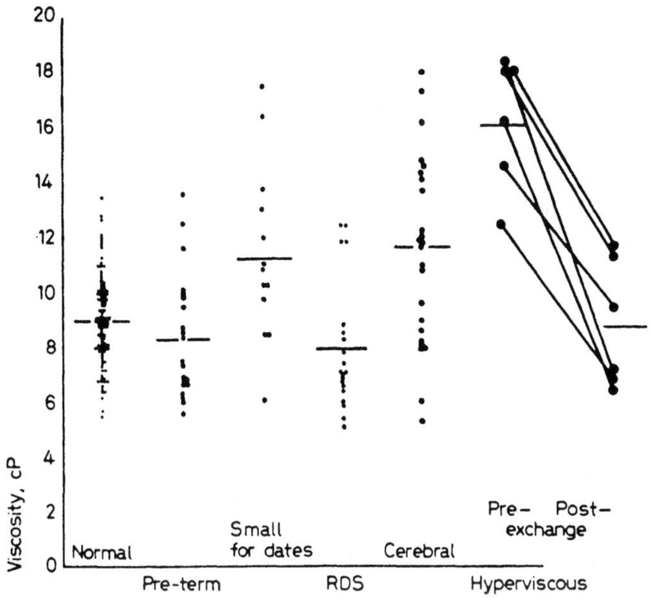

Fig.17.5. Viscosity values (shear rate 23 s^{-1}) in various groups of infants (RDS: Respiratory distress syndrome); and effects of haemodilution on viscosity. 1 centipoise (cP) = 1 mPa.s

1974); and with the genesis of innocent cardiac murmurs (Sabbah et al., 1979). Coagulation defects (Rivers, 1975) and even disseminated intravascular coagulation (DIC) (Kontras et al., 1970) have been attributed to low blood flow with "sludging". One wonders how many more related conditions would be found if one were to look.

17.5 The Relationship Between Hyperviscosity and Other Factors Influencing Blood Flow in the Newborn

17.5.1 Cerebral Blood Flow

A great deal of work has been done in recent years in an attempt to establish the relationship between intraventricular cerebral haemorrhage and cerebral blood flow and pressure, blood gases, and the much debated role of bicarbonate therapy. This work has been ably presented by Pape and Wigglesworth (1979) but a striking feature of their publication is the scant reference made to the possible influence of hyperviscosity. However, they do emphasise several points which could either counteract the effects of hyperviscosity or enhance them.

There are those who might argue that the cerebral signs in the newborn which follow asphyxia are the result of the hypoxia itself and that there is no need to invoke hyperviscosity to explain them, but Karlsson et al. (1974) have shown that in states of hypoxia cerebral oxygen consumption is directly dependent on blood flow, which hyperviscosity will surely impair. Moreover, babies who have not been subjected to asphyxia and who have high Hct (and hyperviscosity) also present "cerebral" signs which often resolve quite dramatically following haemodilution (Gross et al., 1973; Ramamurthy et al., 1975; Walker and MacIntosh, 1972).

If asphyxia has been severe enough to impair myocardial function to the point of cerebral hypotension then any accompanying hyperviscosity must further jeopardize microcirculation. Moreover, when there is a tendency to red-cell aggregation, such as occurs in polycythaemia, it would appear that satisfactory maintenance of blood volume and blood pressure does not guarantee adequate perfusion of tissues (Bicher et al., 1971). On the other hand the hypercapnea and acidosis of mild to moderate hypoxia is known to cause an increase in cerebral blood flow (Fig. 17.6) (Betz, 1972; Cooke et al., 1979; Lassen, 1974), thus possibly counteracting the effects of hyperviscosity. No evidence could be found to indicate whether or not very high Hct blood would actually prevent this increase in cerebral blood flow but this must be a distinct possibility since, as mentioned above, hypoxia and acidosis have been shown to reduce the deformability of foetal red cells (Gross and Hathaway, 1972).

It is interesting to speculate as to which factors are responsible for the quite consistent rise in intracranial pressure which occurs in pre-term babies in the first 24 h of life (Donn and Philip, 1978) just when the haematocrit is rising. Even in term infants the apparent cerebral blood flow falls during the first few hours of life as the Hct is rising (Fig. 17.7) (Cooke et al., 1979). Since not all areas of the brain receive the same proportion of total blood flow at all times (Weiss and Edelman, 1976) and since auto-regulation of cerebral flow is defective or absent in the ill newborn (Lou et al., 1979; Volpe, 1976), there is no way of knowing whether or not flow to the vulnerable areas is adequate.

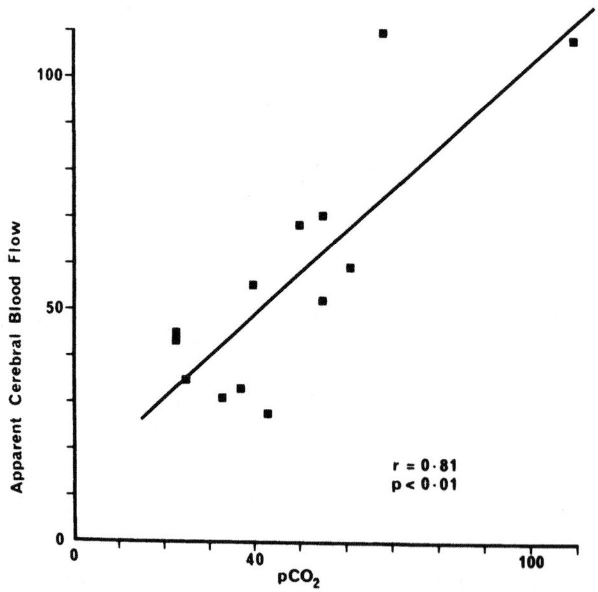

Fig.17.6. Effect of pCO_2 on apparent cerebral blood flow in 13 pre-term infants with hyaline membrane disease. (Courtesy of Cooke et al., 1979)

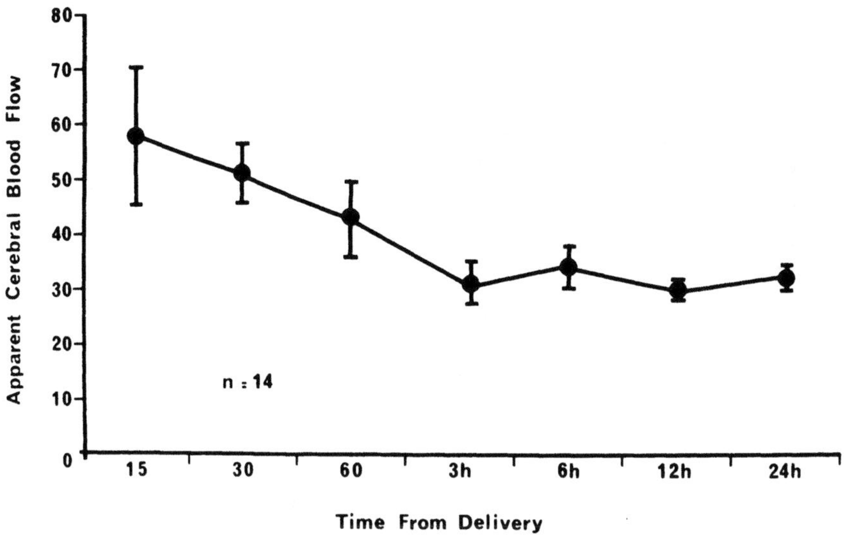

Fig.17.7. Mean apparent cerebral blood flow following delivery in 14 term newborn infants. (Courtesy of Cooke et al., 1979)

The hypoxic stimulus to erythropoiesis (Finne, 1966) and the accompanying tendency to thrombosis seen in infants with cyanotic heart defects (Kontras et al., 1970) is an excellent example of how polycythaemia as such can encourage vascular occlusion in infants and young children. There is some doubt as to whether the lower levels of coagulation factors seen in these conditions are technical artefacts (Pearson, 1976) but in any event cerebral thrombosis is a recognised complication in which hyperviscosity must again surely play a part.

Where hyperviscosity may also be of significance is in the production of less dramatic microcirculatory lesions in the brain in the severely "small-for-gestational-age" baby. Although extremely difficult to prove, it is not impossible that the long-term morbidity seen in these babies (Gross et al., 1973) is due in part to ischaemia caused by occlusion of the microcirculation in the neonatal period or even *in utero*.

17.5.2 Pulmonary Blood Flow

Marked impairment of respiratory function with distress (Fig. 17.8) can occur in normal full-term infants who have no aetiological factor other than a very high Hct (Danks and Stevens, 1964; Oh et al., 1967) and many regard pulmonary hypotension as an aetiological factor just as important as deficiency of surfactant in the early stages of RDS (Chu et al., 1965).

Once again asphyxia, with its accompanying higher Hct, is known to be a primary aetiological factor in RDS in which intravascular microaggregation is a recognised histological finding. This in turn causes diminished effective pulmonary blood flow, even in term infants, and could be of considerable importance as a cause of intra-alveolar transudation of fluid and secondary surfactant deficiency (Hutchison and Russell, 1976) in the development of the syndrome.

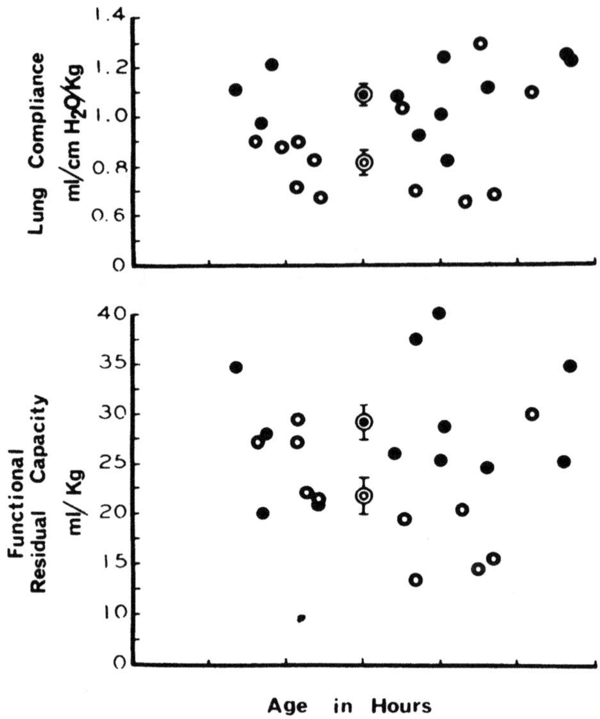

Fig. 17.8. Lung compliance and functional residual capacity in 12 infants whose cords were clamped late (○) and 11 clamped early (●) studied in the first 6 h of life. (Courtesy of Oh et al., 1967)

17.5.3 Renal Blood Flow

It might be expected that renal function would be impaired by reduced blood flow associated with high Hct and this has been well documented by Aperia et al. (1974). The effect of this upon an already overloaded circulation in a baby with, for example, a late clamped cord could be considerable, but any fluid retention could, paradoxically, be helpful.

17.5.4 Haemostasis

Blood flow in the umbilical cord is, in a sense, "protected" by the presence, especially in the artery, of high levels of activity of both plasminogen activator and plasmin, and of low levels of antiplasmin (Foley et al., 1977; Jurgens et al., 1979). Thus the chances of fibrin occlusion of flow during birth are reduced. These initially high levels of activator are apparently matched by appropriate levels of inhibitor (Ekelund et al., 1970) so it is uncertain to what extent there is increased fibrinolytic activity at birth. However, after birth the tendency for the blood to resist clotting does not last long since antiplasmin levels are normal or slightly elevated and plasminogen levels themselves are low (42% if adult values) (Ekelund et al., 1970) especially in the pre-term baby. It has been suggested that hyaline membrane disease might be prevented by the early administration of plasminogen (Ambrus et al., 1972).

Significant changes in coagulation factors have been found in the newborn with high Hct (Rivers, 1975) and it has been suggested that low blood flow and "sludging" may actually cause DIC (Kontras et al., 1970). It has also been suggested that some infants exhibit a frankly hyper-coagulable state immediately after birth (von Kaulla et al., 1965; Markarian et al., 1971), and proceed rapidly to a consumption coagulopathy. Whether or not this is so, the evidence is in favour of there being increased fibrinolytic activity in most babies at and immediately after birth (Foley et al., 1977), which is less evident in the pre-term, and least in pre-term infants with RDS. This would favour coagulation in those babies who are perhaps already prone to vascular occlusion due to hypotension and to the effects of asphyxia, acidosis and haemoconcentration already discussed. These additional factors would perhaps explain in part why giving coagulation factor concentrate to low-birth-weight babies fails to prevent intraventricular haemorrhage (Waltl et al., 1973).

It remains uncertain, but seems likely, that there is a relationship between cerebral thrombosis and haemorrhage and these coagulation disorders, and equally likely that the situation is greatly modified by the many other factors influencing blood flow concurrently. The timing of their interaction could be all-important, since it is when the excessive fibrinolytic activity is over that the haematocrit and viscosity are at their highest and the cerebral vascular dynamics and blood gas values in the low-birth-weight baby are likely to be swinging excessively due to the respiratory problems. This is precisely the period during which cerebral haemorrhage in the newborn is thought to occur.

17.6 Treatment

Of the three main mechanisms which have been discussed (viscosity, cardiovascular dynamics and coagulation), disorders of the first are the easiest to evaluate and treat. There has been sufficient clinical experience to confirm that cerebral, pulmonary and renal (Fig. 17.9) disorders can be alleviated rapidly with the improvement of blood flow brought about by partial exchange haemodilution. It must be said, however, that there is still some doubt as to whether lowering the viscosity will always improve oxygen extraction and brain function (Karlsson et al., 1974) and as to whether it will prevent the long-term neurological sequelae (Goldberg et al., 1979) described by Gross et al. (1973). Nevertheless, most of the experience recorded to date supports the view that if the venous Hct is above 0.65 in a baby with symptoms, this should be reduced, though there are no firm data to indicate how far. Arbitrarily, one would choose to obtain a haematocrit of around 0.50 at the end of the procedure and certainly care must be taken not to reduce the oxygen-carrying power to the point where hypoxia durther impairs metabolic activity. While in adults the cardiac output and oxygen-carrying capacity increase with dilution to an Hct of 0.30 (Messmer et al., 1972) a minimum figure of 0.45-0.50 might be more appropriate for the newborn (Fig. 17.4) (Linderkamp et al., 1978c; Walker and MacIntosh, 1971). The dangers of this procedure in skilled hands are minimal, and albumin has the advantage over plasma of being free from potential icterogenic factors.

It is probably wiser to perform such a partial exchange transfusion to ensure adequate pulmonary and cerebral microcirculatory flow even in the face of the unlikely event of having to replace some blood later.

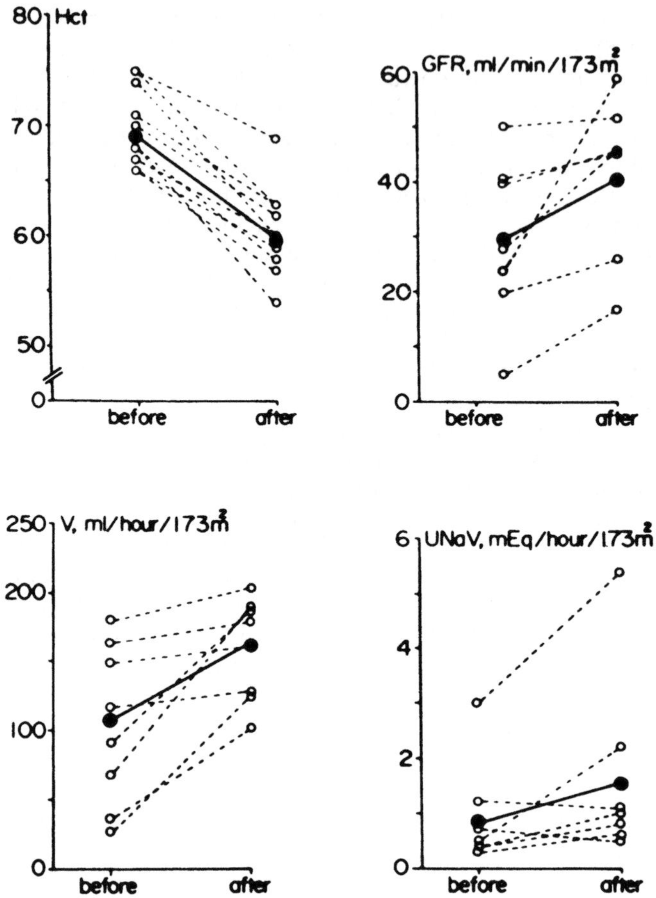

Fig. 17.9. The effect of haemodilution on haematocrit (*Hct*), glomerular filtration rate (*GFR*), urine volume (*V*) and sodium excretion (*UNaV*). Individual (○) and mean (●) values. (Courtesy of Aperia et al., 1974)

17.7 Conclusion

Since the incidence of hyperviscosity in the newborn is estimated at between 3% and 5% of all newborns and in view of its probable importance in the development of neonatal symptoms and long-term sequelae (Gross et al., 1973) the estimation of Hct should be routine in the ill newborn. Haemodilution should be considered in the treatment of any such baby when the venous haematocrit exceeds 0.65. The need for Hct evaluation as a screening procedure for all babies (Wessenberg, 1978) has yet to be confirmed, as has the efficacy of early partial exchange transfusion in the prevention of long-term neurological sequelae.

Acknowledgements. I wish to thank Mr. Anthony Whytock for his assistance in the preparation of illustrations and Mrs. Forsyth, Mrs. Donaldson and Mrs. Mathers for their secretarial help.

17.8 References

Ambrus C A, Choi T S, Cunnanan E, Eisenberg B, Staub H P, Weintraub D H, Courey N G, Patterson R J, Jockin H, Pickren H W, Bross I D, Jung O S, Ambrus J L (1977) Prevention of hyaline membrane disease with plasminogen. JAMA *237:* 1837-1841

Anderson J M, Bain A D, Brown J K, Cockburn F, Forfar J O, Machin G A, Turner T L (1976) Hyaline membrane disease, alkaline buffer treatment, and cerebral intraventricular haemorrhage. Lancet *i:* 117-119

Aperia A, Bergqvist G, Broberger O, Thodenius K, Zetterström R (1974) Renal function in newborn infants with high hematocrit values before and after isovolemic haemodilution. Acta Paediatr Scand *63:* 878-884

Baum R S (1966) Viscous forces in neonatal polycythemia. In: Proceedings of the Society for Pediatric Research. J. Pediat *69:* 975

Bergqvist G (1974) Viscosity of the blood in the newborn infant. Acta Paediatr Scand *63:* 858-864

Bergqvist G, Zetterström R (1974) Blood viscosity and peripheral circulation in newborn infants. Acta Paediatr Scand *63:* 865-868

Bergqvist G, Zetterström R (1975) Submaximal blood flow and blood viscosity in newborn infants. Acta Paediatr Scand *64:* 253-256

Betz E (1972) Cerebral blood flow: Its measurement and regulation. Physiol Rev *52:* 595-630

Bicher H I, Bruley D, Knisely M H, Reneau D D (1971) Effects of microcirculation changes on brain tissue oxygenation. J Physiol *217:* 689-707

Bland R D, Clarke T L, Harden L B, Meyer J L, Ries J P, Madden W A, Crast F W, Coyer W F, Bass J W (1973) Early albumin infusion to infants at risk for respiratory distress. Arch Dis Child *48:* 800-805

Bratteby L, Garby L, Wadman B (1968) Studies on erythrokinetics in infancy XII. Survival in adult recipients of cord blood red cells labelled in vitro with $DF^{32}P$. Acta Paediatr Scand *57:* 403-310

Bussman Y L, Tillman M L, Pagliara A S (1977) Neonatal thyrotoxicosis associated with the hyperviscosity syndrome. J Pediatr *90:* 266-268

Chou P J, Ackerman B D (1973) Perinatal acidosis and placental transfusion. Acta Paediatr Scand *62:* 417-422

Chu J, Clements J A, Cotton E, Klaus M H, Sweet A Y, Thomas M R, Tooley W H (1965) The pulmonary hypoperfusion syndrome. Preliminary report. Pediatrics *35:* 733-742

Cooke R W I, Rolfe P, Howat P (1979) Apparent cerebral blood-flow in newborns with respiratory disease. Dev Med Child Neurol *21:* 154-160

Cort R L, Pribylova H (1964) Placental transfusion and fluid metabolism on the first day of life. Arch Dis Child *39:* 363-370

Danks D M, Stevens L H (1964) Neonatal respiratory distress associated with a high haematocrit reading. Lancet *ii:* 499-500

Donn S M, Philip A G S (1978) Early increase in intracranial pressure in preterm infants. Pediatrics *61:* 904-907

Dunn P M (1970) Neonatal polycythaemia. In: Proceedings of the Paediatric Research Society. Arch Dis Child *45:* 273

Ekelund H, Hedner U, Nilsson I M (1970) Fibrinolysis in newborns. Acta Paediatr Scand *59:* 33-43

Finne P H (1966) Erythropoietin levels in cord blood as an indicator of intrauterine hypoxia. Acta Paediatr Scand *55:* 478-488

Flod N E, Ackerman B D (1971) Perinatal asphyxia and residual placental blood volume. Acta Paediatr Scand *60:* 433-436

Foley M E, Clayton J K, McNicol G P (1977) Haemostatic mechanisms in maternal umbilical vein and umbilical artery blood at the time of delivery. Br J Obstet Gynaecol *84:* 81-87

Foley M E, Isherwood D M, McNicol G P (1978) Viscosity, haematocrit, fibrinogen and plasma proteins in maternal and cord blood. Br J Obstet Gynaecol *85:* 500-504

Fouron J-C, Hebert F (1973) The circulatory effects of hematocrit variations in normovolemic newborn lambs. J Pediatr *82:* 995-1003

Friedman J J (1976) Transcapillary protein leakage and fluid movement. Microvasc Res *12:* 275-290

Gaehtgens P, Schickendantz S (1975) Rheological properties of maternal and neonatal blood. 8th Europ Conf Microcirculation, Le Touquet 1974. Bibl Anat *13:* 107-108

Gatti R A, Muster A J, Cole R B, Paul M H (1966) Neonatal polycythemia with transient cyanosis and cardiorespiratory abnormalities. J Pediatr *69:* 1063-1072

Goldberg K, Wirth F, Guggenheim M A, Lubchenco L O (1979) (Personal communication) Neonatal blood hyperviscosity: effect of partial plasma exchange transfusion. University of Colorado Medical Center, Denver.

Greene R, Hughes J M B, Iliff L D, Pineo G F (1973) Red cell flexibility and pressure-flow relations in isolated lungs. J Appl Physiol *34:* 169-175

Gross G P, Hathaway W E (1972) Fetal erythrocyte deformability. Pediat Res *6:* 593-599

Gross G P, Hathaway W E, McGaughey H R (1973) Hyperviscosity in the neonate. J Pediatr *82:* 1004-1012

Hakanson D O, Oh W (1977) Necrotizing enterocolitis and hyperviscosity in the newborn infant. J Pediatr *90:* 458-461

Humbert J R, Abelson H, Hathaway W E, Battaglia F C (1969) Polycythemia in small for gestational age infants. J Pediatr *75:* 812-819

Hutchison A A, Russell G (1976) Effective pulmonary capillary blood flow in infants with birth asphyxia. Acta Paediatr Scand *65:* 669-672

Ingomar C J, Klebe J G (1973) The influence of the placental transfusion on the capillary blood gas and acid-base balance in the newborn infant. Acta Paediatr Scand *62:* 121-124

Ingomar C J, Klebe J G (1974) The transcapillary escape rate of T-1824 in newborn infants of diabetic mothers and newborn infants with respiratory distress or birth asphyxia. Acta Paediatr Scand *63:* 565-570

Jensen A H-B, Josso F, Zamet P, Monset-Couchard M, Minkowski A (1973) Evolution of blood clotting factor levels in premature infants during the first 10 days of life: A study of 96 cases with comparison between clinical status and blood clotting factor levels. Pediatr Res *7:* 638-644

Jurgens H, Gobel U, Bokelmann J, Von Voss H, Wahn V (1979) Coagulation studies on umbilical arterial and venous blood from normal newborn babies. Europ J Pediatr *131:* 199-204

Karlsson K, Hrbek A, Olsson T, Riha M, Rosen K-G, Kjellmar I (1974) Cerebral reactions to intrauterine asphyxia in sheep III. Effects of alterations of hematocrit and viscosity. J Perinat Med *2:* 161-169

Kaulla K N von, Kaulla E von, Butterfield J (1965) Fibrinolytic activity, thrombin inhibitor and kinetics of clot formation in premature infants with respiratory distress syndrome. Acta Paediatr Scand *54:* 587-592

Klebe J G, Ingomar C J (1974) Placental transfusion in infants of diabetic mothers elucidated by placental residual blood volume. Acta Paediatr Scand *63:* 59-64

Kontras S B (1972) Polycythemia and hyperviscosity syndromes in infants and children. Pediatr Clin N Am *19:* 919-933

Kontras S B, Bodenbender J G, Craenen J, Hosier D M (1970) Hyperviscosity in congenital heart disease. J Pediatr *76:* 214-220

Lanning P, Koivisto M, Suramo I, Lanning M (1977) Neonatal polycythaemia and chest roentgenograms. Europ J Pediatr *126:* 103-108

Lassen N A (1974) Control of cerebral circulation in health and disease. Circ Res *34:* 749-760

Leake R D, Thanopoulos B, Nieberg R (1975) Hyperviscosity syndrome associated with necrotizing enterocolitis. Am J Dis Child *129:* 1192-1194

Linderkamp O, Strohhacker I, Versmold H T, Klose H, Riegel K P, Betke K (1978a) Peripheral circulation in the newborn: Interaction of peripheral blood flow, blood pressure, blood volume, and blood viscosity. Europ J Pediatr *129:* 73-81

Linderkamp O, Versmold H T, Messow-Zahn K, Muller-Holve W, Riegel K P, Betke K (1978b) The effect of intra-partum and intra-uterine asphyxia on placental transfusion in premature and full-term infants. Europ J Pediatr *127:* 91-99

Linderkamp O, Versmold H T, Fendel H, Riegel K P, Betke K (1978c) Association of neonatal respiratory distress with birth asphyxia and deficiency of red cell mass in premature infants. Europ J Pediatr *129:* 167-173

Lou H C, Lassen NA , Friis-Hansen B (1979) Impaired autoregulation of cerebral blood flow in the distressed newborn infant. J Pediatr *94:* 118-121

MacIntosh T F, Walker C H M (1973) Blood viscosity in the newborn. Arch Dis Child *48:* 547-533

McWhirter W R (1975) Tocopherol and anaemia of prematurity. Abs Arch Dis Child *50:* 404

Markarian M, Githens J H, Rosenblut E, Fernandez F, Jackson J J, Bannon A E, Lindley A, Lubchenco L O, Martorell R (1971) Hypercoagulability in premature infants with special reference to the respiratory distress syndrome and hemorrhage I. Coagulation studies. Biol Neonate *17:* 84-97

Messmer K, Sundfer-Plassmann L, Klövekorn W P, Holper K (1972) Circulatory significance of hemodilution: Rheological changes and limitations. Adv Microcirc *4:* 1-77

Messow-Zahn K, Sarafoff M, Riegel K P (1978) Stress at birth: Noradrenaline concentrations of women in labour and in cord blood. Klin Wochenschr *56:* 311-312

Moss A J, Monset-Couchard M (1967) Placental transfusion: Early versus late clamping of the umbilical cord. Pediatrcs *40:* 109-126

Oh W, Lind J, Gessner I H (1966) The circulatory and respiratory adaptation to early and late cord clamping in newborn infants. Acta Paediatr Scand *55:* 17-25
Oh W, Wallgren G, Hanson J S, Lind J (1967) The effects of placental transfusion on respiratory mechanics of normal term newborn infants. Pediatrs. *40:* 6-12
Ongkie A S (1974) Syndrome of hyperviscosity in the neonate. Paediatr Indones *14:* 232-248
Oski F A, Barness L A (1967) Vitamin E deficiency: A previously unrecognized cause of hemolytic anemia in the premature infant. J Pediatr *70:* 211-220
Oski F A, Naiman J L (1965) Red cell metabolism in the premature infant I. Adenosine triphosphate levels, adenosine triphosphate stability and glucose consumption. Pediatrics *46:* 104-112
Papageorgiou A, Stern L (1972) Polycythemia and gangrene of an extremity in a newborn infant. J Pediatr *81:* 985-987
Pape K E, Wigglesworth J S (1979) Haemorrhage, Ischaemia and the perinatal brain. Clinics in Developmental Medicine Nos. 69/70. W. Heinemann Medical Books. Spastics International Medical Publications, London
Pearson H A (1976) The blood. In: Smith C A, Nelson N M (eds). The physiology of the newborn infant, 4th ed. Charles C Thomas, Illinois, pp 263-311
Philip A G S, Yee A B, Rosy M, Surti N, Tsamtsouris A, Ingall D (1969) Placental transfusion as an intrauterine phenomenon in deliveries complicated by foetal distress. Br Med J *1:* 11-13
Pietra G G, D'Amodio M D, Leventhal M M, Oh W, Braudo J L (1968) Electron microscopy of cutaneous capillaries of newborn infants: Effects of placental transfusion. Pediatrics *42:* 678-683
Powell L W, Dunnicliff M A, Billing B H (1968) Red cell survival in experimental cholestatic jaundice. Br J Haematol *15:* 429-435
Ramamurthy R S, Royal J E, Pyati S P, Seeler R A (1975) Neonatal polycythemia hyperviscosity syndrome. IMJ *148:* 48-53
Rivers R P (1975) Coagulation changes associated with a high haematocrit in the newborn infant. Acta Paediatr Scand *64:* 449-456
Sabbah H N, Lee T G, Stein P D (1979) Role of blood viscosity in the production of innocent ejection murmurs. Am J Cardiol *43:* 753-756
Secher O, Karlberg P (1962) Placental blood transfusion. Lancet *i:* 1203-1205
Strang L B (1966) The pulmonary circulation in the respiratory distress syndrome. In: James L S (ed) Pediatric clinics of North America. Saunders, Philadelphia-London, pp 693-701
Tillmann W, Lenard H G, Wagner D, Dönges H, Schröter W (1974) Rheologic properties of erythrocytes in Duchenne muscular dystrophy. Pediatr Res *13:* 152-155
Turnbull E P N, Walker J (1955) Haemoglobin and red cells in the human fetus II. The red cells. Arch Dis Child *30:* 102-110
Usher R, Shepherd M, Lind J (1963) The blood volume of the newborn infant and placental transfusion. Acta Paediatr Scand *52:* 497-512
Volpe J J (1976) Perinatal hypoxic-ioschemic brain injury. Pediatr Clin N Am *23:* 383-397
Walker C H M, Balf C L (1954) Capillary resistance in the newborn infant and in the mother in late pregnancy, labour and early puerperium. J Obstet Gynaecol Br Emp *61:* 1-16 and 17-30
Walker C H M, Mackintosh T F (1972) The treatment of hyperviscosity syndromes of the newborn with hemodilution. In: Messmer K, Schmid-Schönbein H (eds). Hemodilution. Int Symp Rottach-Egern 1971. Karger, Basel, pp 271-288
Walker J L, Turnbull E P N (1953) Haemoglobin and red cells in the human fetus and their relation to the oxygen content of the blood in the vessels of the umbilical cord. Lancet *ii:* 312-318
Watl H, Kurz R, Mitterstieler G, Födisch H J, Hohenauer L, Rössler H (1973) Intracranial haemorrhage in low-birth-weight infants and prophylactic administration of coagulation-factor concentrate. Lancet *i:* 1284-1286
Weed R I, LaCelle P L, Merill W E (1969) Metabolic dependence of red cell deformability. J Clin Invest *48:* 795-809
Weinberger M M, Oleinick A (1970) Congenital marrow dysfunction in Down's syndrome. J Pediatr *77:* 273-279
Weiss H R, Edelman N H (1976) Effect of hypoxia on small vessel blood content of rabbit brain. Microvasc Res *12:* 305-315
Wessenberg R L (1978) Neonatal 'thick blood' syndrome. Hosp Pract *13:* 137-140, 145
Yao A C, Lind J (1974) Placental transfusion. Am J Dis Child *127:* 128-141

Section E

Therapeutic Aspects of Blood Rheology

18 Isovolaemic Haemodilution

H. Schmid-Schönbein and H. Rieger

18.1 Introduction

Historically, the concept that slowly flowing blood becomes very viscous in certain diseases can be traced back to the dawn of written medicine. The classic Hippocratic and Galenic theory of humoural pathology attributed malfunction of organs to changes in the blood, which was accepted as a *flowing* medium and mixture of elements. Especially after the discovery of the function of the blood as the *circulating* source of all life in the body, it was widely accepted that "running cold of blood" was the cause of disease processes. In the 17th and 18th centuries, the "iatrophysical" or "iatromechanical" school (Boerhave, 1703) was firmly convinced that abnormal mixture of the elements of blood led to a condensation and congealment of the humours and subsequent obstruction of organs.

In the second half of the 18th and the first half of the 19th century, more modern empirical schools discovered that microscopic stasis of blood was a prominent feature in post-mortem appearances of tissues and organs, objective observations that reaffirmed these archaic beliefs. Despite the fact that the doctrine of "humoural pathology" was almost entirely abandoned during the second half of the 19th and in the 20th century, its therapeutic consequence, phlebotomy, was never entirely excluded from medical practice. In the 20th century, extensive intravital microscopic research in experimental animals as well as in man reinstituted the belief that changes in blood viscosity were instrumental in disease. Knisely's and Bloch's extensive subjective observations of "blood sludging" (Knisely, 1965) and Gelin's discovery that infusions with low-molecular-weight dextran solutions ameliorated observable changes in the micro-circulation (Gelin, 1961) are further milestones in the long history of haemo-rheological research.

Objective measurement of the effects of low-molecular-weight dextran infusions in experimental animals and in man later confirmed the subjective impressions obtained by microscopic observation. These objective measurements in turn greatly augmented the interest in haemorheological research, which also profited from the general availability of rotational viscometers from the early 1960s. Very soon, however, the original claim of a "specific disaggregating action" of low-molecular-weight dextrans was challenged by workers who claimed that the effect of low-molecular-weight dextran infusion was primarily one of short-lasting plasma expansion and — rheologically — of simple haemodilution. For a review the reader is referred to two symposia on this subject (Messmer and Schmid-Schönbein, 1972; 1975).

Not only the striking flow improvements obtained by this kind of therapy, but also the advent of thrombin-like enzymes derived from snake venom, which are capable of reducing the fibrinogen level, opened the door for rheological therapy. Therapeutic defibrinogenation not only interferes with the coagulation system, but also reduces plasma viscosity and red-cell aggregation (Ehringer et al., 1971; Ehrly, 1973). (For a review of clinical results see Lowe et al., Chapter 21 of this volume.) The effect of these agents in producing a significant reduction of apparent *blood* viscosity *in vitro* remains in dispute (Schmid-Schönbein et al., 1978). While clearly blood viscosity reduction at high shear rates can be attributed entirely to reduction in plasma viscosity, at low shear rates the fall in apparent blood viscosity following defibrinogenation was neither consistent nor dramatic. This may be the consequence of the presence in the patients of red-cell-aggregating serum proteins which are not affected by the defibrinogenating enzymes. Alpha$_2$-macroglobulin and IgM have previously been identified as strongly aggregating macromolecules: others, including lipoproteins, may well have the same action (Schmid-Schönbein et al., 1973).

Reduction of the haematocrit by whatever means lowers apparent blood viscosity, but in terms of oxygen transport rate, there is no advantage to be expected from such therapy, since the *gain* in flow rate is offset by a *loss* in oxygen transport capacity. Indeed, it has been clearly shown that the red-cell *flux* per unit time is roughly constant *in vivo* at haematocrit levels between about 27% and 55% in resting experimental animals (Hint, 1964; Guyton and Richardson, 1961). The precise mechanism which causes this wide plateau of haematocrit levels is not known. It is likely to be a result of compensatory mechanisms such as peripheral vasodilation, increased venous return and consequently an increased stroke volume. Gaehtgens and colleagues have recently shown that a similar plateau of "optimal haematocrit" is seen in working skeletal muscle. As was to be expected under these conditions, the maximum work capacity as well as the oxygen consumption fell as the haematocrit was reduced to 30%; in absolute terms, the effect is surprisingly small (Fig. 18.1). It is noteworthy that elevation of the haematocrit above 55% was often associated with an exceedingly slow onset of both perfusion and work capacity of the muscle (Gaehtgens, personal communication; Kreutz, 1978).

No significant gain in normal perfusion is to be expected from a reduction of blood viscosity by an induced anaemia for the following physiological reasons:

1) The haematocrit in the microvessels is far lower than the central haematocrit as measured in venous (or arterial) blood vessels. In other words, not much is to be gained by an attempt to exaggerate the *physiological* dynamic haemodilution which results from strong axial migration and the subsequent Fahråeus and Fahråeus-Lindqvist effects (Fahråeus, 1929; Fahråeus and Lingdqvist, 1931).

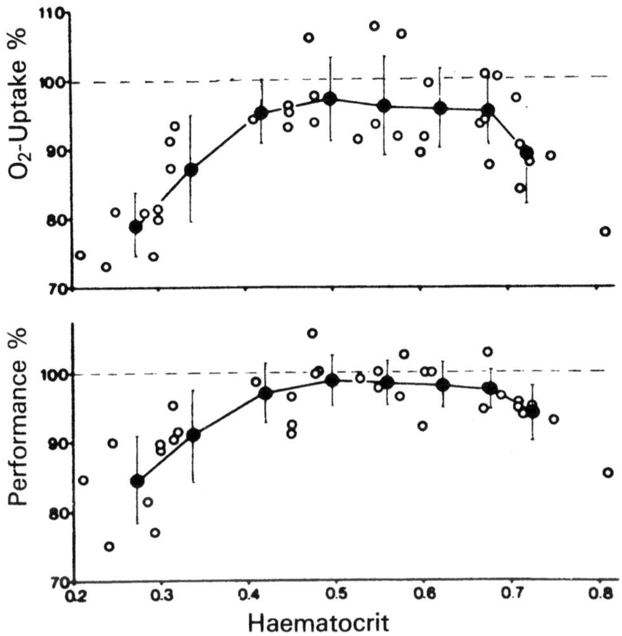

Fig.18.1. Influence of haematocrit value on oxygen uptake and work capacity in maximally working isolated dog muscle (from Gaehtgens and Kreutz, 1979)

2) In conditions of normal circulation, overall viscous energy dissipation is far less affected by the haematocrit level than is the measured apparent viscosity of blood in glass tubes. The surprisingly small effect of haematocrit changes on resistance in maximally dilated vascular beds, as originally discovered by Whittaker and Winton (1933), has since been confirmed by many workers.

However, in rheological terms, fast- and slow-flowing blood differ radically in their flow behaviour. Therefore, the above arguments do not apply in the consideration of pathophysiological conditions. Instead, in the absence of the physiologically high flow forces, the so-called anomalous, pseudoplastic or thixotropic properties of the blood have to be considered — and then the role of the red-cell volume fraction in blood fluidity is entirely different.

18.2 Rheology and Creeping Blood

As detailed elsewhere (Chapter 6), the non-Newtonian flow behaviour of blood is the consequence of an extremely variable flow behaviour of mammalian erythrocytes, which are in essence not cells but fluid-filled cell fragments with a surplus of highly flexible membrane area for their given volume. All mechanisms

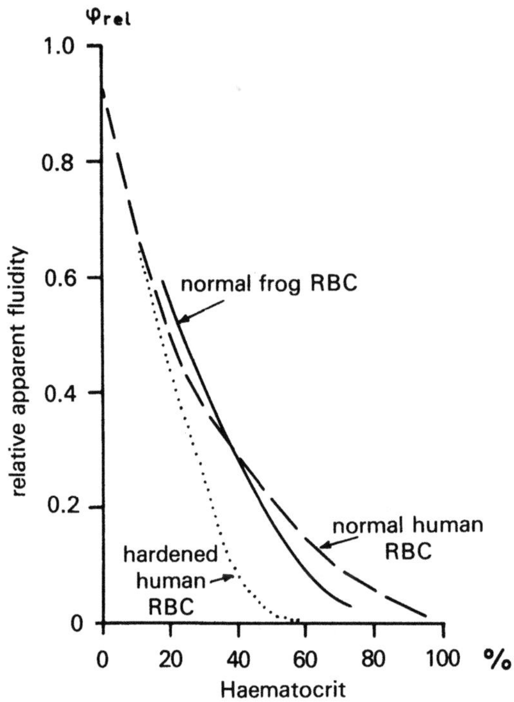

Fig.18.2. Influence of haematocrit on relative apparent fluidity (φ) of blood. (φ rel = φ blood/φ plasma). Note strong curtailment of apparent blood fluidity as haematocrit in rigidified mammalian and frog red cells is elevated above 60%. Normal mammalian red cell suspensions maintain their fluidity even at Hct 90%. (From Chien et al., 1971)

allowing an almost ideal passive adaptation of these fluid-drop-like entities to the flow of plasma can be attributed to these characteristics of mammalian blood, as can the ability to sustain high haematocrits and to flow in narrow capillaries. All favourable flow conditions of blood require the presence of *high shear stresses* (i.e. high blood pressure gradients). In their absence, blood degenerates from a highly fluid, strongly shear-thinning emulsion to an ordinary, often reticulated, suspension. The reticulation is the consequence of rouleaux formation and the build-up of three-dimensional rouleaux networks. It is clear that the integrity, the density and thence the mechanical strength of such structures is highly dependent upon small changes in haematocrit once a level of about 30% is exceeded. A simple calculation may serve to illustrate this argument. Irrespective of shape and orientation in flow, at a haematocrit of 33% each cell is surrounded by a plasma layer of twice its own volume. At 66% haematocrit the cells are surrounded by a volume of plasma only half their volume. What is more important, their "elbow room" has gone to zero, since all cells are in physical contact with each other.

It is therefore not surprising that the fluidity (the inverse value of their viscosity) of all suspensions is abolished entirely once the volume fraction exceeds 55%. While rapidly flowing mammalian blood is the unique exception to this rule, most suspensions of nucleated red cells comply (Fig. 18.2), as do the slowly flowing suspensions of aggregated human red cells at low shear rates (Fig. 18.3).

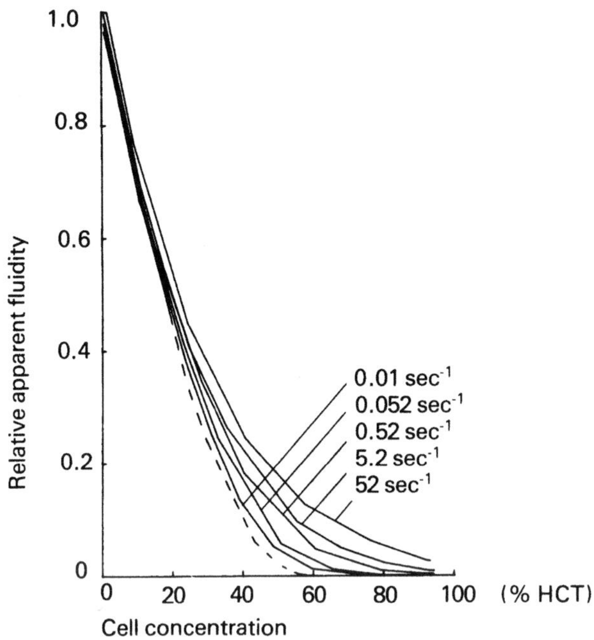

Fig.18.3. Influence of haematocrit and shear rate on apparent relative fluidity (φ) of normal human blood. (φ rel. = φ blood/φ plasma.) As shear rate is reduced the behaviour of normal blood approaches that of suspensions of artificially hardened cells (broken line). (modified after Chien, 1970)

The important message of Figs. 18.2 and 18.3 is reemphasised in Fig. 18.4: when the haematocrit is lowered to 30%, the fluidity of the blood is maintained despite the fact that the shear stresses are lowered substantially. At a haematocrit level of 30% the significant microrheological differences between nucleated and non-nucleated red cells of animal species as well as between normally flexible and artificially hardened cells do not significantly affect the macroscopic fluidity of the suspension, provided that the tube diameter is larger than the cell diameter.

It has long been suspected that standing blood behaves like a visco-elastic solid below its yield point. In the haemorheological literature of the last two decades there is an unsettled argument as to whether blood can indeed lose its fluidity altogether in the presence of finite shear stresses. The first-generation rotational viscometers, which are popular at present, in which the shear rate is established and the resulting shear stress is measured for computing apparent viscosity, are in principle *inappropriate* to settle this argument. However, now that a rheometer applying pre-set and very low shear stresses has become available (Deer constant shear stress rheometer, Deer Rheometers Ltd., c/o Rheometer Marketing Ltd., 15 Greenacre Park Rise, Rawson, Leeds LS19 6RU, Yorkshire, England), we have found that a true yield shear stress could only be measured at haematocrit levels above 70% (Kiesewetter et al., 1979). However, in models of microscopic tube networks (diameter 30 -150 μm), the same group of authors, using the same

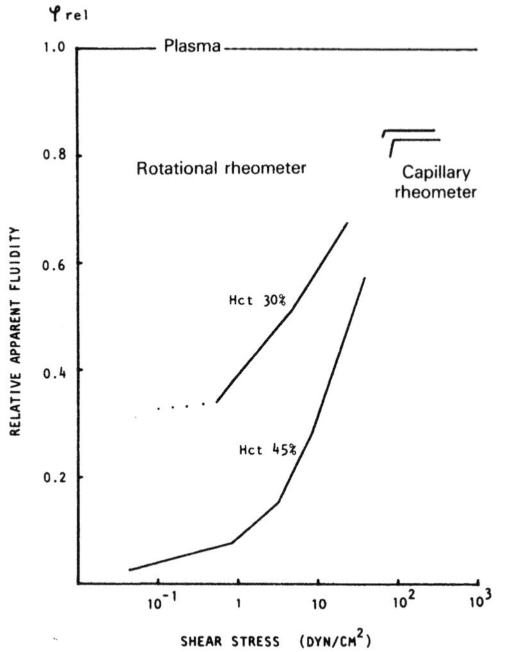

Fig.18.4. Relative apparent fluidity (φ rel) of blood plotted as a function of shear stress (1 dyn/cm^2 = 0.1 Pa). Note strong decrease in apparent fluidity as shear stress falls, but only at haematocrit values of 45% — not at 30% (φ rel = φ blood/φ plasma). Measurements in rotational and capillary rheometers. (from Devendran et al.)

blood samples, clearly observed cessation of flow at finite pressure gradients (Figs. 18.5 and 18.6). In these models of the microcirculation, yield shear stresses as high as 0.1 Pa could be measured directly, values two orders of magnitude higher than the yield shear stress values previously extrapolated from Casson-equation plots. Table 18.1 shows preliminary data; it is evident that the yield values depend not only on the tendency to aggregation but also markedly on the haematocrit. At haematocrit levels of 35% we found in most cases no yield shear stress at all. As the haematocrit was raised from 45% to 55% (by removing plasma), the yield shear stress was increased tenfold.

Judging from measurements of *in vivo* flow behaviour, it is probably justified to state that the presently measured high yield shear stress values *underestimate* rather than overestimate the yield values in still smaller blood vessels *in vivo*. Furthermore, there are strong hysteresis effects and time-dependent increases of the measured yield shear streses (Fig. 18.6). As the blood sample was left standing, structural rearrangements of the stagnant rouleaux occurred, the cause of which is unclear. This resulted in a progressively increased yield shear stress in the plastic tubes used. Similar time effects occur in stagnant, blood-filled microvessels *in vivo*. Again we feel that in our *in vitro* experiments we underestimate such effects *in vivo*, which are likely to be complicated by increased vascular permeability with resulting local hyperconcentration of high-molecular-weight plasma proteins, haemoconcentration with locally increased haematocrit, and possibly endothelial swelling.

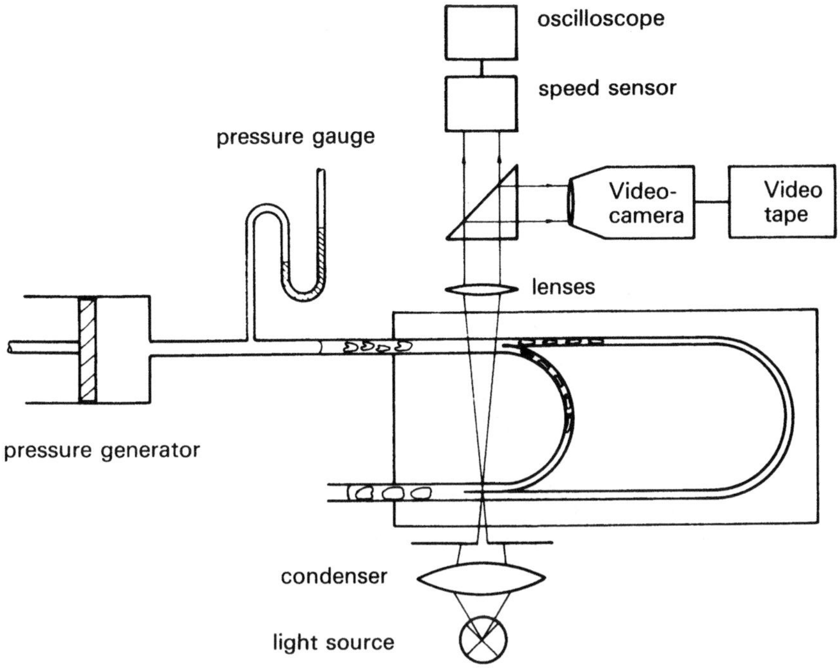

Fig. 18.5. Schematic representation of the apparatus used to measure the yield shear stress of human blood in branched models of a microvascular network. A long (150-mm) and a short (15-mm) channel (I.D. 150 μm) are positioned in parallel). For any given pressure gradient, the shear stresses are 10 times *lower* in the *long* channel. Flow rates are measured electronically from the projected microscopic image of the two channels; pressures are recorded by strain-gauges

The data presented are in keeping with the hypothesis that slow-flowing blood behaves like a pseudoplastic and thixotropic material; in other words, that it can lose its fluidity once the locally effective shear stresses are lowered below a critical threshold. The magnitude of this threshold is extremely haematocrit dependent: at 35% it is undetectably low and blood maintains its high fluidity no matter how low the incident shear stresses. Between 45% and 55% haematocrit, the yield shear stress is elevated by at least a factor of 10. In other words, at a haematocrit level of 55% the critical pressure threshold for rheological occlusion is 10 times higher than at a haematocrit of 45%.

18.3 Rheological Rationale for Haemodilution

The data obtained by these extensive experiments *in vitro* suggested a hypothesis that supplied the basis for proposing intentional haemodilution as a therapeutic intervention in chronic low-flow states.

1) Decompensated states of obstructive peripheral arterial disease are characterised by the absence of sufficient shear stresses to keep the blood in its physiological state of high fluidity.

2) In adaptation to the normal flexibility of the red cells, the haematocrit is comparatively high, and favourable for the normal circulation, but turns into a

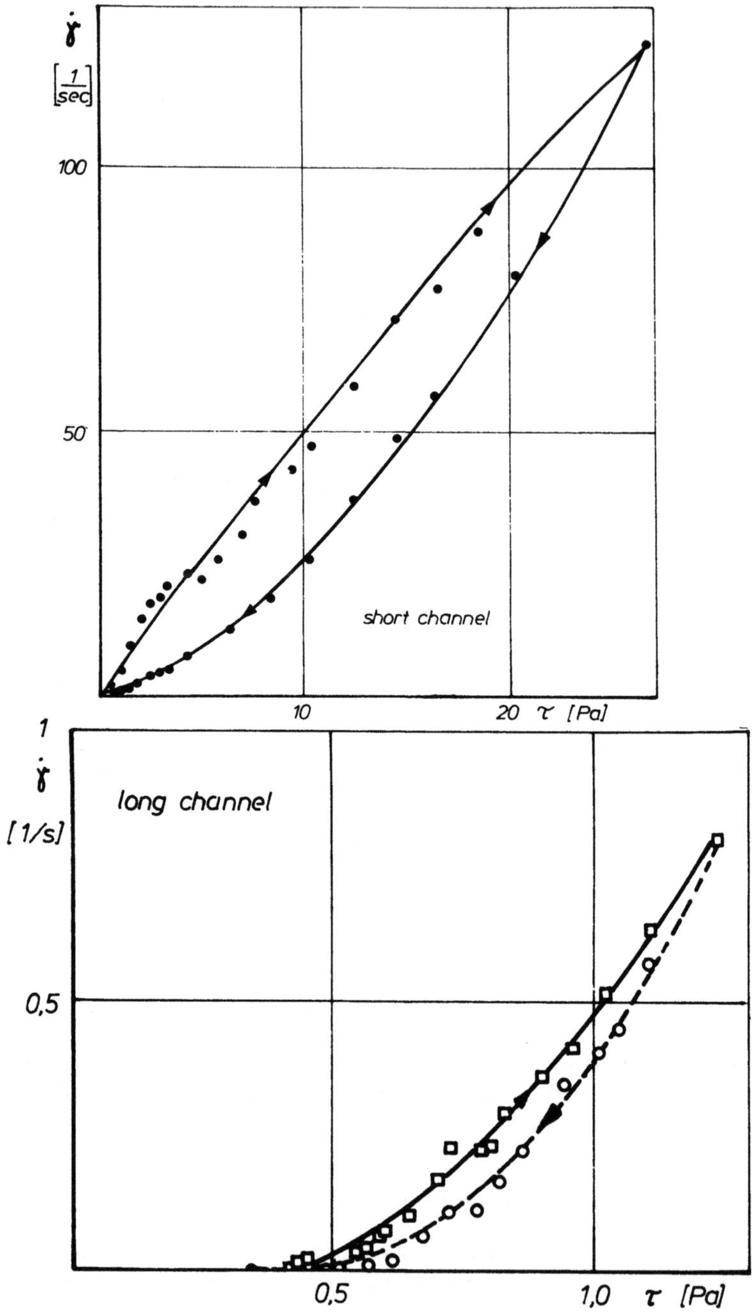

Fig.18.6. Shear rate ($\dot{\gamma}$) (calculated from the centre line flow) plotted as a function of shear stress (τ) in the short channel (upper graph) and the long channel (lower graph). Despite drop in driving pressure, flow is maintained in the short channel, but comes to full stop ($\dot{\gamma}$ = 0 shear rate) in the long channel. Mean value of six measurements in the blood of a diabetic patient with haematocrit 55%; calculated yield shear stress 0.5 ± 0.1 Pa. (From Kiesewetter et al., 1979)

dangerous risk factor for the lack of maintenance of residual flow in post-stenotic regions.

3) Under these conditions, maintenance of a haematocrit level of 30%-35% should restore the fluidity of blood even if the locally effective shear stresses cannot be therapeutically improved.

Extensive animal experiments (for reviews, see Messmer and Schmid-Schönbein, 1972; 1975) have proven the efficacy of haemodilution and, in man, extensive use in open heart surgery as well as general surgery has been safe. Despite the fact that there have been to date only a limited number of *in vivo* experimental investigations into the specific role of the non-Newtonian properties of blood (Lipowski and Zweifach, 1974; Driessen et al., 1979) we felt that their results justified the attempt to institute haemodilutional therapy in patients with chronic arterial disease.

Table 18.1. Influence of the tendency to aggregation (which is elevated in patient groups) and haematocrit value on directly measured yield shear stress of blood (Pa) (from Kiesewetter et al., 1979). Note that at Hct value of 35% the yield shear stress vanishes: residual flow remains despite very low shear stresses.

Haematocrit	Normals	Essential hypertensives	Diabetics
35%	0	0	0 - 0.002
45%	0 - 0.002	0.002 - 0.01	0.002 - 0.1
55%	0 - 0.01	0.01 - 0.1	0.01 - 0.3

18.4 Patient Selection, Procedures and Measurements

Patients with advanced ischaemic ulcers of the leg or hand were studied. Reduction of the haematocrit value was achieved by the withdrawal of 500 ml whole blood, which was replaced with 300-500 ml low-molecular-weight dextran (Rheomacrodex). The blood was withdrawn into flexible plastic containers (two bag system, Travenol), centrifuged, and the supernatant plasma was reinfused immediately, in order to maintain constant blood volume (isovolaemic haemodilution). In the present series of cases, the blood volume was not actually measured, but none of the patients showed any symptoms of hypovolaemia. In future studies, to find the ideal plasma substitute, plasma volumes will be measured at timed intervals.

This procedure was repeated until a mean haematocrit value of 0.31 ± 0.02 was reached (before haemodilution mean haematocrit was 0.44 ± 0.3). During the procedure, continuous measurements of the heart rate and blood pressure were recorded. In contrast to the acute (pre-operative) method of haemodilution in surgery, where a rapid lowering of the haematocrit value is achieved, we performed the haemodilution stepwise over 3-4 days. The level of dilution achieved was maintained over 8-10 weeks by repeating the procedure according to the rise in the haematocrit value.

Measurement of haematocrit, plasma viscosity, apparent blood viscosity at various shear rates, red-cell aggregation and other haematological parameters are described in detail elsewhere (Rieger et al., 1979). *In vivo,* the cardiac output, and the flow velocity in the femoral artery (using an cine-angiographic method) were measured, and the effect of haemodilution on skin temperature was monitored by thermography.

18.5 Results

In all cases involving the *upper limbs* (n = 7), the trophic lesions of the hands and fingers showed a tendency to re-epithelialisation or complete healing (Table 18.2). An example is shown in Fig. 18.7. Figure 18.7a shows the left hand of a patient suffering from Buerger's disease before haemodilution. After a previous amputation of the 4th finger of the left hand, a deep ulcer extending down to the tendons remained. The entire back of the hand was covered with multiple ulcers. All the lesions had been present for at least 3 months. After 8 weeks of haemodilution, the left hand was hyperaemic and the ulcers had healed completely (Fig. 18.7b). In addition to the disappearance of the ulcers, there were remarkable changes in the appearance of nails of those fingers that had not been gangrenous.

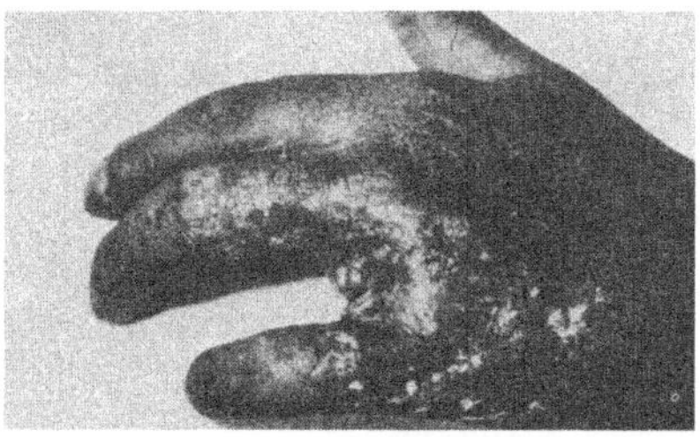
(a)

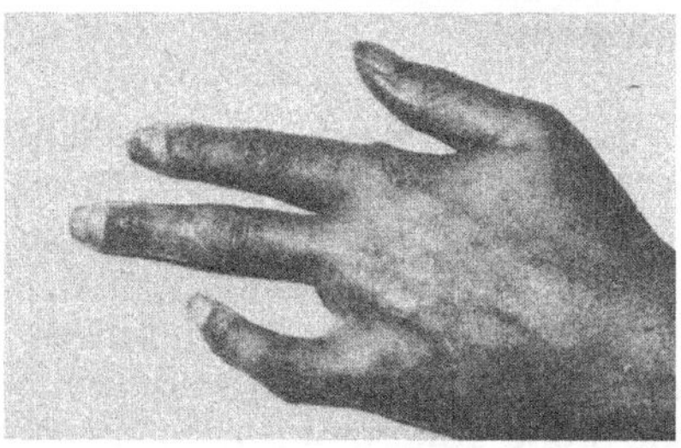
(b)

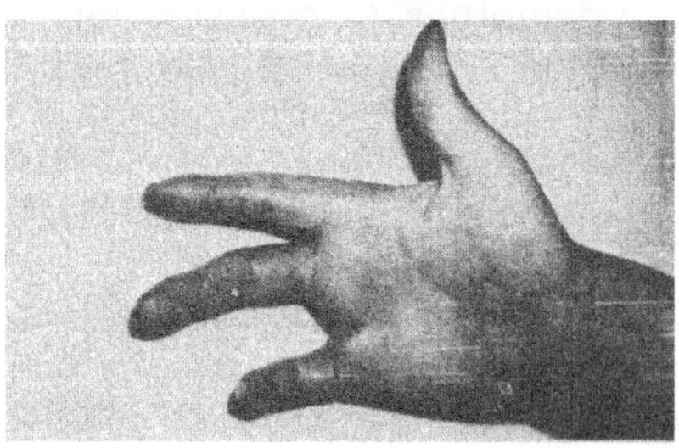
(c)

Fig.18.7. Hand of a patient with Buerger's disease, refractory to all medical and surgical treatment, before (a) and 8 weeks after the start of isovolaemic haemodilution (b). The haematocrit was lowered to 30% by weekly exchange of blood for the patient's platelet-rich plasma and isotonic dextran solution. One year after the end of haemodilution, the favourable result persisted despite normalisation of the haematocrit value (c)

Table 18.2. Clinical effect of haemodilution on upper-limb lesions. *BRA,* brachial artery. *RAD,* radial artery. *UL,* ulnar artery. *DA,* digital arteries. (+) clear tendency of healing. (+ +) complete healing

		Localisation of occlusion				
No.	Med. Report	BRA	RAD	UL	DA	Result
1	176/78				●	+
2	2522/77	●			●	+
3	645/77		●	●	●	+ +
4	2749/77		●	●		+ +
5	2105/76			●	●	+ +
6	1135/78				●	+
7	2055/78				●	+ +

Table 18.3. Clinical effect of haemodilution on lower-limb lesions. *IL,* iliac artery. *FEM,* femoral artery. *PRO,* profunda artery. *POP,* popliteal artery. *TA,* anterior tibial artery. *TP,* posterior tibial artery. *FIB,* fibular artery. *DA,* digital arteries. P_{sys}, systolic ankle blood pressure. (●) arterial obliteration. (△) arterial stenosis. (—) no clinical effect. (+) clear tendency of healing. (+ +) complete healing

		Localisation of occlusion									
No.	Med. Report	IL	FEM	PRO	POP	TA	TP	FIB	DA	Result	P_{sys}
1	312/78		●		●	●	●	●		—	50
2	2522/77		●		●	●	●	●		—	?
3	892/77		●			●	●	●		—	50
4	1222/76		●	●	●	●				—	30
5	1150/77			●	●	●		●		—	42
6	2488/76				●	●	●	●		+	60
7	986/77		△			●	●	●		—	52
8	2749/77					●	●	●		—	72
9	865/76		●			●	●			—	?
10	1827/76					●	●		●	—	?
11	2105/76		●			●	●			—	50
12	398/78		●		●	●				+ +	70
13	1320/77					●	●	●		+	50
14	297/78					●	●	●		+	80
15	1531/77		●	●	●					+	30
16	888/76						●	●	●	+ +	70
17	1684/76				●		●	●		+	70
18	678/77		△			●	●			+ +	100
19	818/76		●							+ +	104
20	1979/76		●							+	50
21	2423/77							●		+	150
22	720/76		●							+	60

The previously hyperplastic, thickened nails started to grow thin; the transition zone between the previously hyperplastic and the presently eutrophic nail area grew toward the periphery.

In the lower limbs, the clinical effect was absent or less marked and depended mainly upon the individual number of arterial occlusions and on the level of the post-occlusive systolic ankle blood pressure. There were patients with recent ulcers that did not respond at all to this therapy, while others showed a clear healing of toe ulcers that had resisted all previous therapeutic attempts (including sympathectomy, defibrinogenating and thrombolytic therapy) for 9 years. The clinical results are summarised in Table 18.3, which lists the site occlusion, systolic ankle pressure, and the therapeutic results in the 22 patients treated to date. The table shows that patients with angiographically documented in-series occlusions of five consecutive vessel segments did not respond to haemodilution (Nos. 1 and 2). Obstruction in four vessels was treated successfully in only 2 of 5 cases (Nos. 3-6). In 11 patients with three vessel occlusion, 6 responded (Nos. 12-17), and 5 did not (Nos. 7-11). All 5 patients with two or one sites of obstruction showed a good response (Nos. 18-22). The heart rate remained constant during single episodes of haemodilution as well as in the periods between

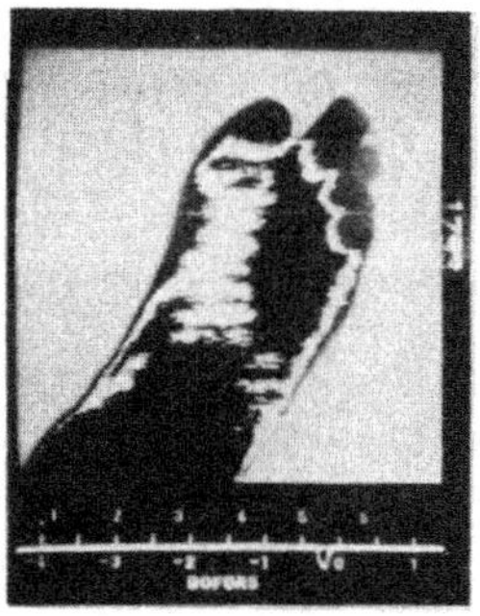

(a) 19.5.76, 9 a.m.
Hct 45%

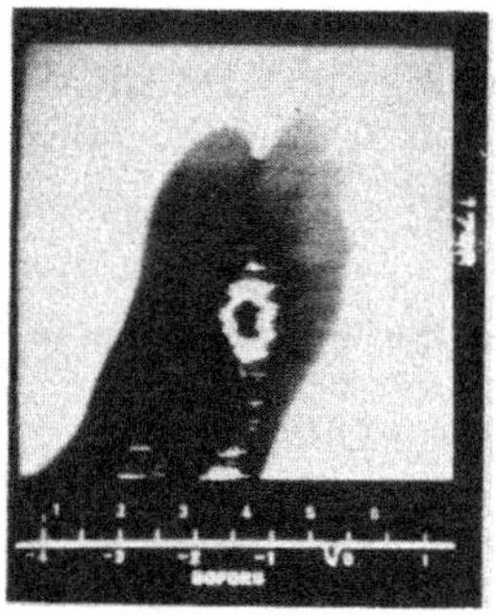

(b) 20.5.76, 9 a.m.
Hct 37%

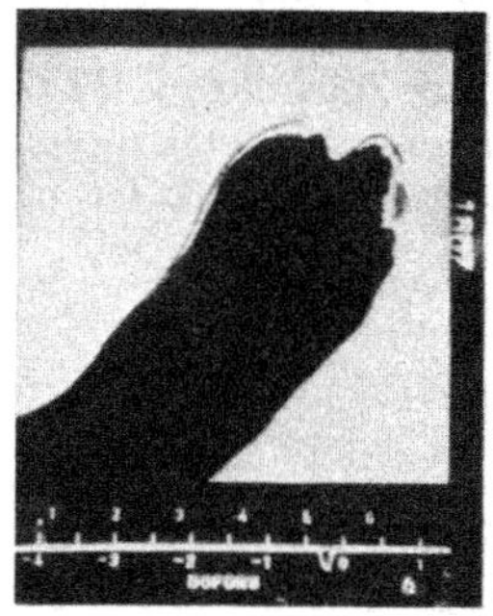

(c) 21.5.76, 8.30 a.m.
Hct 30%

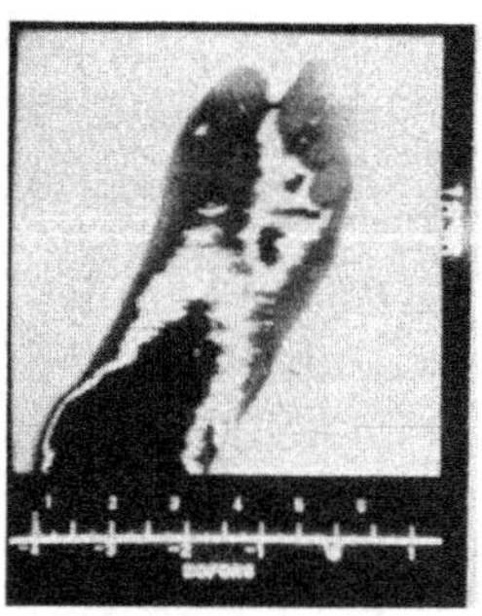

(d) 24.5.76, 11 a.m.
Hct 31%

(e) 28.5.76, 9 a.m.
Hct 32%

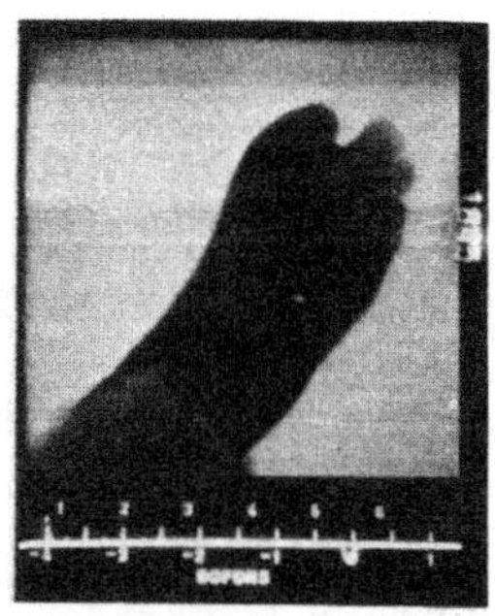

(f) 10.6.76, 9 a.m.
Hct 35%

Fig.18.8. Serial thermograms of the foot of a patient (female, age 54) suffering from atherosclerosis obliterans. Note immediate migration of isotherm line to the periphery after 6 h of treatment (b) and further warming of foot 20 h after end of haemodilution to Hct 30% (c) them. Cardiac output (Kohler and Rieger, 1979) and the mean velocity of femoral artery blood flow increased significantly by approximately 24% and 32%, respectively. In 10 patients studied by thermography, the skin temperature increased in the diseased as well as the intact limbs. A typical example is shown in Fig. 18.8. As can be seen, the black isotherm (indicative of high temperature) moves centrifugally toward the periphery and concentrically toward the lesion during the maintenance of the low haematocrit.

The haemorheological changes before and during the therapy were indicative of conversion of the blood into a more Newtonian fluid with lower overall blood viscosity, with greatly reduced velocity of red-cell-aggregate formation, and a slight thrombocytosis. The reticulocyte count went up and the serum iron levels fell. Mean cellular haemoglobin concentration remained unchanged. Plasma viscosity, fibrinogen level, and all coagulation parameters were not significantly different from levels before haemodilution (Rieger et al., 1979).

18.6 Discussion

Since the carefully controlled studies of Gottstein et al. (1972) and Heiss and Prosenz (1971), it has been established that hypervolaemic haemodilution (using dextran infusion), or isovolaemic haemodilution, is the only reliable way to improve cerebral perfusion pharmacologically (for details see Pearson et al., Chapter 10 of this volume). The present results show that in selected cases of peripheral hypoperfusion which have not responded to therapy it is not only possible to improve flow, cutaneous perfusion and skin temperature, but also to enhance the healing process. Since 1977, a large number of groups have included isovolaemic haemodilution in their therapeutic programmes for peripheral arterial and venous ulceration. The majority have reported positive results.

Much remains to be learned about the actual role of action of haemodilution, and, for that matter, on the role of "pseudoplastic" properties of creeping blood in the pathophysiology of cutaneous hypoperfusion. One has to stress that for rheological reasons it is probable that the pathophysiology of intermittent claudication is entirely different from that of cutaneous necrosis. We do not expect an immediate improvement of walking performance in patients with stage II arterial obstructive disease. If one accepts that a reduction of structural viscosity (or pseudoplasticity) of the blood is responsible for the beneficial effects of haemodilution, the explanation of the haemodynamic changes is rather complex; obviously we have not only changed the "flow properties" of the blood, but also flow conditions.

Aside from the haemorheological effects of haemodilution, its haemodynamic consequences must also be taken into consideration. In the present study we observed an increase of cardiac output but constant arterial pressure, which is in line with the findings of Sunder-Plassman et al. (1972), seen in acute haemodilution. We believe it most likely that the general decrease of the total peripheral resistance, followed by an increase in the venous return to the heart, represents the cause of the increased cardiac output. Since our haemodiluted patients do not show any significant increase in heart rate, the increased cardiac output shows that the reduction in red-cell mass is compensated by an increased stroke volume, which in turn might be the consequence of increased venous return without enhanced sympathetic nervous activity. The reduction of the oxygen-transport capacity cannot be the real cause for the increased cardiac output for two reasons: following haemodilution down to a haematocrit value of 31% the oxygenation of all vital organs is not decreased but increased (Kessler and Messmer, 1975; Messmer et al., 1975). Moreover the increase of cardiac output produced by haemodilution cannot be blocked by simultaneous hyperbaric oxygenation (Dedichen et al., 1967). Thus, there are two synergistic effects following haemodilution: the first is reduction of the red-cell mass and hence blood viscosity; the second is augmentation of cardiac output and hence mean blood flow viscosity. Both are able to reduce the probability of red-cell aggregation, which would occur initially in the pre- and post-capillary vessels. Since apparent blood viscosity, especially in the microcirculation, depends so strongly on the incident forces, blood viscosity is kept more in the range of its normal low value.

In the nutrient capillaries themselves, the situation is even more complex: when rapidly perfused, apparent blood viscosity is probably not significantly different from the plasma viscosity (Fahråeus-Lindqvist effect). For geometrical reasons aggregates are unlikely to form here, but pre- or post-capillary aggregates can nevertheless have a very pronounced indirect effect on capillary perfusion. In the light of the paramount effect of plasma viscosity on capillary perfusion, a reduction of plasma viscosity in vascular patients, which is frequently elevated (Rieger et al., unpublished observations), might result in enhanced capillary perfusion. Studies are presently under way to investigate this possibility, using defibrinogenating enzymes in combination with haemodilution.

The rheological and haemodynamic changes following haemodilution must be regarded as prerequisites for the improvement of both the collateral perfusion and the overall flow within the disturbed microcirculation, especially in those gangrenous areas where the flow had previously come to a complete standstill. However, whether or not the most peripheral localised hypoperfused marginal tissues of the ulcers really profit from this is at present difficult to demonstrate objectively, for obvious methodological reasons. The clinical observations and the thermographic findings, however, seem to indicate an improved flow (erythrocytes or plasma?) in close proximity to the skin ulcers. It seems that the nearer the lesion to the heart, the better the tendency of healing. A curative effect was seen in all cases affecting the upper limbs. Nevertheless, it must also be considered that ulcers of the upper limbs show a more pronounced spontaneous tendency to heal than ulcers of the lower limbs. In this connection the study of Bailey et al. (1979) is of interest: they found in 59 successive diabetics who underwent amputations in the feet that the haemoglobin before operation was significantly lower in amputations that healed than in amputations that failed.

As was to be expected, the healing tendency of lesions of the lower limbs seems to depend on the number of arterial occlusions in several vessel regions and the level of the post-occlusive systolic blood pressure, respectively (Table 18.2). At this point, the theoretical as well as the practical limitations of this procedure become evident. If, for a maximally dilated vessel with fixed geometry the flow rate is the product of the driving pressure and the blood fluidity ($\dot{V} = \triangle P . \emptyset$) even the most dramatic improvement of the fluidity is useless if the post-stenotic pressure (and hence the pressure gradient over the microvascular exchange network) is insufficient. At best, therefore, we can expect a beneficial effect of haemodilution in the narrow range (in time and space) between a residual perfusion compatible with survival and total standstill of the perfusion. In this context the possibility must be seriously discussed that the beneficial effects in the necrotic and the perifocal regions are not primarily attributable to an improved supply of oxygen, but rather a reinstitution of the removal of catabolites. This mechanism might explain the subjective improvement of the resting pain (by better clearance of pain-producing metabolites such as potassium) as well as the often dramatic clearance of the perifocal oedema. Experiments to clarify this possibility are in preparation.

Since the first successful application of haemodilution in peripheral and cerebral arterial insufficiency, a pilot study of haemodilution in ocular vascular disorders has been instituted (Wiederholt, 1980). In "central venous occlusion", which is now regarded as a consequence of gradually developing arterial obstruction, the immediate results were positive, i.e. vision and/or visual acuity were improved significantly. Despite the fact that laser photocoagulation was still necessary (to prevent haemorrhage on the basis of neovascularisation), the beneficial effects of haemodilution over photocoagulation alone warrant more haemorheological research in this important field of degenerative vascular disease. Since the consequences of arterial retinal occlusion become irreversible before therapeutic haemodilution could be started, it is not surprising that so far there are no successful treatments of this disorder by haemodilution. Nevertheless, even in these desperate cases there is hope if preventive measures could be started in high-risk patients.

18.7 Summary and Conclusions

In 25 patients with arterial occlusive disease in the clinical stage IV (Fontaine), haemodilution down to an average haematocrit value of 0.31 ± 0.02 was performed. Eighteen patients had peripheral ulcers of the lower limbs and 3 patients had ulcers of the upper limbs. Four patients had tissue lesions both of the feet and toes and hand and fingers. The average age of the patients was 43 years. Haemodilution was achieved by repeated withdrawals of approximately 500 ml blood, followed by reinfusion of the autologous plasma along with infusion of solutions of low-molecular-weight dextran to avoid hypovolaemia. In all patients with peripheral ulcers of the upper limbs, a good to very good clinical effect was obtained ranging from partial to complete healing. The healing tendency of lesions in the feet and toe region was on the whole less marked and depended basically on the total number of vascular blockages found. The rheological and haemodynamic mechanisms of the therapeutic effect of haemodilution, the preliminary criteria for indicating this therapy, and its observed complications, are discussed.

18.8 References

Bailey M J, Johnston C L W, Yates C J P, Somerville P G, Dormandy J A (1979) Preoperative haemoglobin as predictor of outcome of diabetic amputations. Lancet *ii:* 168-170

Boerhave H (1703) De usu rationicinii mechanici in medicina. Leyden

Dedichen H, Race D, Schenk W G (1967) Hemodilution and concomitant hyperbaric oxygenation. J Thor Cardiovasc Surg *53:* 341-347

Driessen G K, Heidtmann H, Schmid-Schönbein H (1979) Effect of hematocrit on red cell velocity in the capillaries of rat mesentery during hemodilution and hemoconcentration. Pfluegers Arch *380:* 1-6

Ehringer H, Dudczak R, Kleinberger G, Lechner K, Reiterer W (1971) Arvin: Schlangengift als neue Therapiemöglichkeit bei Durchblutungsstörungen. Wien Klin Wochenschr *83:* 411-416

Ehrly A M (1973) Verbesserung der Fließeigenschaften des Blutes: Ein neues Prinzip zur medikamentösen Therapie chronisch arterieller Durchblutungsstörungen. Vasa, Suppl. No. 1

Fahråeus R (1929) The suspension stability of blood. Physiol Rev *9:* 241-274

Fahråeus R, Lindqvist T (1931) The viscosity of blood in narrow capillary tubes. Am J Physiol *96:* 562-568

Gelin L E (1961) Disturbance of the flow properties of blood and its counteraction in surgery. Acta Chir Scand *122:* 287-293

Gottstein H, Held K, Sedlmeyer I (1972) In: Messmer K, Schmid-Schönbein H, (eds). Hemodilution-theoretical basis and clinical application. Karger, Basel, New York, pp. 247-257

Guyton A C, Richardson T Q (1961) Effect of hematocrit on venous return. Circ Res *9:* 157-165

Heiss W D, Prosenz P (1971) The influence of low molecular weight dextran on total and regional cerebral blood flow. Abstracts of Cerebral Blood Flow Symposium, Rome

Hint H D (1964) The flow properties of erythrocyte suspensions in isolated rabbit ear, the effect of erythrocyte aggregation, hematocrit and perfusion pressure. Bibl Anat *4:* 112-118

Kessler M, Messmer K (1975) Tissue oxygenation during hemodilution. In: Messmer K, Schmid-Schönbein H. (eds). Intentional hemodilution. Proceedings 2nd Int Sump. Bibl Hematol *41:* Karger, Basel, New York, pp. 16-20

Kiesewetter H, Schmid-Schönbein H, Radtke H, Stolwerk G (1979) In vitro demonstration of collateral blood viscidation: flow measurement in a model of vascular networks. Microvasc Res *17:* 72

Knisely M H (1965) Intravascular erythrocyte aggregation (blood sludge). In: Hamilton W F, Dow P. (eds). Handbook of physiology, Sect. 2, vol. III, 2249-2292. Amer Physiol Soc, Washington D.C.

Kohler M, Rieger H (1979) Über das Verhalten des Herzzeitvolumens und anderer Kreislaufparameter während subakuter Hemodilution bei Kranken mit Arterienverschluß. Vortrag 44. Tagung der Deutschen Gesellschaft für Kreislaufforschung, Bad Nauheim

Kreutz F (1978) Der optimale Hämatokrit für den arbeitenden Skelettmuskel. Dissertation. Köln

Lipowsky H H, Zweifach B W (1974) Network analysis of microcirculation of cat mesentery. Microvasc Res *7:* 73-83

Messmer K, Schmid-Schönbein H (eds) (1972) Hemodilution, theoretical basis and clinical application. Karger, Basel

Messmer K, Schmid-Schönbein H (eds) (1975) Intentional hemodilution. Proceedings 2nd Int Symp. Bibl Hematol *41,* Karger, Basel, New York

Messmer K, Gornandt L, Jesch F, Sinagowith E, Sunder-Plassmann L, Kessler M (1975) Oxygen transport and tissue oxygenation during hemodilution with dextran. In: Bruley D F, Bicher H I (eds). Oxygen transport to tissue. Plenum-Press, New York, pp. 669-680

Rieger H, Kohler M, Schoop W, Schmid-Schönbein H, Roth F J, Leyher A (1979) Hemodilution (HD) in patients with ischemic skin ulcers. Klin Wochenschr *57:* 1153-1161

Schmid-Schönbein H, Gallasch G, Volger E, Klose H J (1973) Microrheology and protein chemistry of pathological red cell aggregation (blood sludge) studied in vitro. Biorheology *10:* 213-227

Schmid-Schönbein H, Weiss J, Volger E, Klose H J, Malotta H (1978) Mikro-Hamorheologie und Defibrinogenisierung. ZFA (Stuttgart) *54:* 1635-1643

Sunder-Plassmann L, Klovekorn W P, Messmer K (1972) Hemodynamic and rheological changes induced by hemodilution with colloids. In: Messmer K, Schmid-Schönbein H (eds). Hemodilution — theoretical basis and clinical application. Karger, Basel, New York, 184-198

Wiederholt M (1980) Retinal vascular occlusions. Hemodilution, a new therapeutic concept? Microvasc Res (in press)

Whittaker S R F, Winton F R (1933) The apparent viscosity of blood flowing in the isolated hindlimb of the dog, and its variation with corpuscular concentration. J Physiol (London) *78:* 339-369

19 Plasma Exchange

A. J. Dodds

19.1 Introduction

Plasma exchange has been used as a therapeutic tool since the beginning of the century (Abel et al., 1914). However its use has increased markedly since the development of efficient machines to perform the manoeuvre. A distinction is sometimes drawn between the terms plasmapheresis and plasma exchange. Plasmapheresis (plasma + Gr. *aphairesis* or removal) strictly means the removal of plasma (Buskard, 1978). Plasma exchange implies the removal of plasma and replacement with whole or fractionated plasma or plasma substitutes. Large-volume plasmapheresis is not possible without replacement, so plasma exchange is the more appropriate term.

The first major application of the principle was to control and treat blood hyperviscosity in the malignant paraproteinaemias (Solomon and Fahey, 1963; Powles et al., 1971). The benefits of plasma exchange in these conditions have now been established (Russell et al., 1977).

More recently the procedure has been tried in a wide variety of conditions with varying success. It has been used to remove both autoantibodies and immune complexes, to control hyperlipidaemia and to remove drugs from the plasma (Pinchling, 1978). Because of the nature of the procedure it may be difficult to distinguish between effects produced by the removal of an abnormal plasma component and depletion of normal plasm proteins. Indeed the removal of the mediators of inflammation such as complement and fibrinogen may provide another benefit in inflammatory diseases (Kincaid-Smith and d'Apice, 1978).

Finally it is logical to assume that alteration of the components of bulk blood will lead to alteration in its physical properties. Thus removal of abnormal plasma proteins may result in a reduction of blood viscosity. The potential of using this form of therapy in other conditions such as vascular disease has not been fully explored. Here the intention would be to improve blood flow by altering its physical properties.

19.2 Techniques of Plasma Exchange

A manual method for plasmapheresis was described in 1914 (Abel et al., 1914). It involved removal of a unit of blood, allowing the red cells to sediment, removing the plasma and then reinfusing the cells. Although this method is slow and cumbersome it can still be used in an emergency.

The first mechanical method was developed in 1951 (Cohn et al., 1951), but only provided satisfactory yields of plasma or red cells. Subsequent development and simplification have led to modern, efficient cell separators.

Two main systems are used for plasma exchange. The first type uses continuous flow centrifugation (CFC) and was developed by IBM (Judson et al., 1968). Blood is removed by a venous line and fed into a rapidly rotating bowl which separates the blood into red-cell, buffy-coat and plasma layers. Any layer can be extracted and the remainder re-infused into the patient. This instrument is rather large and requires considerable expertise to operate. It also suffers from the disadvantage of needing to be cleaned and sterilized after each use. The Aminco Celltrifuge is a similar instrument. IBM have now developed a simplified version which has a disposable bowl and is currently undergoing clinical trial.

The second type works on an intermittent system (IFC). The Haemonetics cell separator is such a machine. Blood is removed into a bowl and centrifuged to separate the components. The required fraction is removed, the remainder returned to the patient and another cycle can then take place. The system is a closed one and all the components which come into contact with blood are disposable. This system has been cheaper, more portable and easier to operate. However the volume of plasma removed per cycle is dependent on the patient's haematocrit.

In general the procedure is well tolerated irrespective of the machine used although a number of side-effects may occur. The commonest are chills, nausea, vomiting and syncope. These have a variety of causes but the exact mechanism in an individual patient can be difficult to determine. Mild hypocalcaemia due to too rapid re-infusion of citrated red cells is common. In addition some of the replacement solutions are low in calcium. Syncope can be caused by hypovolaemia if a large volume of blood is removed too rapidly, especially in an older person. Allergic reactions may occur to some replacement solutions, particularly dextran. Hypotension is a poorly recognised complication of rapid infusion of Plasma Protein Fraction (PPF). It is thought to be due to a bradykinin-like substance in the PPF (Isbister and Biggs, 1976; Culliver and Pennington, 1979; Izaka et al., 1974). Fluid overload may also occur if replacement solutions are given too enthusiastically and hypothermia may occur if they are at too low a temperature. Major complications such as air embolism or haemolysis are very rare.

More long-term effects are due to depletion of normal plasma components during intensive plasma exchange. This will depend on the replacement solutions used. Frequent plasma exchange can reduce immunoglobin levels (Friedman et al., 1975) and complement (Keller and Urbaniak, 1978). Although depletion of coagulation factors and mild thrombocytopenia develops bleeding does not seem to be a problem (Bayer et al., 1975; Keller et al., 1979).

19.3 Haemorheological Effects of Plasma Exchange

The apparent viscosity of blood is not only dependent on the red-cell concentration but also the phenomena of red-cell aggregation and deformation. Progressive aggregation occurs at shear rates below 100 s^{-1} even in normal blood and leads to an increase in apparent viscosity (Charm et al., 1964). Aggregation requires a flexible membrane (Chien et al., 1970) and the presence of aggregating macromolecules (Chien, 1970). Fibrinogen is the most important of these in the plasma. At low shear rates a direct relationship between apparent blood viscosity and plasma fibrinogen concentration can be demonstrated, although this may not apply to very low fibrinogen levels (Blättler et al., 1979). The relationship between plasma fibrinogen concentration and red-cell deformability is not as yet settled (Rampling and Sirs, 1972; Dodds et al., 1979; Drummond et al., 1979).

Abnormal components of plasma can influence blood viscosity. Cryoglobulin and paraprotein may both be associated with an elevation of viscosity (McGrath and Penny, 1976; McGrath and Penny, 1978). Plasma factors, apart from fibrinogen, may also affect red-cell deformability (Braasch, 1974; Dodds et al., 1979).

The effects of plasma exchange on blood rheology are likely to be complex but in general will depend on the extent of fibrinogen depletion as well as the removal of abnormal plasma components. Changes in haematocrit may also occur if fluid balance is not maintained. There is some debate as to the specific rheological effects of the synthetic plasma substitutes. Schmid-Schönbein et al. (1972) have investigated the effect of some of them *in vitro* on the microrheology of human blood. They found that if a 1 in 10 haemodilution was made *in vitro*, then all agents reduced apparent viscosity. However, low-molecular-weight dextran reduced relative apparent viscosity at high shear rates, implying better cell deformation and orientation. High-molecular-weight dextran and Haemaccel slightly increased relative apparent viscosity at low shear rates, which implied that they

Table 19.1. Haemorheological variables before and immediately after plasma exchange in 10 patients (mean ± SD)

	Before	After	*p* value[a]
Blood viscosity 0.77 s^{-1} (mPa.s)	25.6 ±4.0	12.5 ±2.6	<0.002
Blood viscosity 91 s^{-1} (mPa.s)	4.5 ±0.2	3.9 ±0.3	<0.002
Red cell deformability index	0.5 ±0.06	0.41 ±0.11	NS
Plasma viscosity (mPa.s)	1.41 ±0.08	1.24 ±0.09	<0.002
Plasma fibrinogen (g/litre^{-1})	2.6 ±0.4	1.4 ±0.2	<0.002
Haematocrit (%)	44.0 ±0.4	43.5 ±0.3	NS

[a]paired Student's *t* test

had a slight aggregating ability in their own right. All the agents increased plasma viscosity. Dormandy (1971) found that low-molecular-weight dextran was no different from Hartman's solution in ability to lower blood viscosity at a high shear rate. It seems probable that if these agents do have any specific rheological properties it is not pronounced and that the effect of fibrinogen depletion is more significant in the context of plasma exchange.

Kilpatrick et al. (1979) have investigated the reduction of blood viscosity in seven hyperlipidaemic patients following a 2.4-4.0-litre exchange. They found a 83% fall in apparent blood viscosity at 0.07 s^{-1} immediately following exchange and a 30% fall at 3.4 s^{-1}; Plasma fibrinogen fell by 59%, plasma viscosity by 21% and haematocrit by 4.5%. They also followed the sequential values in two patients. Blood viscosity had almost returned to pre-exchange levels by the 6th day, although plasma fibrinogen and plasma viscosity had not. They speculated that the procedure may result in short-term enhancement of blood flow. However they found no objective evidence of this.

We have also investigated the haemorheological changes in 10 patients with vascular disease following a single 2-litre plasma exchange (Table 19.1). The mean viscosity at 0.77 s^{-1} fell by 49% following exchange, and by 13% at 91 s^{-1}. Plasma fibrinogen fell by 47% and plasma viscosity by 12%. Red-cell filtration rate, however, did not change in this group. PPF and Haemaccel were used as replacement solutions.

Reduction in blood viscosity may also be responsible for the reported anti-hypertensive effect of plasma exchange (Whitworth et al., 1978).

19.4 Results of Plasma Exchange in Raynaud's Syndrome

To investigate the possible use of plasma exchange in the treatment of patients with Raynaud's syndrome and severe digital ischaemia, a controlled clinical trial has been carried out (O'Reilly et al., 1979). Plasma exchange weekly for 4 weeks was found superior to both placebo and intermittent heparinisation. In eight patients the haemorheological changes were studied during this treatment and related to clinical results (Dodds et al., 1979). A Doppler ultrasound probe was used to quantify the presence of blood flow in digital arterial segments. The primary disease was scleroderma in four and four were idiopathic. No cryoprotein, paraprotein or cold agglutinins were detected in any of the cases.

Apparent whole-blood viscosity was measured at 0.77 s^{-1} and 91 s^{-1} and adjusted to a standard haematocrit (Dormandy, 1974). Plasma fibrinogen and microhaematocrit were also measured. Red-cell deformability was assessed using a filtration technique in which a suspension of red cells was passed through Nuclepore polycarbonate sieves (Dodds et al., 1979). To assess blood flow in patent arterial segments, each finger was divided into proximal, middle and distal thirds, each having a radial and ulnar segment. Plasma exchange was carried out using a Haemonetics Model 30 Blood Processor. The mean volume exchanged per week was 1.9 litre and the mean volumes of replacement solutions used were 1.2 litre PPF, 0.5 litre Haemaccel and 0.5 litre physiological saline. Measurements were carried out before treatment and 6 weeks after the last exchange. Four of the patients had more detailed tests during the treatment.

Whole-blood viscosity at 0.77 s^{-1} was significantly lower at follow-up ($p < 0.001$; paired Student's t test) (Fig. 19.1). However there was no difference in viscosity at the high shear rate or in haematocrit. Plasma fibrinogen was lower

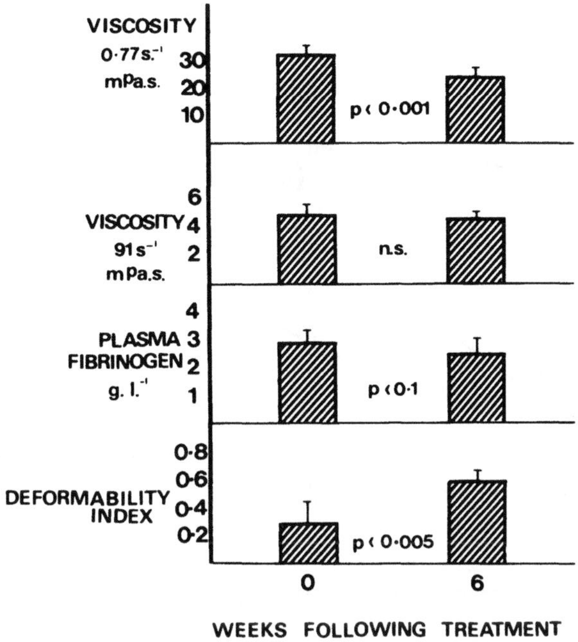

Fig.19.1. Haemorheological values before and 6 weeks after four weekly plasma exchanges in eight patients (viscosity results are adjusted to a standard haematocrit (45%) (mean ± SD)

but not significantly so. Red-cell deformability improved following treatment ($p<0.005$; paired Student's t test). Serial changes in four patients (Fig. 19.2) revealed an acute fall in viscosity at low shear following each exchange and a progressive fall over the time of treatment. Haematocrit fell on average by 5% after each exchange but was no different at follow-up. Red-cell deformability, only measured before each exchange, improved particularly in the first 2 weeks. All the patients showed an increase in the number of digital artery segments containing detectable blood flow (Fig. 19.3). This was significant at both 15°C and 21°C (Wilcoxon rank sum test).

The acute changes in blood viscosity seen in this study were largely brought about by a reduction in plasma fibrinogen. Thus the major change occurred at the low shear rate. The blood flow in the digits of such patients is known to be sluggish (McGrath et al., 1977; Coffman and Cohen, 1971). Therefore a reduction in viscosity at low shear may be particularly important. Hyperviscosity is well documented in Raynaud's syndrome (Dintenfass, 1977; Goyle and Dormandy, 1976). This rise would be accentuated by the low temperatures in the digits.

The sustained reduction in blood viscosity at follow-up did not appear to be associated with a fall in plasma fibrinogen. It is possible that the improvement in red-cell deformability is a factor. A reduction in deformability in Raynaud's syndrome has been described (Dintenfass, 1977). Acidosis associated with tissue ischaemia may amplify this reduction and lead to a further reduction in blood flow (Leading article, Lancet, 1978). The improvement in deformability following plasma exchange could be explained by the removal of an abnormal plasma factor interfering with red-cell behaviour (Braasch, 1974).

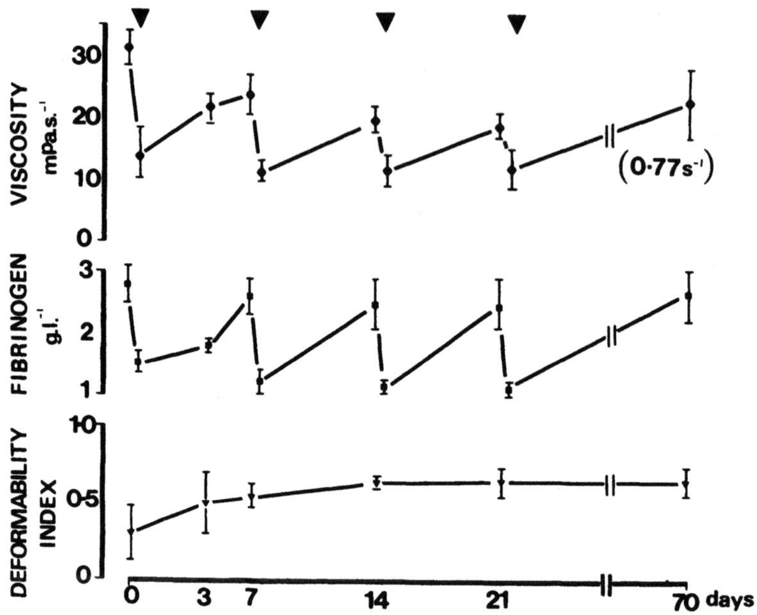

Fig.19.2. Serial haemorheological values before, immediately after, and 6 weeks following four weekly plasma exchanges in four patients. Viscosity results are adjusted to a standard haematocrit (45%). RCD was not measured immediately after a plasma exchange (mean ± SD)

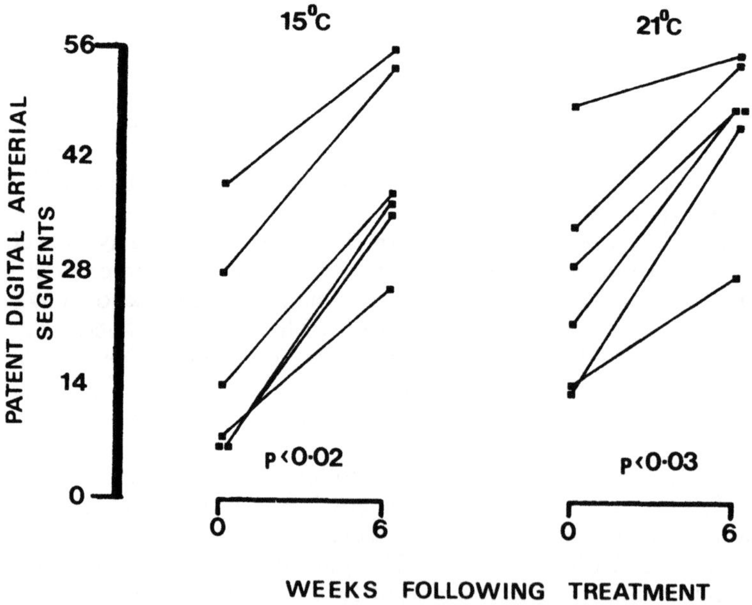

Fig.19.3. Digital artery patency rate before and 6 weeks after four weekly plasma exchanges in six patients at 15°C and 21°C

19.5 Conclusion

Although plasma exchange is both novel and dramatic, it is rather crude in concept because it removes all the components of plasma rather than selectively removing what is required. In the future more selectivity may be obtained by combining plasma exchange with affinity chromatography. This would then reduce the dependence on limited supplies of PPF.

Plasma exchange is also expensive and requires continuous supervision by highly trained personnel. It therefore must remain, in many conditions, an experimental tool and it has proved an adjunct to, rather than a substitute for, more conventional therapy. However it has still had a considerable impact on medicine. In the rheological area it has an established place in the treatment of the malignant paraproteinaemias and it may also find a place in the management of patients with vascular insufficiency, particularly digital ischaemia.

Acknowledgements. The author is most grateful for the advice and encouragement of Mr. J. Dormandy and Professor P.T. Flute. He is supported by Berk Pharmaceuticals and the Pathology Research Fund of St. George's Hospital. Mrs. P. Hopgood and Mrs. J. Allen provided technical assistance. The manuscript was typed by Mrs. C. Velarde.

19.6 References

Abel J J, Rowntree L G, Turner B B (1914) Plasma removal with return of the corpuscles. J Pharmacol Exp Ther *5:* 625-641

Bayer W L, Farrales F B, Summers T, Belcher C (1975) Coagulation studies after Plama Exchange with plasma protein fraction and lactated Ringer's solution. In: Goldman J M, Lowenthal R M (eds). Leucocytes: Separation, collection and transfusion. Academic Press, London

Blättler W, Straub P W, Jeanneret C, Horaic G S (1979) Effect of low fibrinogen concentrations on the rheology of human blood in vitro. Am J Physiol *236:* H447-H450

Braasch D (1974) Annotation: Red cell elasticity. Br J Haematol *24:* 405-407

Buskard N A (1978) Plasma exchange and plasmapheresis (editorial). Can Med Assoc J *119:* 681-683

Charm S E, McComis W, Kurland G (1964) Rheology and structure of blood suspensions. J Appl Physiol *19:* 127-133

Chien S (1970) Shear dependence of effective cell volume as a determinant of blood viscosity. Science *168:* 977-978

Chien S, Usami S, Dellenback R J, Gregersen M I (1970) Shear dependent deformation of erythrocytes in rheology of human blood. Am J Physiol *219:* 136-142

Coffman J D, Cohen A S (1971) Total and capillary blood flow in Raynaud's Phenomenon. N Engl J Med *285:* 259-263

Cohn E J, Tullis J L, Surgentor D M, Batchelor W H, D'Hont M D (1951) Biochemistry and biomechanics of blood collection, processing and analysing of human blood and other tissues. Address presented before the National Academy of Science, Yale University, New Haven

Culliver H, Pennington D G (1979) Mechanisms of vascometer reactions in the use of SPPS. Vox Sang *36:* 201-207

Dintenfass L (1977) Haemorheological factors in Raynaud's Syndrome. Angiology *28:* 472-481

Dodds A J, O'Reilly M J G, Yates C J P, Cotton L T, Flute P T, Dormandy J A (1979) Haemorheological response to Plasma Exchange in Raynaud's Syndrome. Br Med J *4:* 1186-1187

Dormandy J A (1971) Influence of blood viscosity on blood flow and the effect of low molecular weight Dextran. Br Med J *4:* 716-719

Dormandy J A (1974) Mechanical and engineering problems of blood viscosity. Biomed Eng *9:* 284-303

Drummond M M, Lowe G D O, Belch J J F, Forbes C D, Barbenel J C (1979) An assessment of red cell deformability using a simple filtration method. Thromb Haemostas *42:* 106

Friedman B A, Schork M A, Mocniak J L, Oberman H A (1975) Short-term and long-term effects of plasmapheresis on serum proteins and immunoglobulins. Transfusion *15:* 467-472

Goyle K B, Dormandy J A (1974) Abnormal blood viscosity in Raynaud's Phenomenon. Lancet *i:* 1317-1318

Isbister J P, Biggs J L (1976) Reactions to rapid infusion of stable plasma protein solution during large volume plasma exchange. Anaesth Intensive Care *4:* 105-107

Izaka K, Tsutsui E, Mima Y, Hasegawa E (1974) A bradykinin-like substance in heat-treated human plasma protein solution. Transfusion *14:* 242-248

Judson G, Jones A, Kellogg R, Buckner D, Eisel R, Perry S, Greenough W (1968) Closed continuous flow centrifuge. Nature (London) *217:* 816-818

Keller A J, Urbaniak S J (1978) Intensive plasma exchange on the Cell Separator: Effects on serum immunoglobulins and complement components. Br J Haematol *38:* 531-540

Keller A J, Chirnside A, Urbaniak S J (1979) Coagulation abnormalities produced by plasma exchange on the cell separator with special reference to fibrinogen and platelet levels. Br J Haematol *42:* 593-603

Kilpatrick D, Fleming J, Clyne C, Thompson G R (1979) Reduction of blood viscosity following plasma exchange. Atherosclerosis *32:* 301-306

Kincaid-Smith P, d'Apice A J F (1978) Plasmapheresis in rapidly progressive glomerulonephritis. Am J Med *65:* 564-566

Leading Article (1978) Red cell deformability Lancet *ii:* 1348-1349

McGrath M A, Penny R (1976) Paraproteinaemia: blood hyperviscosity and clinical manifestations. J Clin Invest *58:* 1155-1162

McGrath M A, Peek R, Penny R (1977) Blood hyperviscosity with reduced skin blood flow in scleroderma. Ann Rheum Dis *36:* 569-574

McGrath M A, Penny R (1978) Blood hyperviscosity in cryoglobulinaemia: temperature sensitivity and correlation with reduced skin blood flow. Aust J Exp Biol Med Sci *56:* 127-137

O'Reilly M J G, Talpos G, Roberts V C, White J M, Cotton L T (1979) Controlled trial of plasma exchange in the treatment of Raynaud's Syndrome. Br Med J *1:* 1113-1115

Pinchling A J (1978) Plasma exchange. Br J Hosp Med *20:* 552-559

Powles R, Smith C, Kohn J, Hamilton Fairley G (1971) Method of removing abnormal protein rapidly from patients with malignant paraproteinaemias. Br Med J *3:* 664-667

Rampling M, Sirs J A (1972) The interactions of fibrinogen and dextrans with erythrocytes. J Physiol (London) *223:* 199-212

Russell J A, Toy J L, Powles R L (1977) Plasma exchange in malignant paraproteinaemias. Exp Haematol *5:* 100-116

Schmid-Schönbein H, Klose H J, Volger E (1972) Effect of colloidal plasma substitutes on microrheology of human blood. In: Messmer K, Schmid-Schönbein H (eds). Haemodilution: Theoretical basis and clinical application. Karger, Basel, New York, pp. 66-83

Solomon A, Fahey J L (1963) Plasmapheresis therapy in macroglobulinaemia. Ann Intern Med *58:* 789-800

Whitworth J A, d'Apice A J F, Kincaid-Smith P, Shulkes A A, Skinner S L (1978) Antihypertensive effect of plasma exchange. Lancet *i:* 1205

20 Plasma Exchange in Macroglobulinaemia

T. L. Allan

20.1 Introduction

In 1944, Waldenström described a new syndrome in three patients who presented with a variety of haemorrhagic and ocular disorders associated with marked increases in both serum viscosity and in a serum globulin which had a very high molecular weight on ultracentrifugation; an underlying lymphoproliferative disorder involving the bone marrow was subsequently found in all three patients. In a recently published study of 40 cases of Waldenström's macroglobulinaemia (WM), one-third of the patients had clinical evidence of this hyperviscosity syndrome, (Mackenzie and Fudenberg, 1972). Some authors (Fahey et al., 1965; Mackenzie et al., 1970) have claimed that individual patients have a threshold value for blood viscosity above which they become symptomatic, while others (Pronk et al., 1969) could find no clear relation between either serum viscosity or paraprotein level and the clinical symptomatology. There is a similar divergence of opinion on the relation between IgM level and viscosity. Some workers have found a highly significant linear correlation between serum viscosity and IgM (Somer, 1966; 1975; Rosenblum and Asofsky, 1968), yet others have found a steep non-linear increase in serum viscosity, particularly with IgM levels above 40 g/litre (Preston et al., 1978). This discrepancy may in part be explained by the observation of Mackenzie and Babcock (1975) that a steep non-linear relationship is found in patients whose monoclonal paraprotein is markedly asymmetrical in configuration, while those with a more normal spheroidal IgM tend to exhibit a linear relationship. In our experience, there is a highly significant linear correlation between whole-blood viscosity and IgM level, at least up to 40 g/litre paraprotein (Fig. 20.1).

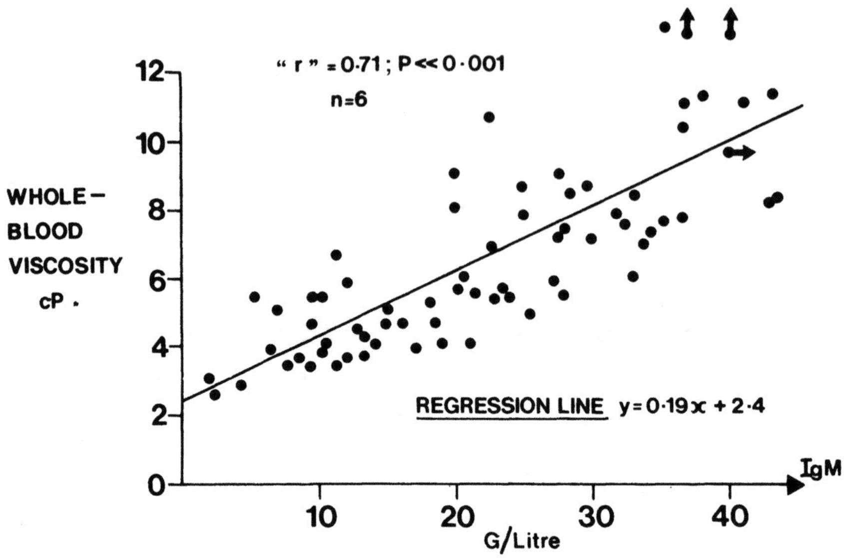

Fig.20.1. Relation between whole-blood viscosity and IgM level in six patients with Waldenström's macroglobulinaemia (results pooled). 1 centipoise (cP) = 1 mPa.s

20.2 Methods

We have treated six patients with WM using an Aminco cell separator. There were three male and three female patients ranging in age from 62 to 80 years (mean 69.3) and on each occasion, 2.5-4.0 litres plasma was exchanged for stable Plasma Protein Solution (43 g/litre albumin solution). These patients all satisfied the criteria for WM in having serum IgM levels greater than 10 g/litre at presentation, along with characteristic marrow findings. The indications for plasmapheresis were symptoms referrable to hyperviscosity, e.g. tiredness, exertional dyspnoea, early congestive cardiac failure and characteristic fundal changes.

Before and after each plasma exchange, measurements were made of immunoglobulins, by radial immunodiffusion, and of plasma and whole-blood viscosity at 37°C, using a Wells-Brookfield microviscometer. As fibrinogen levels are usually normal or slightly low in WM (Waldrenström, 1944) and as physiological variations in fibrinogen have a negligible effect on whole-blood viscosity unless at ultra-low shear rates (Begg and Hearns, 1966; Weaver et al., 1969) fibrinogen was not routinely assayed. All whole-blood viscosity measurements were corrected to a haematocrit of 0.30 (the mean Hct. value for the series) using Cokelet's (1963) equation for red-cell suspensions,

$$\frac{\mu_s}{\mu_p} = \frac{1}{(1-H)2.5}$$

where μ_s = viscosity of suspension; μ_p = viscosity of plasma; and H = haematocrit.

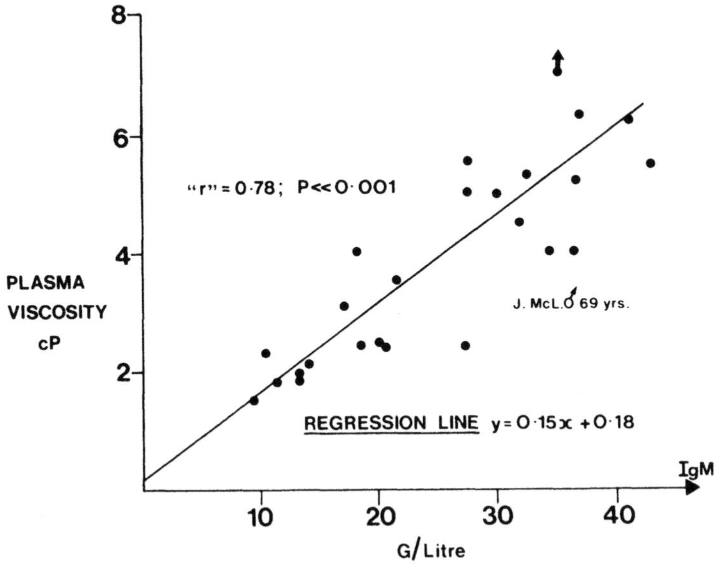

Fig.20.2. Relation between plasma viscosity and IgM level in one case of Waldenström's macroglobulinaemia. 1 centipoise (cP) = 1 mPa.s

20.3 Results and Discussion

Figures 20.2 and 20.3 show the linear relationship between IgM level and plasma and whole-blood viscosities respectively from serial observations on one patient, while Fig. 20.1 shows a similar relationship obtained by pooling 70 observations on all six patients. The effect of five plasma exhanges on IgM level and viscosity in one patient is shown in Fig. 20.4.

These results show a clear-cut linear relationship between whole-blood viscosity and IgM level, at least up to 40 g/litre. Higher levels of paraprotein were not encountered in this series. In the early stages of WM the serum IgM level slowly rises, with a doubling time of about 3-24 months (Hobbs, 1969), reflecting the gradual proliferation of abnormal lymphoid cells in the marrow. Eventually, however, there is a tendency for a steady state to be reached, and the markedly elevated IgM level then reflects the rate of IgM synthesis which has been shown by means of radio-labelled IgM studies (Barth et al., 1964) to be as high as 8.0 to 40.0 g/day (or approximately 5000 times normal rate). A fixed percentage of IgM (approx. 10%-14%) is catabolized daily. This helps to explain why the IgM level often rises rapidly to its former value following plasmapheresis. In fact, on some occasions there even appears to have been a rebound phenomenon, resulting in the IgM level climbing to a new high within a few days of plasmapheresis, despite a reduction in the patient's IgM of 20 g/litre or more during the exchange. It may be that in removing the paraprotein a chalone inhibitor of "B" cells is also removed. The main value of plasmapheresis in WM is in the rapid and effective treatment of the hyperviscosity syndrome. It has also been advocated as the treatment of choice in patients who are precluded from having chemotherapy by concomitant illnesses, or whose tumour has proved unresponsive to chemotherapy (Buskard et al., 1977).

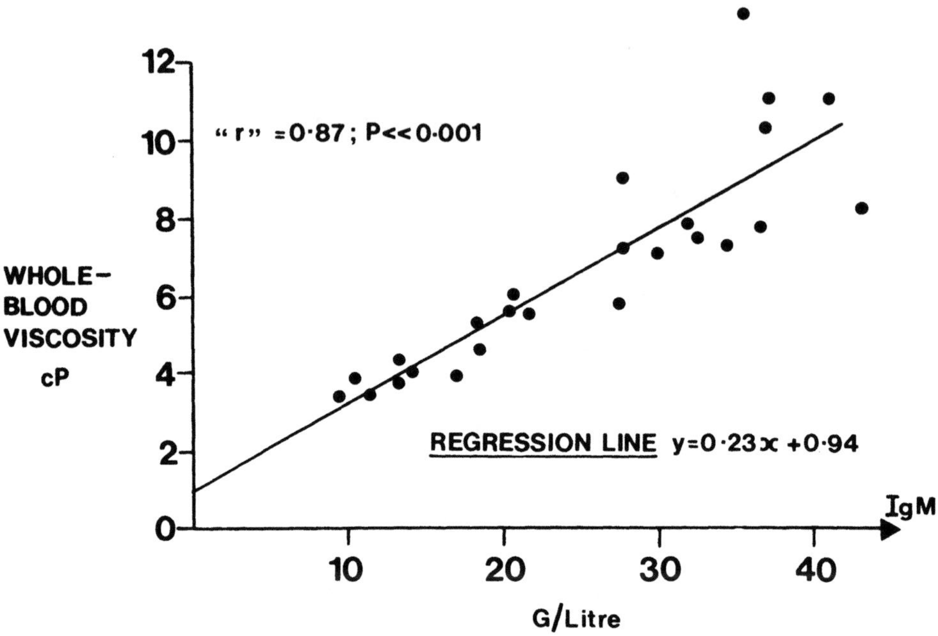

Fig.20.3. Relation between whole-blood viscosity and IgM level in one case of Waldenströms macroglobulinaemia. 1 centipoise (cP) = 1 mPa.s

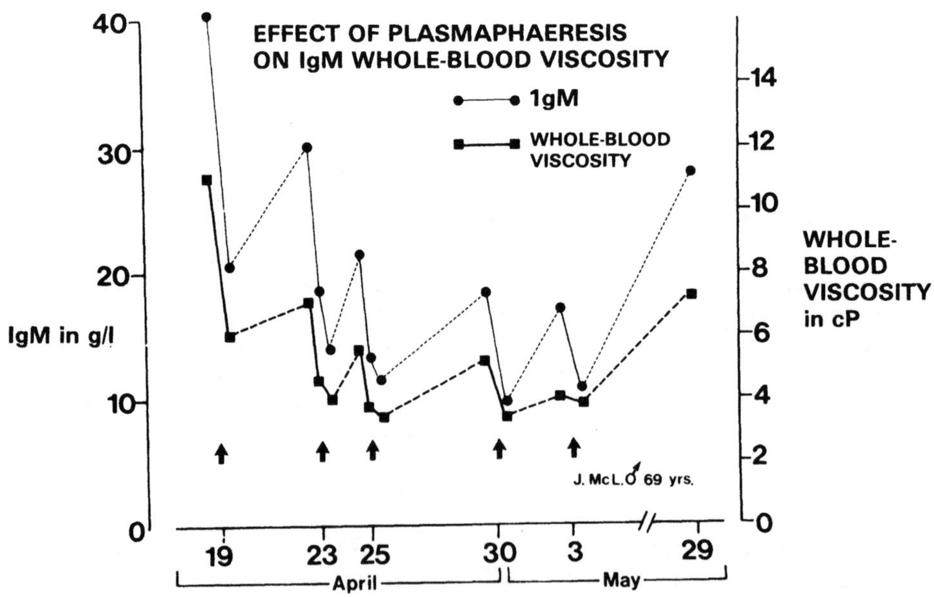

Fig.20.4. Effect of repeated plasma exchange on IgM level and whole-blood viscosity in one patient with Waldenström's macroglobulinaemia. 1 centipoise (cP) = 1 mPa.s

20.4 References

Barth W F, Wochner R D, Waldmann T A, Fahey J L (1964) Metabolism of human gamma macroglobulins. J Clin Invest *43:* 1036-1048
Begg T B, Hearns J B (1966) Components in blood viscosity. The relative contribution of haematocrit, plasma fibrinogen and other proteins. Clin Sci *31:* 87-93
Buskard N A, Galton D A, Goldman J M, Kohner E M, Grindle C F, Newman D L, Twinn K W, Lowenthal R M (1977) Plasma exchange in the long term management of Waldenström's macroglobulinaemia. Can Med Assoc J *117:* 135-137
Cokelet G R (1963) The rheology of human blood. Doctoral dissertation. M.I.T., Cambridge, Mass.
Fahey J L, Barth W F, Solomon A L (1965) Serum hypersensitivity syndrome. JAMA *192:* 464-467
Hobbs J R (1969) Immunochemical classes of myelomatosis. Br J Haemat *16:* 599-606
Mackenzie M R, Babcock J (1975) Studies of the hyperviscosity syndrome. 2. Macroglobulinaemia. J Lab Clin Med *85:* 227-234
Mackenzie M R, Brown E, Fudenberg H H, Goodenday L (1970) Waldenström's macroglobulinaemia. Correlation between expanded plasma volume and increased serum viscosity. Blood *35:* 394-408
Mackenzie M R, Fudenberg H H (1972) Waldenström's macroglobulinaemia. An analysis of forty patients. Blood *39:* 874-889
Preston F E, Cooke K B, Foster M E, Winfield D A, Lee D (1978) Myelomatosis and the hyperviscosity syndrome. Br J Haematol *38:* 517-530
Pronk E A S, Klein F, Elkerbout F, Radema H, Cleton F J (1969) Therapy of macroglobulinaemia. Acta Med Scand *186:* 273-281
Rosenblum W I, Asofsky R M (1968) Factors affecting blood viscosity in macroglobulinaemic mice. J Lab Clin Med *71:* 201-211
Somer T (1966) The viscosity of blood, plasma and serum in dys- and paraproteinaemias. Acta Med Scand *Suppl 456:* 1-97
Somer T (1975) Hyperviscosity syndrome in plasma cell dyscrasias. Adv Microcirc *6:* 1-55
Waldenström J (1944) Incipient myelomatosis or "essential" hyperglobulinaemia with fibrinogenopenia — a new syndrome? Acta Med Scand *117:* 216-247
Weaver J P A, Evans A, Walder D N (1969) The effect of increased fibrinogen content on the viscosity of blood. Clin Sci *36:* 1-10

21 Defibrinating Agents

G.D.O. Lowe, C.D. Forbes and C.R.M. Prentice

Two defibrinating agents are currently commercially available for use in the management of certain vascular diseases: both are proteolytic enzymes purified from snake venoms. Ancrod (Arvin) is derived from the venom of *Agkistrodon rhodostoma,* while batroxobin (Defibrase) is derived from *Bothrops atrox:* these two snakes are pit vipers from Southeast Asia and from Central-South America, respectively. A detailed review of the pharmacology of these agents has been made by Stocker (1978).

21.1 Effects on Blood Coagulation and Fibrinolysis

Unlike thrombin, which converts fibrinogen to fibrin by splitting off the two fibrinopeptides A and B, ancrod and batroxobin release only fibrinopeptide A from fibrinogen. Fibrin micro-clots and soluble fibrin complexes are formed, which are rapidly cleared from the circulation, resulting in a fall in circulating fibrinogen. Fibrinogen/fibrin degradation products (FDP) appear in the circulation at this time. The breakdown of fibrin is probably mediated partly by the reticulo-endothelial system and partly by plasminogen activation. Infusion of ancrod or batroxobin does not usually result in increased systemic fibrinolytic activity, but plasma plasminogen levels fall (by about 50%) and plasmin-antiplasmin complexes appear, confirming plasminogen activation, which is presumably secondary to intravascular fibrin formation and which may occur in the microcirculation (Collen and Verstraete, 1978). Levels of clotting factors other than fibrinogen are unaffected.

21.2 Effects on Platelets

Neither ancrod nor batroxobin appear to affect platelets *in vitro,* in contrast to thrombin, which induces platelet aggregation and the release reaction. Following infusion in humans and animals, reduced aggregation of platelets in response to ADP is observed: whether this is due to low fibrinogen levels, or to high levels of FDP which have platelet-suppressant effects, is uncertain (Prentice et al., 1969; Egberg and Johnsson, 1975). Patients with vascular diseases have been found to have abnormal platelet electrophoretic mobility responses to ADP, and increased "circulating" platelet aggregates detected by formalin fixation in freshly drawn blood: these abnormalities are normalised following defibrination with ancrod (Turpie et al., 1972; Lowe et al., 1979a). Moderate, transient falls in the platelet count and decreased levels of platelet factors 3 and 4 have also been described during the initial infusion of ancrod, and may reflect platelet trapping and damage in micro-clots (Vinazzer, 1975). Bleeding times and platelet adhesiveness are unaffected by infusion of defibrinating agents.

21.3 Effects on Blood Rheology

The inclusion of a review of defibrinating agents in this volume is due to the well-recognised fall in blood and plasma viscosity observed during defibrination therapy. Indeed, ancrod and batroxobin are at present the only drugs which are consistently found to affect blood rheology. The plasma fibrinogen level is well known to be a determinant of blood viscosity, by virtue of (a) its effect on plasma viscosity and (b) its effect on red-cell aggregation. Defibrination with ancrod reduces high-shear blood viscosity and plasma viscosity by 5%-10% (Ehrly, 1972), and it is likely that the fall in blood viscosity at high shear reflects the fall in plasma viscosity. With decreasing shear rate, defibrination causes a greater fall in blood viscosity (Dormandy et al., 1977), probably due to the additional effect of decreased red-cell aggregation. Low-shear viscosity measurements underestimate the effects of plasma fibrinogen on red-cell aggregation, because with increasing aggregation (causing increased apparent viscosity) there is also increasing separation of plasma from red cells at the viscometer sample chamber wall (causing decreased apparent viscosity). Measurement of red-cell aggregates by rheoscopy (Schmid-Schönbein et al., 1969) or photometry (Schmid-Schönbein et al., 1978) has shown more marked effects of defibrination therapy than those observed in low-shear viscometers. Decreased aggregates, decreased resistance of aggregates to dispersal by shearing forces, and decreased speed of reaggregation, have all been observed. It should be noted that, even after near-total defibrination, red-cell aggregates still exist, due to the presence of other plasma proteins which aggregate erythrocytes, such as α_2-macroglobulin and IgM (Schmid-Schönbein et al., 1978).

Following the observations that soluble fibrinogen-fibrin complexes, generated *in vitro* by thrombin, increased red-cell aggregation, blood viscosity, and "yield stress", the possibility of decreased blood fluidity during initial infusions of ancrod or batroxobin has been raised. Such decreased fluidity might theoretically result in impaired blood flow and increasing ischaemia. Blättler et al. (1974) infused volunteers with batroxobin and found no increase in plasma or blood viscosity at 37°C. These workers did find increases in plasma and blood viscosity at the non-physiological temperature of 22°C, and suggested that soluble fibrinogen-fibrin complexes and their rheological effects are low-temperature artefacts and do not apply *in vivo*. We have serially measured blood and plasma viscosity in 10 patients during a 6-h induction infusion of ancrod, and likewise found no significant increase in viscosity at 37°C (unpublished observations).

Red-cell deformability, measured by filtration methods, does not appear to be altered after defibrination. Defibrination has been reported to decrease erythrocyte flexibility as measured by the centrifuge technique of Sirs, which correlates highly with plasma fibrinogen levels (Rampling and Sirs, 1972). Further work is required to clarify these phenomena.

21.4 Effects on Blood Flow and Blood Pressure

Reduction of blood viscosity should in theory lower the resistance of the peripheral vascular bed and hence increase blood flow (Strandness and Sumner, 1975). The effect of defibrination therapy on blood flow has so far only been studied in the lower limbs of patients with peripheral arterial disease. Increase in resting and reactive leg flow, measured by calf plethysmography, was described by Ehringer et al. (1974). Meissner et al. (1978) confirmed an increase in resting leg muscle blood flow, using a ^{133}Xenon clearance technique. Barrie and Schenk (1976) described increased blood flow through an arterial stenosis in an animal model. Because this increased flow could represent arteriovenous shunting, rather than increased nutritive flow in tissue capillaries, other methods are required to demonstrate changes in capillary flow. In severe peripheral arterial disease, causing rest pain and pre-gangrene, skin nutrition in particular is decreased. We have used an atraumatic, epicutaneously applied ^{133}Xenon clearance technique to measure nutritive skin blood flow in the feet of patients with peripheral arterial disease. Mean resting blood flow was normal in this group of patients, as reported previously (McEwan et al., 1970). An increase of 50% in resting skin flow was observed after defibrination, suggesting limitation of flow rate by increased basal viscosity values, which were associated with high fibrinogen levels (Lowe et al., 1979b). Further evidence that nutritional blood flow is increased by defibrination is the increase in leg muscle tissue oxygen pressure described by Ehrly and Schroeder (1977). While most workers have ascribed the improvement in blood flow to viscosity reduction, another possibility is dispersal of "circulating" platelet-aggregates following defibrination (Lowe et al., 1979a).

Increase in the ratio of leg arterial pressure to arterial pressure in the arm (an index of large-vessel obstruction) following defibrination has been reported by some workers (Dormandy et al., 1977; Tonnesen et al., 1978; Lowe et al., 1979b) but not by others (Martin et al., 1976; Humphreys et al., 1977).

21.5 Administration, Control and Adverse Effects

Ancrod (Arvin) is the only defibrinating agent available at present in the United Kingdom. It is usually given intravenously: an induction dose of 2 units/kg body weight is given by slow infusion (diluted in saline) over at least 4 h to avoid excessively fast formation of intravascular fibrin. Maintenance doses (1-2 units/kg) are then given by slow intravenous injection every 12-24 h. Plasma fibrinogen can thus be maintained at about 0.5 g/litre (normal range 1.5-4 g/litre): the clot grade test of Reid et al. (1963) or the rapid fibrinogen assay of Clauss (1957) are suitable for control of therapy. Subcutaneous injections have also been used: an initial dose of 4 units/kg is followed by daily maintenance doses of 1-2 units/kg. After 4-6 weeks of therapy, ancrod and batroxobin are inactivated by formation of antibodies (Vinazzer, 1975). Clinical reactions resulting from these antibodies are rare, but they do limit the duration of treatment.

Bleeding during defibrination therapy can occur, particularly during the induction period when high levels of anti-platelet and anti-coagulant FDP co-exist with very low fibrinogen levels (Vinazzer, 1975). It is usual to observe conventional contraindications to anticoagulant therapy. However, bleeding appears to be less common than with heparin therapy. Adverse reactions other than bleeding are uncommon.

21.6 Peripheral Arterial Disease

It is well recognised that the natural history of symptomatic peripheral arterial disease is generally benign, improvement in claudication — or even rest pain and pre-gangrene — commonly occurring without (or in spite of) therapeutic manoeuvres. There is also a marked placebo response. It is therefore essential to perform double-blind, randomised controlled trials to establish the efficacy of any treatment.

Intermittent claudication. Dormandy et al. (1977), in an uncontrolled study, found a significant increase in walking ability (59%) in 15 patients treated with subcutaneous ancrod for 5 weeks. On the other hand, Martin et al. (1976) found no benefit in a randomised controlled trial of intravenous batroxobin, given for 3 weeks to 10 patients, compared to intravenous saline which was given to another 10 patients. There was only a small (c. 15%) increase in walking ability in either group — possibly because all patients in this trial had already achieved maximal walking ability from regular exercise. This failure of defibrination to improve claudication may be due to insufficient reduction in blood viscosity to decrease significantly the segmental resistance of obstructed arteries and their collaterals, and hence increase blood flow to exercising muscles (Strandness and Sumner, 1975). Furthermore, the limitation of defibrination therapy to a few weeks by antibody formation limits its application to the management of the chronic affliction of claudication.

Rest pain and pre-gangrene. These manifestations of more severe degrees of peripheral ischaemia appear more attractive indications for defibrination therapy. Any reduction of arterial or microcirculatory peripheral resistance due to a decrease in viscosity would theoretically be of value (Strandness and Sumner, 1975) and we have already referred to observations of normalisation of platelet-aggregates and increased foot skin blood flow in these patients (Lowe et al., 1979a,b). Anecdotal evidence suggests that critical ischaemia may follow thrombus formation on atherosclerotic plaques, and a few weeks of defibrination might increase tissue perfusion sufficiently to salvage ischaemic tissue while thrombi lyse or organise and collateral vessels develop — and hence avoid, or minimise, amputation in those patients who are unsuitable for reconstructive arterial surgery. Repeated courses of defibrination may be given at intervals of several months for recurrent symptoms (Vinazzer, 1975; Lowe et al., 1977).

Several groups of workers have reported uncontrolled series of patients with rest pain and pre-gangrene. In most series, over 50% of patients experienced relief of pain within a few days of defibrination, and a majority of these remained free of rest pain for several months thereafter (Vinazzer, 1975; Gerber and Safer, 1978). Some workers, however, could not confirm such benefit (Humphreys et al., 1977).

Wolf (1976) reported two sequentially analysed controlled trials of ancrod in patients with rest pain. In the first study, intravenous ancrod or subcutaneous heparin was given for 7 days; in the second, subcutaneous ancrod or intravenous then oral vasodilators plus oral anticoagulants were given for 4 weeks. Both studies showed marked preference for ancrod therapy, but these results must be interpreted cautiously for two reasons. Firstly, neither study was double-blind. Secondly, we are not aware of good evidence for beneficial effects of anticoagulants or vasodilators in rest pain. It is quite possible that vasodilators are deleterious — by stealing blood away from areas of critical ischaemia (Strandness and Sumner, 1975). Heparin also may be harmful, as it induces platelet aggregation and, occasionally, arterial thrombosis. We therefore doubt that these agents should be used as "controls".

To date, two double-blind, placebo-controlled trials of ancrod in ischaemic rest pain have been reported. In our own study (Lowe et al., 1980) patients were randomised to 8 days' intravenous infusion of either ancrod or saline. After 27 patients had completed treatment the study was terminated, as two-thirds of patients in each group claimed improvement in rest pain. There was no difference between groups in analgesic consumption, nor in the rate of surgical intervention in the subsequent 6 months. We conclude that previous reports of improvement in pain after defibrination therapy were quite consistent with a placebo response. Tonnesen et al. (1978) randomised 42 patients to 3 weeks' subcutaneous ancrod or saline: one-quarter of the patients in each group claimed improvement in rest pain. There was no difference in healing of ulcers, nor in the rate of amputation in the 3-month follow-up period.

At present, therefore, defibrination therapy does not have an established place in the management of peripheral arterial disease. It may be that selected subgroups of patients may benefit, or that results might be improved by combination with other measures, such as haemodilution (Bohme, 1978; Tonnesen et al. 1978). Nevertheless we believe strongly that double-blind, placebo-controlled trials are mandatory for demonstration of true therapeutic efficacy.

Dormandy and Reid (1978) have suggested two further possible applications of defibrination therapy in arterial disease — to promote healing of skin flaps after leg amputations; and to prevent early thrombosis of arterial grafts. Controlled trials are required to examine these interesting possibilities.

21.7 Coronary Arterial Disease

Spottl et al. (1978) have reported improvement in patients with unstable angina pectoris and increased plasma fibrinogen, following ancrod therapy in an uncontrolled study; controlled trials have not been reported. The widespread use of beta-adrenergic blocking drugs and coronary artery surgery in the management of angina, and the difficulties of defining "unstable angina", would appear to limit the application of defibrination in this condition. A recent multicentre study of streptokinase in acute myocardial infarction reported reduced mortality, and reduced pulmonary and systemic blood pressures, in streptokinase-treated patients; the authors suggested that reduction of plasma fibrinogen, and hence blood viscosity, during streptokinase therapy, might account for these effects (European Collaborative Study Group, 1979). If so, defibrination therapy after myocardial infarction might be worthy of study: possible benefits might include increased coronary and systemic perfusion, decreased cardiac work load, and reduced venous thrombo-embolism (see Section 21.8).

21.8 Venous Thrombo-embolism

Intravenous ancrod therapy produces effective anticoagulation by reducing the plasma fibrinogen level to around 0.5 g/litre, and three studies have shown that ancrod is an effective alternative to heparin for the treatment of deep vein thrombosis (Kakkar et al., 1969; Davies et al., 1972; Tibbutt et al., 1974). Like heparin, ancrod prevents extension of venous thrombosis but does not promote lysis. Ancrod may have advantages over heparin in that its action is more predictable, it is easier to control, and it may be less likely to cause bleeding. In addition we, like others, have been impressed by the rapid decrease in leg oedema following defibrination, which may be due to increased flow in partially obstructed or collateral veins, consequent upon decreased blood viscosity. There is some evidence that ancrod and batroxobin potentiate the thrombolytic action of streptokinase and urokinase (Kwaan and Grumet, 1975), and pre-treatment with a defibrinating agent prior to administration of plasminogen activator might also promote lysis by reducing circulating fibrinogen (which competes with fibrin for the action of plasmin), by reducing circulating antiplasmin (Griguer et al., 1979) and by increasing perfusion of the thrombus due to reduced viscosity. Combined treatment appears reasonably safe (Forbes et al. 1976) and preliminary clinical results are encouraging (Griguer et al., 1979).

Ancrod has also been used for the prevention of venous thrombo-embolism following surgery. Post-operative deep vein thrombosis is associated with increased pre-operative levels of plasma fibrinogen (Clayton et al., 1976) and blood viscosity (Dormandy and Edelman, 1973). Peri-operative ancrod might prevent formation or extension of thrombi by removal of a large part of the circulating fibrinogen (the substrate for fibrin formation) and by reducing blood viscosity, which might increase leg blood flow (see 21.4) and hence reduce stasis in leg veins. Two preliminary studies showed that peri-operative ancrod was feasible and safe, and that once-daily subcutaneous ancrod could maintain the plasma fibrinogen concentration at 0.5-1.5 g/litre, reducing post-operative plasma and blood viscosity by about 10% (Barrie et al., 1974: Lowe et al., 1978a). Such partial defibrination has been shown to prevent thrombosis of femoral vein grafts in dogs (Ishimaru et al., 1979); interestingly, *total* defibrination led to thrombosis as well as bleeding in this animal study.

Low-dose subcutaneous heparin has been widely adopted for thrombotic prophylaxis in general and gynaecological surgery; however it is not effective in preventing thrombo-embolism after hip surgery. In this group of patients only full-dose oral anticoagulants are of proven antithrombotic efficacy. As orthopaedic surgeons do not use oral anticoagulants because of the incidence of bleeding and the need for daily control tests, we evaluated a fixed regime of subcutaneous ancrod in patients undergoing surgery for fractured neck of femur in a randomised, placebo-controlled trial (Lowe et al., 1978b). Ancrod reduced both the incidence and the extent of deep vein thrombosis, detected by bilateral venography; the efficacy of the regime appeared comparable to that of oral anticoagulants. No increased incidence of bleeding, wound complications or fracture complications was observed. In view of these encouraging preliminary results, further studies to determine whether defibrination prevents thromboembolism after elective hip surgery, and whether mortality from pulmonary embolism is reduced, would appear worthwhile.

21.9 Other Disorders

There have been anecdotal reports of successful treatment of retinal vein thrombosis and priapism by defibrination. Naish and Peters (1974) found that pre-treatment with ancrod reduced the severity of experimental glomerulonephritis in rabbits, but there is little information on treatment of nephritis in humans. Defibrination may also affect metastasis in malignant disease (Donati et al., 1977), but again trials in patients with cancer have yet to be reported.

21.10 Conclusions

There is good evidence that defibrinating agents reduce blood and plasma viscosity, and some evidence that thereby leg blood flow in increased. They are safe and effective agents in the treatment of venous thrombosis, and possibly also in its prevention in high-risk patients. As yet, controlled trials do not support the routine use of defibrination in patients with arterial disease.

21.11 References

Barrie W W, Schenk W G (1976) Improvement in blood flow through a critical arterial stenosis by defibrination with ancrod. Arch Surg *111:* 561-563

Barrie W W, Wood E H, Crumlish P, Forbes C D, Prentice C R M (1974) Low-dosage ancrod for prevention of thrombotic complications after surgery for fractured neck of femur. Br Med J *2:* 130-133

Blättler W, Straub P W, Peyer A (1974) Effect of in vivo produced fibrinogen-fibrin intermediates on viscosity of human blood. Thromb Res *4:* 787-801

Bohme H (1978) Defibrination in the therapy of disorders of arterial blood flow. ZFA (Stüttgart) *54:* 1645-1653

Clauss A (1957) Gerinnungsphysiologische Schnellmethode zur Bestimmung des Fibrinogens. Acta Haematol *17:* 237-246

Clayton J K, Anderson J A, McNicol G P (1976) Pre-operative prediction of post-operative deep vein thrombosis. Br Med J *2:* 910-912

Collen D, Verstraete M (1978) Activation of the fibrinolytic system during reptilase therapy and following strenuous exercise in man. In: Davidson J F, Rowan R M, Samama M M, Desnoyers P C (eds). Progress in chemical fibrinolysis and thrombolysis, Vol. 3. Raven Press, New York, pp 155-157

Davies J A, Merrick M V, Sharp A A, Holt J M (1972) Controlled trial of ancrod and heparin in treatment of deep vein thrombosis of lower limb. Lancet *i:* 113-115

Donati M B, Poggi A, Mussoni L, Kornblihtt L, de Gaetano G, Garattini S (1977) Effect of warfarin, warfarin enantiomers and defibrase on Lewis lung carcinoma metastasis growth. Thromb Haemostas *38:* 158

Dormandy J A, Edelman J B (1973) High blood viscosity: an aetiological factor in venous thrombosis. Br J Surg *60:* 187-190

Dormandy J A, Reid H L (1978) Controlled defibrination in the treatment of peripheral vascular disease. Angiology *29:* 80-88

Dormandy J A, Goyle K B, Reid H L (1977) Treatment of severe intermittent claudication by controlled defibrination. Lancet *i:* 625-626

Egberg N, Johnsson H (1972) Platelet aggregation induced by ADP and thrombin in reptilase defibrinated dogs. Thromb Res *i:* 95-112

Ehringer H, Dudczak R, Lechner K (1974) A new approach in the treatment of peripheral arterial occlusions: defibrination with arvin. Angiology *25:* 279-289

Ehrly A M (1972) Influence of arvin on the flow properties of blood. Biorheology: *9:* 151

Ehrly A M, Schroeder W (1977) Oxygen pressure in ischaemic muscle tissue of patients with chronic occlusive arterial diseases. Angiology *28:* 101-108

European Collaborative Study Group for Streptokinase Treatment in Acute Myocardial Infarction (1979) Streptokinase in acute myocardial infarction. N Engl J Med *301:* 797-802

Forbes C D, Barbenel J, Prentice C R M (1976) Treatment of thrombosis by sequential therapy with ancrod followed by streptokinase: clinical pharmacology and rheology. Haemostas *5:* 348-354

Gerber R, Safer A (1978) Retrospective follow-up study on the duration of the therapeutic effect of arvin therapy in chronic arterial occlusive diseases. Folia angiol *26:* 22-27

Griguer P, Dusson P, Brochier M (1979) Biological findings during sequential thrombolytic treatments with defibrase, lys-plasminogen and urokinase. Thromb Haemostas *42:* 281

Humphreys W V, Walker A, Charlesworth D (1977) A report on the use of arvin in patients with pre-gangrene of the lower limb. Br J Surg *64:* 31-33

Ishimaru S, Hirayama T, Furakawa K, Takahashi M (1979) Antithrombotic effect of defibrinogenating enzyme on experimental small caliber vein replacement using expanded PTFE graft. Thromb Haemostas *42:* 231

Kakkar V V, Flanc C, Howe C T, O'Shea M, Flute P T (1969) Treatment of deep vein thrombosis. A trial of heparin, streptokinase and Arvin. Br Med J *1:* 806-810

Kwaan H C, Grumet G N (1975) The place of thrombolytic and defibrinating agents in the treatment of venous thromboembolism. In: Nicolaides A N (ed). Thromboembolism. Medical and Technical Publishing, Lancaster, pp 251-267

Lowe G D O, Forbes C D, Prentice C R M (1977) Multiple course of ancrod (Arvin) therapy. Br Med J *1:* 508

Lowe G D O, Morrice J J, Fulton A, Forbes C D, Prentice C R M, Barbenel J C (1978a) Subcutaneous ancrod after operation for fractured hip — a dose-ranging and feasibility study. Thromb Haemostas *40:* 134-143

Lowe G D O, Campbell A F, Meek D R, Forbes C D, Prentice C R M, Cummings S (1978b) Subcutaneous ancrod in prevention of deep vein thrombosis after operation for fractured neck or femur. Lancet *ii:* 698-700

Lowe G D O, Reavey M M, Johnston R V, Forbes C D, Prentice C R M (1979a) Increased platelet aggregates in vascular and non-vascular illness: correlation with plasma fibrinogen and effect of ancrod. Thromb Res *14:* 377-386

Lowe G D O, Morrice J J, Forbes C D, Prentice C R M, Fulton A J, Barbenel J C (1979b) Subcutaneous ancrod therapy in peripheral arterial disease: improvement in blood viscosity and nutritional blood flow. Angiology *30:* 594-599

Lowe G D O, Dunlop D, Lawson D H, Pollock J G, Watt J K, Forbes C D, Prentice C R M, Drummond M M (1980) Double-blind controlled clinical trial of ancrod in the relief of ischaemic rest pain of the leg. Angiology, in press

McEwan A J, Stalker C G, Ledingham I M (1970) Foot skin ischaemia in atherosclerotic peripheral vascular disease. Br Med J *3:* 612-615

Martin M, Hirdes E, Auel H (1976) Defibrinogenation treatment in patients suffering from severe intermittent claudication — a controlled study. Thromb Res *9:* 47-57

Meissner J A, Karpowicz M, Konopka L, Lopaciuk S (1978) Changes in the muscle rest blood flow associated with therapeutic defibrinogenation. Abstracts of 17th Congress of International Society of Haematology, Paris, p 326

Naish P F, Peters D K (1974) Mediators of intraglomerular fibrin deposition in nephrotoxic serum nephritis. Clin Sci Mol Med *46:* 16

Prentice C R M, Hassanein A A, Turpie A G G, McNicol G P, Douglas A S (1969) Changes in platelet behaviour during arvin therapy. Lancet *i:* 644-647

Rampling M, Sirs J A (1972) The interactions of fibrinogen and dextrans with erythrocytes. J Physiol *223:* 199-212

Reid H A, Chan K E, Thean P C (1963) Prolonged coagulation defect (defibrination syndrome) in Malayan viper bite. Lancet *i:* 621-626

Schmid-Schönbein H, Wells R, Schildkraut R (1969) Microscopy and viscometry of blood under uniform shear rate (rheoscopy). J Appl Physiol *26:* 674-678

Schmid-Schönbein H, Weiss J, Volger E, Klose H J, Malotta H (1978) Microhaemorheology and defibrination. ZFA (Stüttgart) *54:* 1635-1643

Spottl F von, Pimmingstorfen E, Froschauer J (1978) Therapeutic defibrination by Arvin in unstable angina pectoris combined with hyperfibrinogenaemia. Wein Klin Wochenschr *22:* 792-796

Stocker K (1978) Defibrinogenation with thrombin-like snake venom enzymes. In: Markwardt F (ed). Fibrinolytics and anti-fibrinolytics. Springer-Verlag, Berlin, pp 451-484

Strandness D E, Sumner D S (1975) Haemodynamics for surgeons. Grune and Stratton, New York pp 290-341

Tibbutt D A, Williams E W, Walker M N, Chesterman C N, Holt J M, Sharp A A (1974) Controlled trial of ancrod and streptokinase in the treatment of deep vein thrombosis of the lower limb. Br J Haematol *27:* 407-414

Tonnesen K H, Sager P, Gormsen J (1978) Treatment of severe foot ischaemia by defibrination with ancrod: a randomised blind study. Scand J Clin Lab Invest *38:* 431-435

Turpie A G G, McNicol G P, Douglas A S (1972) Platelet electrophoresis: effect of defibrination by ancrod (Arvin). Cardiovasc Res *6:* 101-108

Vinazzer H (1975) Clinical experience with Arvin. Folia Angiol *23:* 402-406

Wolf G K (1976) Arvin in peripheral arterial circulatory disorders: controlled multicentre trials. Europ J Clin Pharmacol *9:* 387-392

22 Drug Modification of Erythrocyte Deformability

J.A. Dormandy

22.1 Introduction

There has been no significant advance in the treatment of peripheral circulatory insufficiency since reconstructive arterial surgery became generally accepted over 20 years ago. The next breakthrough, as well as the final solution, is almost certainly going to be non-surgical. Less than 20% of patients with peripheral ischaemia, whether in the limbs or the brain, are suitable for surgical treatment and even in this highly selected group the results are discouraging. Nor is there any indication that either the scope or the success of this approach is likely to improve in the near future. Medical attempts at treatment have concentrated on the removal of risk factors and the prevention of thrombosis by anti-platelet drugs. But another avenue which, in the short term at least, may well prove more profitable to explore is haemorheological manipulation, by which it is hoped to improve the flow properties of blood. Although this would leave the vessel disease untouched, it could probably produce more rapid and dramatic results in terms of improving tissue perfusion. Haemorheological therapy may rather artificially be divided into treatment aimed at improving blood viscosity or improving the flow properties of the individual red cells. This distinction is purely artificial as undoubtedly one of the most important determinants of whole-blood viscosity, particularly at high shear rates, is the ability of individual red cells to deform. This chapter will, however, only consider the action of drugs specifically on erythrocyte deformability.

The fact that red cells have to undergo considerable deformation in order to negotiate the microcirculation, where most passages are narrower than the diameter of the resting red cells, has been appreciated for centuries. But it is only recently that evidence (considered in Chapter 13) has begun to accumulate suggesting that tissue ischaemia may be due to a decreased red-cell deformability. From this has arisen the concept of making the red cells more flexible and thereby improving the microcirculation. Understandably, the pharmaceutical industry has not been slow to appreciate the potential of this idea although at the moment there is insufficient solid evidence to confirm this hypothesis. We have to go at least some way towards solving the following problems before beginning to accept the improvement of red-cell deformability as a practical therapeutic aim:

1) By contrast to the measurement of whole-blood viscosity, there is still no general agreement on the clinically most valid technique for assessing red-cell deformability. (The available techniques are discussed in Chapter 5.)

2) Although abnormalities in red-cell deformability, usually using a filtration technique, have now been described in a number of circulatory disorders (see Chapter 13), we have no idea whether these abnormalities are the cause or the consequence of the diseases. Possibly they are both; the hypoxia and acidosis associated with ischaemia making the red cells more rigid, which may in turn add to the degree or extent of the ischaemia. In other words a vicious cycle may be set up.

3) There is considerable disagreement about the determinants of red-cell deformability, occasionally amounting to claims of directly opposing influences, as in the case of plasma fibrinogen.

4) So far nobody has been able to correlate abnormalities in red-cell deformability measured *in vitro* with physiological measurements relating to tissue perfusion *in vivo*.

5) Similarly, there is so far insufficient evidence that improvement of red-cell deformability *in vitro* can be correlated with any clinical improvement associated with improved tissue perfusion.

Despite the existence of all these unknowns, attempts have already been made to investigate the possible practical potential of drugs to improve blood flow by an effect on the physical properties of red cells; and it could be argued that in the process we have to some degree increased our knowledge of the more fundamental problems mentioned above.

Before considering individual drugs, the theoretically possible mechanisms which may achieve an improvement in red-cell deformability will be reviewed.

22.2 Possible Mechanisms for Improving Red-cell Deformability

At this early stage of the art, there are virtually limitless possibilities, none of which have been fully explored. Only a few examples will be briefly mentioned here.

22.2.1 *Role of Adenosine Triphosphate (ATP) and Calcium*

ATP is involved in most energy transfer processes within the red cell and is, therefore, vital for the maintenance of its integrity (Wilkinson and Robinson, 1974). It has been demonstrated that artificially ATP-depleted red cells become rigid and as a result of extensive studies by Weed and LaCelle the hypothesis has been suggested that ATP has a chelating effect at the inner membrane surface. They have further postulated that a sol-to-gel transformation occurs as local membrane calcium rises, resulting in a thickened membrane, stiffened by the accumulated protein gel (Weed et al., 1969; LaCelle, 1970). On the basis of this hypothesis, one may expect an improvement in red-cell deformability by an increase in available ATP or by a calcium antagonistic agent. It should, however, be added that recently Meiselman et al. (1978) have thrown some doubt on the importance of ATP in this context.

22.2.2 *Importance of Membrane–Plasma Surface Tension*

Braasch (1969) suggested the hypothesis that the surface tension between the cell membrane and the surrounding plasma is an important determinant of red-cell shape, and it is generally accepted that alterations in the shape of the red cells will inevitably influence its deformability. This would suggest a line of investigation into surface active agents. Changes in osmolarity would have a similar effect on shape, but would hardly be a practical therapeutic proposition (Bessis amd Mohandas, 1975).

22.2.3 *Surface Red-cell Charge*

On purely theoretical grounds the magnitude of the negative red-cell surface charge (or zeta potential) may also be important, but this does not appear to have been investigated experimentally.

22.2.4 *Prostaglandins*

Allen (1974) has suggested that prostaglandin PGE_1 increases the deformability of red cells, while PGE_2 has the opposite effect. These changes take place within seconds using physiological doses of prostaglandins, which exert this effect by changing the levels of cyclic adenosine monophosphate.

22.2.5 *Polymerisation of Red-cell Proteins*

Spectrine, which is the principal skeletal protein of the red cell, is present in both polymerised and unpolymerised forms, the ratio depending on specific phosphorylating enzymes. Alterations in this balance, with possible improvement in the deformability of the cell, may be brought about by changes in the concentration of the specific enzymes concerned with phosphorylation of proteins like spectrine.

22.3 Specific Drugs Said to Improve Red-cell Deformability

22.3.1 General Considerations

Despite the many large gaps in our current knowledge about red-cell deformability, its determinants and its clinical relevance, there are now a number of drugs marketed claiming to improve red-cell flexibility and hence blood flow. These drugs were almost invariably originally developed for their other properties, usually as vasodilators. (At this early stage, there has not yet been time to develop a new generation of drugs specifically for their effect on red-cell deformability.)

The evidence about a possible effect on the physical properties of red cells by the existing drugs to be considered falls into three main groups: firstly, what will be called *in vitro* evidence, the effect of the drug when added to red cells after withdrawal. Often the red cells have first been rendered artificially rigid by storage or incubation with various solutions. Secondly, so-called *in vivo* experience, when red-cell deformability was assessed before and after administration of the drug intravenously or orally to patients. Lastly, the correlation of these results with some measure of the clinical change in the patients' tissue perfusion, whether subjective or objective, in open or blind clinical trials. Some of the commoner drugs, marketed in this country, and said to have a specific effect on red-cell deformability, will be considered and the evidence discussed under these three headings. Clinical evidence will only be considered where an improvement has been directly linked to measurements of improved red-cell deformability.

22.3.2 Pentoxifylline (Trental)

Apart from its action on red-cell deformability, this drug is also a vasodilator and is said to have an effect on platelet behaviour and whole-blood viscosity. Its effect on red-cell deformability is thought to be due to increasing intracellular ATP levels (Stefanovitch, 1975; Vigneron and Stoltz, 1977). Ehrly (1975) used an *in vitro* system for making red cells abnormally rigid in a hyperosmolar suspension and then showed that pentoxifylline had a protective action against this challenge. Werner (1975) showed a similar result using abnormally rigid stored red cells.

In vivo studies on rabbits have shown an improvement in red-cell deformability (Vigneron and Stoltz, 1977). But in humans Volker (1976) could not show any change after 10 days' treatment with 900 mg/day, although the whole-blood viscosity was improved. On the other hand, more recently Martin and Vives (1979) gave pentoxifylline to 28 patients recovering from cerebrovascular accidents and showed a significant improvement in red-cell filterability compared with 21 similar patients who did not receive the drug. A clinical improvement, associated with an improvement in red-cell deformability, had also been reported in a smaller uncontrolled study of patients with chronic cerebral ischaemia (Pecker et al., 1978).

22.3.3 Isoxsuprine (Duvadilan)

This is also a vasodilator, with possible other effects on platelets and plasma fibrinogen. Although its action on the physical properties of the red cells themselves has not been assessed directly, Di Perri et al. (1978), from findings based on whole-blood viscosity measurements, suggested that isoxsuprine may have a beneficial effect on red-cell deformability.

22.3.4 Cinnarizine (Stugeron Forte)

Cinnarizine and its more recent difluorinated derivative flunarizine are calcium antagonists which have a specific vasodilator effect by decreasing calcium influx into depolarised muscle cells (Godfraind et al., 1973). They also have an antihistaminic action. It has been suggested (Emanuel, 1979) that these substances may improve red-cell deformability by a similar action on the calcium-ATP-mediated mechanism controlling the flexibility of the red-cell membrane as originally postulated by Weed et al. (1969). This suggestion has been supported by *in vitro* work using artificially crenated red cells (Chasin and Scott, 1978). Di Perri et al., whose work on isoxsuprine has already been quoted (1978), drew similar conclusions regarding cinnarizine (1979). Again, their suggestion that red-cell flexibility may be improved is based on viscometric data. Direct measurement of red-cell deformability using a filtration technique was used by De Cree et al. (1979) in 22 patients with various circulatory disorders. The abnormally rigid red cells of the patients were made significantly more flexible by 80 mg flunarizine given over 24 h.

22.3.5 Alprenolol

There has been one unusual report of this beta-blocker having a haemorheological effect, by Dintenfass and Lake (1976), who compared two groups of 12 patients each with myocardial ischaemia. Using their own method of assessing red-cell flexibility, they found both groups had rigid red cells, but that alprenolol seemed to have an effect tending to minimise this. This unexpected finding does not seem to have been looked at by any other investigator using this or any other beta-blocker.

22.4 Conclusions

Unfortunately, none of the evidence relating to the compounds discussed is conclusive. We should not, however, allow this to disillusion us about the basic concepts involved. It is perfectly plausible that a therapeutic improvement in red-cell deformability can be achieved and that this would result in improved tissue perfusion. Even if red-cell rigidity is not a common primary cause of ischaemia, it may be a secondary effect contributing to the progression of the ischaemia, and therapeutic intervention at this stage may well be clinically beneficial. However, this still remains to be convincingly proved. Whilst continuing to study the properties of the existing compounds mentioned, it is essential also to continue our attempts to clarify the more basic problems mentioned earlier, concerned with the whole field of clinical microrheology.

22.5 References

Allen J E (1974) Prostaglandins in haematology. Arch Intern Med *133:* 86-96

Bessis M, Mohandas W (1975) Deformability of normal, shape-altered and pathological red cells. Blood Cells *1:* 315-321

Braasch D (1969) The relation between erythrocyte deformability, cell shape and membrane surface tension. Pluegers Arch *313:* 316-370

Chasin N, Scott T (1978) Effect of flunarazine on human red cells. Paper presented at 20th Annual meeting of American Society of Angiology, Marbella

De Cree J, De Cock W, Geukens H, De Clerck F, Beerens M, Verhaegen H (1979) The rheological effects of cinnarizine and flunarizine in normal and pathological conditions. Angiology *30:* 505-525

Dintenfass L, Lake B (1976) Beta-blockers and blood viscosity. Lancet *i:* 1026

Di Perri T, Forconi S, Agnusdei D, Guerrini M, Laghi Pasini F (1978) The effect of intravenous isoxsuprine on blood viscosity in patients with occlusive arterial disease. Brit J Clin Pharmacol *5:* 255-260

Di Perri T, Foroni S, Guerrini M, Passini F L, Del Cippola R, Rossi C, Agnusdei D (1979) Action of cinnarizine on the hyperviscosity of blood in patients with peripheral obliterative arterial disease. Angiology *30:* 13-20

Ehrly A M (1975) The effect of pentoxifylline on red cell flexibility. Med Welt *51:* 2300-2301

Emanuel M B (1979) Specific calcium antagonists in the treatment of peripheral vascular disease. Angiology *30:* 454-469

Godfraind T, Kaba A, Rojas R (1973) Inhibition by cinnarizine of calcium channels opening in depolarised smooth muscle. Br J Clin Pharmacol *49:* 164-165

LaCelle P L (1970) Alteration of membrane deformability in haemolytic anaemia. Semin Haematol *7:* 355-371

Martin P, Vives P (1979) Deformabilité des globules rouges et accidents vasculaires cérébraux. Intérêt de la pentoxifylline. Gaz Med France *86:* 2585-2587

Meiselman H J, Evans E A, Hochmuth R M (1978) Membrane mechanical properties of ATP — depleted erythrocytes. Blood *52:* 499-504

Pecker J, Vallee B, Haranger J-M, Guegven M (1978) Torental et accidents ischémiques du système nerveux central. Ouest Med *31:* 1571-1574

Stefanovitch V (1975) The effect of pentoxifylline on erythrocyte adenine nucleotide levels in rats. IRCS Med Sci *3:* 91

Vigneron C, Stolz J-F (1977) Modifications physico-clinique et rhéologique de l'hématie sous l'influence de la pentoxifylline. Médecine Actuelle *4:* 200-204

Volker D (1976) Haemorheological investigations in patients with vascular disease undergoing treatment with pentoxifylline. Pharmatherapeutica *1:* 154-159

Weed R I, LaCelle P L, Merrill E W (1969) Metabolic dependence of red cell deformability. J Clin Invest *48:* 795-809

Werner V (1975) Measurement of the flexibility of erythrocytes incubated with oxpentifylline and low molecular dextran. Med Welt *26:* 2098-2099

Wilkinson J H, Robinson J M (1974) The effect of ATP on release of intracellular enzymes from damaged cells. Nature *249:* 662-663

Subject Index

GPSR Compliance

The European Union's (EU) General Product Safety Regulation (GPSR) is a set of rules that requires consumer products to be safe and our obligations to ensure this.

If you have any concerns about our products, you can contact us on ProductSafety@springernature.com

In case Publisher is established outside the EU, the EU authorized representative is:

Springer Nature Customer Service Center GmbH
Europaplatz 3
69115 Heidelberg, Germany

Batch number: 10620881

Printed by Printforce, the Netherlands